Study Gui

to accompany

Microbiology

Fourth Edition

Study Guide

to accompany

Microbiology

Fourth Edition

Lansing M. Prescott
Augustana College

John P. Harley
Eastern Kentucky University

Donald A. Klein
Colorado State University

Prepared by
Ralph Rascati
Kennesaw State University

Boston   Burr Ridge, IL   Dubuque, IA   Madison, WI   New York   San Francisco   St. Louis
Bangkok   Bogotá   Caracas   Lisbon   London   Madrid
Mexico City   Milan   New Delhi   Seoul   Singapore   Sydney   Taipei   Toronto

WCB/McGraw-Hill

A Division of The McGraw-Hill Companies

Study Guide to accompany
MICROBIOLOGY, FOURTH EDITION

1 2 3 4 5 6 7 8 9 0 QPD/QPD 9 0 9 8

ISBN 0-697-35445-8

www.mhhe.com

Contents

PREFACE

This edition of the Study Guide was designed to accompany and supplement the fourth edition of the textbook *Microbiology* by Prescott, Harley, and Klein, and correlates with the modifications in the textbook. If used properly, it should enhance the comprehension gained from the text, but it is not a substitute for it. The text was designed as a comprehensive discussion of all major areas of microbiology in a manner that is appropriate for an introductory course for biology or microbiology students who are already somewhat familiar with general biological and chemical concepts. The 44 chapters in this Study Guide correspond to the 44 chapters in the text. Most chapters consist of the following sections:

Chapter Overview: A brief summary of the major points of discussion in the chapter.

Chapter Objectives: A list of concepts addressed in the chapter that students should be able to discuss when they have mastered the chapter material.

Chapter Outline: A detailed study outline of the chapter that reinforces the material learned.

Terms and Definitions: An exercise in which students are asked to match the terms (taken primarily from the terms in boldface type in the text) with appropriate definitions.

Fill in the Blank: A completion exercise in which students fill in key terms omitted from phrases and descriptions from the text.

Multiple Choice: An exercise in which students select the best answer from the choices available.

True/False: An exercise in which students decide if the statement, as written, is true or false.

Critical Thinking: A series of two or more questions that require students to put together several pieces of information, and that may require students to pursue some additional reading or express an opinion as well.

Answer Key

Some chapters have additional exercises that are appropriate for the material covered.

How to Use This Study Guide

This Study Guide is most beneficial if used in conjunction with the text in the following manner:

Students should first read the **Chapter Overview** and the **Chapter Objectives** in the Study Guide. Keeping these objectives in mind, students should thoroughly read the corresponding chapter in the text and should then use the **Chapter Outline** in the Study Guide to reinforce the major concepts learned in the text. After doing all of this, the students are then ready to begin testing their mastery of the material. It should be noted that in each chapter of this Study Guide, there are some questions that can be answered only by reading the textbook; the Chapter Outline of the Study Guide, although reasonably comprehensive, does not cover all areas thoroughly enough to answer all of the questions.

    Terms and Definitions is the first exercise encountered in most chapters and should be attempted first, because mastery of any subject area is possible only if the relevant vocabulary is learned thoroughly. Answers can be checked using the answer key provided. Students should study any terms that were missed by rereading the appropriate sections of the text.

    The students should then complete the remaining objective exercises for that chapter. Most answers can be checked using the key provided. For any questions answered incorrectly, the students should reread the corresponding section of the text.

    Mastery of any subject area requires more than assimilation of facts; an understanding of the interrelationships among various concepts and factual items is necessary as well. To help increase this understanding, at least two **Critical Thinking** questions are provided for each chapter. These should be attempted only when students can answer all of the objective questions correctly. If additional development is needed in this area, students should answer some or all of the excellent Questions that the textbook authors have provided throughout each chapter.

    Conscientious use of this Study Guide in this manner should greatly increase students' understanding and, therefore, enjoyment of the fascinating world of microbiology, and should help develop and sharpen the skills needed for effective mastery of any scientific discipline.

Correlation Guide for *Microbes in Motion, II* to *Microbiology, 4th ed* [1]

This correlation guide was compiled in order to assist you integrating the information from *Microbes in Motion, II*, the tutorial CD-ROM developed by WCB/McGraw-Hill, into the textual material presented in this fourth edition of *Microbiology* by Prescott, Harley, and Klein

You will encounter a CD-ROM icon ⊛ in figure legends throughout the text. When you see a CD-ROM icon, be sure to refer to this correlation guide for corresponding information on *Microbe in Motion, II*. In some cases, there will be a reference to supporting information on Microbes in Motion, II in this correlation guide, but no icon in the text. This CD material is usually supporting textual material in the fourth edition of *Microbiology* and an could not be inserted into the text narrative. Be sure to check this correlation table prior to reading a chapter in case there is supporting information on *Microbes in Motion, II.*

Chapter 2

Fig. 2.13	Bacterial Structure & Function Book Bacteria Groups Chapter / Introduction, Categories pp. 1 (audio)–2 (interactive)
Fig. 2.14	Bacterial Structure & Function Book Bacteria Groups Chapter/Categories Classification/pp.2–4

Chapter 3

Fig. 3.5	Bacterial Structure & Function Book Cell Membrane Chapter/pp. 4–6 (animation)
Fig. 3.7	Bacterial Structure & Function Book Cell Membrane Chapter/pp. 8 & 10 (animation)
Fig. 3.19	Bacterial Structure & Function Book Gram Positive Cell Chapter/pp. 1–5 Gram Negative Cell Chapter/pp. 1–9
Fig. 3.23	Bacterial Structure & Function Book Cell Wall Chapter/Peptidoglycan/pp. 1–6 Antimicrobial Action Book/Cell Wall Inhibitors Chapter/Mechanism of Action p. 7 (Peptidoglycan) Synthesis Animation)
Fig. 3.25	Bacterial Structure & Function Book Gram Positive Cell Chapter/Teichoic Acids/pp. 2–5
Fig. 3.27	Bacterial Structure & Function Book Gram Negative Cell Chapter/pp. 1–9
Fig. 3.30	Antimicrobial Action Book Cell Wall Inhibitors Chapter/Mechanism of Action/p. 7 (animation)
Fig. 3.31	Bacterial Structure & Function Book External Structures Chapter / Glycocalyx/pp. 2–11
Fig. 3.39	Bacterial Structure & Function Book

External Structures Chapter/Flagella/pp. 18–24 (video)

Chapter 4

Fig. 4.7	Mycology—Fungal Structure & Function Book General Eucaryotic StructuresChapter/ Internal Structures/pp. 7–9

Chapter 5

Table 5.1 Table 5.2	Microbial Metabolism & Growth Book Microbial Growth Chapter/Metabolic Nutrition/pp. 2–7 (interactive)
Fig. 5.2	Bacterial Structure & Function Book Cell Membrane Chapter/Transport/p. 11 Microbial Metabolism & Growth Book Microbial Growth Chapter/Nutrient Transport/p. 8–10 (interactive)
Fig. 5.3	Bacterial Structure & Function Book Cell Membrane Chapter/Transport/p. 11 Microbial Metabolism & Growth Book Microbial Growth Chapter/Nutrient Transport/pp. 8–9
Fig. 5.7	Microbial Metabolism & Growth Book Microbial Growth Chapter/Media/p. 35 (video)

Chapter 6

Fig. 6.1	Microbial Growth & Metabolism Book Microbial Growth Chapter/Growth Curves/pp. 26, 33, & 34 (animation)
Fig. 6.3	Microbial Growth & Metabolism Book Microbial Growth Chapter/Growth Curves/pp. 26–27 (interactive)
Fig. 6.5	Microbial Growth & Metabolism Book Microbial Growth Chapter/Growth Curves/pp. 28 & 30 (video)

---

[1]Each entry has a condensed format: chapter/topic/pages.

Part I INTRODUCTION TO MICROBIOLOGY

1 The History and Scope of Microbiology

CHAPTER OVERVIEW

This chapter introduces the field of microbiology and discusses the importance of microorganisms not only as causative agents of disease but also as important contributors to food production, antibiotic manufacture, vaccine development, and environmental management. It presents a brief history of the science of microbiology, an overview of the microbial world, and a discussion of the scope and relevance of microbiology in today's society.

CHAPTER OBJECTIVES

After reading this chapter you should be able to:

- define the science of microbiology and describe some of the general methods used in the study of microorganisms
- discuss the historical concept of spontaneous generation and the experiments that were performed to disprove this erroneous idea
- discuss Koch's postulates, which are used to establish the causal link between a suspected microorganism and a disease
- describe some of the various nonpathological activities of microorganisms
- describe procaryotic and eucaryotic morphology, the two types of cellular anatomy, and also the distribution of microorganisms among the various kingdoms or domains in which living organisms are categorized
- discuss the importance of the field of microbiology to other areas of biology and to general human welfare

CHAPTER OUTLINE

I. The Discovery of Microorganisms
   A. Microbiology is the study of organisms too small to be clearly seen by the unaided eye (i.e., microorganisms)
   B. Types of microorganisms include viruses, bacteria, protozoa, algae, and fungi
      1. Some algae and fungi are large enough to be visible, but are included in the field of microbiology because they have similar properties and because similar techniques are employed to study them (isolation, sterilization, culture in artificial media)

    C.   Early discovery of microorganisms
        1.   Invisible living creatures were thought to exist and were thought to be responsible for disease long before they were observed
        2.   Antony van Leeuwenhoek (1632–1723) constructed microscopes and was the first person to observe and describe microorganisms accurately

II.   The Spontaneous Generation Conflict
    A.   The proponents of the concept of spontaneous generation claimed that living organisms could develop from nonliving or decomposing matter
    B.   Francesco Redi (1626–1697) challenged this concept by showing that maggots on decaying meat came from fly eggs deposited on the meat, and not from the meat itself
    C.   John Needham (1713–1781) showed that mutton broth boiled in flasks and then sealed could still develop microorganisms, which supported the theory of spontaneous generation
    D.   Lazzaro Spallanzani (1729–1799) showed that flasks sealed and then boiled had no growth of microorganisms, and he proposed that air carried germs to the culture medium; he also commented that external air might be needed to support the growth of animals already in the medium; the latter concept was appealing to supporters of spontaneous generation
    E.   Louis Pasteur (1822–1895) trapped airborne organisms in cotton; he also heated the necks of flasks, drawing them out into long curves, sterilized the media, and left the flasks open to the air; no growth was observed because dust particles carrying organisms did not reach the medium, instead they were trapped in the neck of the flask; if the necks were broken, dust would settle and the organisms would grow; in this way Pasteur disproved the theory of spontaneous generation
    F.   John Tyndall (1820–1893) demonstrated that dust did carry microbes and that if dust was absent, the broth remained sterile—even if it was directly exposed to air; Tyndall also provided evidence for the existence of heat-resistant forms of bacteria

III.   The Recognition of the Microbial Role in Disease
    A.   Agostino Bassi (1773–1856) showed that a silkworm disease was caused by a fungus
    B.   M. J. Berkeley (ca. 1845) demonstrated that the Great Potato Blight of Ireland was caused by a fungus
    C.   Louis Pasteur showed that the pébrine disease of silkworms was caused by a protozoan parasite
    D.   Joseph Lister (1872–1912) developed a system of surgery designed to prevent microorganisms from entering wounds; his patients had fewer postoperative infections, thereby providing indirect evidence that microorganisms were the causal agents of human disease; his published findings (1867) transformed the practice of surgery
    E.   Robert Koch (1843–1910), using criteria developed by his teacher, Jacob Henle (1809-1895), established the relationship between *Bacillus anthracis* and anthrax; his criteria became known as Koch's Postulates and are still used to establish the link between a particular microorganism and a particular disease:
        1.   The microorganisms must be present in every case of the disease but absent from healthy individuals
        2.   The suspected microorganisms must be isolated and grown in pure culture
        3.   The same disease must result when the isolated microorganism is inoculated into a healthy host
        4.   The same microorganism must be isolated again from the diseased host
    F.   Koch's work was independently confirmed by Pasteur
    G.   Charles Chamberland (1851–1908) was instrumental in identifying viruses as disease-causing agents
    H.   Edward Jenner (ca. 1798) used a vaccination procedure to protect individuals from smallpox
    I.   Louis Pasteur developed other vaccines including those for chicken cholera, anthrax, and rabies

J.   Emil von Behring (1854–1917) and Shibasaburo Kitasato (1852–1931) induced the formation of diphtheria tetanus antitoxins in rabbits which were effectively used to treat humans thus demonstrating humoral immunity

K.   Elie Metchnikoff (1845–1916) demonstrated the existence of phagocytic cells in the blood, thus demonstrating cell-mediated immunity

IV.   The Discovery of Microbial Effects on Organic and Inorganic Matter

A.   Louis Pasteur demonstrated that alcoholic fermentations were the result of microbial activity, that some organisms could decrease alcohol yield and sour the product, and that some fermentations were aerobic and some anaerobic; he also developed the process of pasteurization (see chapter 7) to preserve wine during storage

B.   Sergei Winogradsky (1856–1953) worked with soil bacteria and discovered that they could oxidize iron, sulfur, and ammonia to obtain energy; he also studied anaerobic nitrogen-fixation and cellulose decomposition

C.   Martinus Beijerinck (1851–1931) isolated aerobic nitrogen-fixing bacteria and sulfate reducing bacteria

D.   Beijerinck and Winogradsky pioneered the use of enrichment cultures and selective media

V.   The Development of Microbiology in This Century

A.   Microbiology established a closer relationship with other disciplines during the 1940s because of its association with genetics and biochemistry

B.   George W. Beadle and Edward L. Tatum (ca. 1941) studied the relationship between genes and enzymes using the bread mold, *Neurospora*

C.   Salvadore Luria and Max Delbruck (ca. 1943) showed that mutations were spontaneous and not directed by the environment

D.   Oswald T. Avery, Colin M. MacLeod, and Maclyn McCarty (1944) provided evidence that deoxyribonucleic acid (DNA) was the genetic material and carried genetic information during transformation (see chapter 13)

E.   More recently, microbiology has been a major contributor to molecular biology and has been deeply involved in the elucidation of the genetic code; in studies on the mechanisms of DNA, ribonucleic acid (RNA), and protein synthesis; and in studies on the regulation of gene expression and the control of enzyme activity

F.   In the 1970s new discoveries in microbiology led to the development of recombinant DNA technology and genetic engineering

VI.   The Composition of the Microbial World

A.   Procaryotes have a relatively simple morphology and lack a true membrane-delimited nucleus

B.   Eucaryotes are morphologically complex and have a true, membrane-enclosed nucleus

C.   Organisms are currently divided in five kingdoms: the *Monera* or *Procaryotae, Protista, Fungi, Animalia,* and *Plantae*

D.   Alternative classification schemes involving several empires or domains with multiple kingdoms contained within have been proposed and are discussed in chapter 19

E.   Microbiologists are concerned primarily with members of the first three kingdoms and also with viruses, which are not classified with living organisms

VII.   The Scope and Relevance of Microbiology

A.   Microorganisms were the first living organisms on the planet, live everywhere life is possible, are more numerous than any other kind of organism, and probably constitute the largest component of the earth's biomass

B.   The entire ecosystem depends on the activities of microorganisms, and microorganisms influence human society in countless ways

C.   Microbiology has an impact on medicine, agriculture, food science, ecology, genetics, biochemistry, and other fields

D.   Microbiologists may be interested in specific types of organisms:

1.   Virologists—viruses

2.   Bacteriologists—bacteria

3. Phycologists or Algologists—algae
4. Mycologists—fungi
5. Protozoologists—protozoa
E. Microbiologists may be interested in various characteristics or activities of microorganisms:
1. Microbial morphology
2. Microbial cytology
3. Microbial physiology
4. Microbial ecology
5. Microbial genetics and molecular biology
6. Microbial taxonomy
F. Microbiologists may have a more applied focus:
1. Medical microbiology, including immunology
2. Food and dairy microbiology
3. Public health microbiology
4. Industrial microbiology
5. Agricultural microbiology

AREAS OF MICROBIOLOGY

From the list below select the area that best matches the description in each of the numbered statements.

F 1. Deals with diseases of humans and animals
J 2. Endeavors to control the spread of communicable diseases
D 3. Deals with the mechanisms by which the human body protects itself from disease-causing organisms
A 4. Deals with microorganisms that cause damage to crops or live in herds of domestic animals; also deals with ways of increasing soil fertility
G 5. Studies the relationship between microorganisms and their habitats
B 6. Investigates the causes of spoilage of products for human consumption and the use of microorganisms in the production of cheese, yogurt, pickles, beer, etc.
E 7. Employs microorganisms to make products such as antibiotics, vaccines, steroids, alcohols, vitamins, amino acids, and enzymes
H 8. Investigates the synthesis of antibiotics and toxins, microbial energy production, the ways in which microorganisms survive harsh environmental conditions, etc.
I 9. Focuses on the nature of genetic information and how genes regulate the development and function of cells and organisms
C 10. Involves the insertion of new genes into organisms in order to investigate the genes' functions or to produce useful organisms with new properties

a. agricultural microbiology
b. food and dairy microbiology
c. genetic engineering
d. immunology
e. industrial microbiology
f. medical microbiology
g. microbial ecology
h. microbial physiology
i. molecular biology
j. public health microbiology

FILL IN THE BLANK

1. The concept that living organisms could develop from nonliving or decomposing matter is referred to as _Spontan. G_ . An Italian physician named _Gruncho_ challenged this by showing that maggots developed from fly eggs, not from decaying meat as had previously been thought. However, even after this, many thought that simple microorganisms could develop from nonliving material, even if more complex organisms could not. This was finally disproved by the work of a French scientist named _Louis Past_ and an English physicist named _John T_ , who both demonstrated that organisms only developed in sterile broth exposed to air if dust particles carrying living organisms dropped into the broth.

2. Indirect evidence that microorganisms were agents of human disease came from the work of an English surgeon named _Joseph Liston_ who developed a procedure for antiseptic surgery in which he _heat sterilize_ _____ his surgical instruments before use and also sprayed the compound _phenol_ , which kills bacteria, over the surgical area. These procedures lowered the incidence of postoperative infections.

3. Pasteur and Roux discovered that older cultures of the bacterium that caused _chicken cholera_ lost their ability to cause disease, or were said to be _attenuated_ . If chickens were injected with these older strains, they remained healthy but developed the ability to resist disease. Pasteur called this culture a(n) _vaccin_ in honor of the English physician _Jenner_ who developed the procedure in order to protect against _Smallpox_ .

4. Inactivated toxin from the microorganism that causes diphtheria was injected into rabbits by _____ and _____, inducing the rabbits to produce a soluble substance which would neutralize the toxin. This is referred to as _____ immunity. Taking a different approach, Elie Metchnikoff discovered that some blood cells were capable of ingesting bacteria and thereby destroying them. He called this process _____ and the cells which can do this _____. This is the basis for cell-mediated immunity.

5. The theory that yeast cells were responsible for converting sugar to alcohol was first proposed by _____ and others. However, it was _____ who did the definitive work which clearly demonstrated this to be true. One of his most important discoveries was that some organisms carried out these processes _____ (in the presence of oxygen) while others carried them out _____ (in the absence of oxygen).

6. Two of the most important contributors in the field of microbial ecology were _____ who discovered that soil bacteria could oxidize iron, sulfur, and ammonia to obtain energy; and _____ _____ who isolated a root-nodule bacterium capable of nitrogen fixation. These individuals also pioneered the use of _____ _____ techniques and _____ media.

MULTIPLE CHOICE

For each of the questions below select the *one best* answer.

1. Which of the following is (are) used to define the field of microbiology?
    a. the size of the organism studied
    b. the techniques employed in the study of organisms regardless of their size
    c. Both (a) and (b) are correct.
    d. Neither (a) nor (b) is correct.

2. Which of the following was one of the first to suggest that microorganisms caused disease in humans?
    a. von Leeuwenhoek
    b. Fracastoro _Roid_
    c. Pasteur
    d. Koch

5

3. Which of the following developed a set of criteria that could be used to establish a causative link between a particular microorganism and a particular disease?
   a. von Leeuwenhoek
   b. Fracastoro
   c. Pasteur
   d. Koch

4. Which of the following was the first to observe and accurately describe microorganisms?
   a. von Leeuwenhoek
   b. Fracastoro
   c. Pasteur
   d. Koch

5. Which of the following provided strong evidence against the concept of spontaneous generation?
   a. von Leeuwenhoek
   b. Fracastoro
   c. Pasteur
   d. Koch

6. Which of the following provided evidence that a microorganism could be responsible for a particular disease?
   a. Bassi
   b. Berkeley
   c. Koch
   d. All of the above are correct.

7. Which of the following groups of people provided strong evidence that DNA was the genetic material and carried genetic information during transformation?
   a. Beadle and Tatum
   b. Luria and Delbruck
   c. Avery, MacLeod, and McCarty
   d. Hershey and Chase

8. Which of the following groups of people showed that mutations in bacteria were random and not directed by the environment?
   a. Beadle and Tatum
   b. Luria and Delbruck
   c. Avery, MacLeod, and McCarty
   d. Hershey and Chase

TRUE/FALSE

_____ 1. Although Robert Koch used and published the criteria for establishing a causative link between a particular microorganism and a particular disease, the criteria were actually first developed by his former teacher, Jacob Henle.

_____ 2. Agar is used instead of gelatin to solidify microbial media because agar is not digested by many bacteria while gelatin is.

_____ 3. The first disease to be identified as being caused by a virus was anthrax.

_____ 4. The discovery of viruses and their role in disease was made possible when Charles Chamberland constructed a porcelain filter that would retain bacteria. Using these filters, others found that some disease-causing organisms were not retained by the filter; these were referred to as viruses.

_____ 5. Louis Pasteur was the first to document the use of a vaccination procedure to prevent disease.

_____ 6. Microorganisms can be used in bioremediation to reduce pollution effects.

_____ 7. Genetically engineered microorganisms are used to produce a variety of products including hormones, antibiotics and vaccines.

CRITICAL THINKING

1. Discuss the spontaneous generation theory. Present the evidence that was used to support it and present the evidence that was used to discredit it. How was this theory finally discredited?

2. Describe in detail at least two ways in which microorganisms have a direct and substantial impact on your life (other than as causative agents of disease). Include a discussion of the role(s) of the microorganisms in these processes.

ANSWER KEY

Areas of Microbiology

1. f, 2. j, 3. d, 4. a, 5. g, 6. b, 7. e, 8. h, 9. i, 10. c

Fill in the Blank

1. spontaneous generation; Francesco Redi; Louis Pasteur; John Tyndall 2. Joseph Lister; heat-sterilized; phenol 3. chicken cholera; attenuated; vaccine; Jenner; smallpox 4. Emil von Behring; Shibasaburo Kitasato; humoral; phagocytosis; phagocytes 5. Theodore Schwann; Louis Pasteur; aerobically; anaerobically 6. Sergius Winogradsky; Martinus Beijerinck; enrichment-culture; selective

Multiple Choice

1. c, 2. b, 3. d, 4. a, 5. c, 6. d, 7. c, 8. b

True/False

1. T, 2. T, 3. F, 4. T, 5. F, 6. T, 7. T

2 The Study of Microbial Structure: Microscopy and Specimen Preparation

CHAPTER OVERVIEW

This chapter provides a relatively detailed description and operation principles of the bright-field microscope. Other common types of light microscopes are also described. Following this the students are introduced to various procedures for the preparation and staining of specimens in order to observe general features and/or specific structures. The chapter concludes with a description of the two major types of electron microscopes and the procedures associated with their use.

CHAPTER OBJECTIVES

After reading this chapter you should be able to:

- describe how lenses bend light rays to produce enlarged images of small objects
- describe the various parts of the light microscope and how each part contributes to the functioning of the microscope
- compare the images obtained by the different types of light microscope
- describe the preparation and simple staining of specimens for observation with the light microscope
- describe the Gram-staining procedure and how it is used to categorize bacteria
- describe the basis for the various staining procedures used to visualize specific structures associated with microorganisms
- compare the operation of the transmission and scanning electron microscopes with each other and with light microscopes

CHAPTER OUTLINE

I. Lenses and the Bending of Light
   A. Light is refracted (bent) when passing from one medium to another
   B. Lenses bend light and focus the image at a specific place known as the focal point; the distance between the center of the lens and the focal point is the focal length
II. The Light Microscope
   A. The bright-field microscope produces a dark image against a brighter background
   B. Microscope resolution refers to the ability of a lens to separate or distinguish small objects that are close together; magnification (total) is the product of the magnification of the objective lens and the magnification of the ocular (eyepiece) lens
   C. The major factor determining resolution is the wavelength of light used
   D. The dark-field microscope produces a bright image of the object against a dark background and is used to observe living, unstained preparations
   E. The phase-contrast microscope enhances the contrast between intracellular structures that have slight differences in refractive index, and is an excellent way to observe living cells

F. The fluorescence microscope exposes a specimen to ultraviolet, violet, or blue light and shows a bright image of the object resulting from the fluorescent light emitted by the specimen

III. Preparation and Staining of Specimens

    A. Fixation refers to the process by which internal and external structures are preserved and fixed in position and by which the organism is killed and firmly attached to the microscope slide

        1. Heat fixing is normally used for bacteria; this preserves overall morphology but not internal structures

        2. Chemical fixing is used to protect fine cellular substructure and the morphology of larger, more delicate microorganisms

    B. Dyes and simple staining are used to make internal and external structures of the cell more visible by increasing the contrast with the background

    C. Differential staining is used to divide bacteria into separate groups based on their different reactions to an identical staining procedure

        1. Gram staining is the most widely used differential staining procedure because it divides bacterial species into two roughly equal groups—gram positive and gram negative

            a. The smear is first stained with crystal violet which stains all cells purple

            b. Iodine is used as a mordant to increase the interaction between the cells and the dye

            c. Ethanol or acetone is used to decolorize; this is the differential step because gram-positive bacteria retain the crystal violet whereas gram-negative bacteria lose the crystal violet and become colorless

            d. Safranin is then added as a counterstain to turn the gram-negative bacteria pink while leaving the gram-positive bacteria purple

        2. Acid-fast staining is a differential staining procedure that identifies two medically important species of bacteria—*Mycobacterium tuberculosis,* the causative agent of tuberculosis, and *Mycobacterium leprae,* the causative agent of leprosy

    D. Staining specific structures

        1. Negative staining is widely used to visualize diffuse capsules surrounding the bacteria; those capsules are unstained by the procedure and appear colorless against a stained background

        2. Spore staining is a double staining technique by which bacterial endospores are left one color and the vegetative cell a different color

        3. Flagella staining is a procedure in which mordants are applied to increase the thickness of flagella to make them easier to see after staining

IV. The electron microscope focuses beams of electrons to produce an image with a resolution roughly 1,000 times better than that of the light microscope

    A. In transmission electron microscopy, electrons scatter when they pass through thin sections of a specimen; the transmitted electrons (those that do not scatter) are used to produce an image of the internal structures of the organism

    B. Specimen preparation for the electron microscope involves procedures for cutting thin sections, chemical fixation, and staining with electron-dense materials (analogous to the procedures used for the preparation of specimens for light microscopy)

    C. Other preparation methods include shadowing or freeze-etching

    D. The Scanning electron microscope uses electrons reflected from the surface of a specimen to produce a three-dimensional image of its surface features

    E. The scanning tunneling electron microscope uses a sharp probe to create an accurate three-dimensional image of the surface atoms of a specimen; the resolution is such that individual atoms can be observed

    F. The atomic force microscope is similar to the scanning tunneling microscope in that it uses a scanning probe; however, in this microscope the probe maintains a constant distance from the specimen and is useful for surfaces that do not conduct electricity well

TERMS AND DEFINITIONS

Place the letter of each term in the space next to the definition or description that best matches it.

q	1.	The bending of light rays at the interface of one medium with another
R	2.	A measure of how greatly a substance changes the velocity of light, and a factor determining the direction and magnitude of the bending of light rays
J	3.	The point at which a lens focuses parallel rays of light
I	4.	The distance between the center of a lens and the focal point
b	5.	Conventional microscope that produces a dark image against a brighter background
O	6.	Describes a microscope whose image remains in focus when the objectives are changed
S	7.	The ability of a lens to separate or distinguish small objects that are close together
L	8.	A colorless liquid with the same refractive index as glass; when used between the specimen and the objective lens, it minimizes loss of light which is normally refracted at air-glass interfaces
Z	9.	The distance between the front surface of the lens and the surface of the cover glass (or the specimen) when the specimen is in sharp focus
C	10.	A microscope that produces a bright image of the specimen against a dark background
P	11.	A microscope that converts slight differences in refractive index and cell density into easily detected variations in light intensity
H	12.	A microscope that exposes specimens to ultraviolet, violet, or blue light and forms an image from the resulting light emitted, which has a different wavelength
F	13.	The process by which the internal and external structures of cells and organisms are preserved and maintained in position
W	14.	A staining process in which a single staining agent is used
D	15.	A staining process that divides organisms into two or more separate groups depending on reaction to the same staining procedure
M	16.	A substance that accelerates the reaction of cell structures with a dye so that the cell is more intensely stained
N	17.	A staining procedure in which the background is dark and the organism remains unstained

a. atomic force microscope
b. bright-field microscope
c. dark-field microscope
d. differential staining
e. electron microscope
f. fixation
g. flagella staining
h. fluorescence microscope
i. focal length
j. focal point
k. freeze-etching
l. immersion oil
m. mordant
n. negative staining
o. parfocal
p. phase-contrast microscope
q. refraction
r. refractive index
s. resolution
t. scanning electron microscope (SEM)
u. scanning tunneling electron microscope
v. shadowing
w. simple staining
x. spore staining
y. transmission electron microscope (TEM)
z. working distance

X 18. A staining process in which heat is used to increase the affinity of bacterial endospores to dye; endospores are usually resistant to simple staining procedures

G 19. A staining process that enables the observation of thin, threadlike flagella by increasing their thickness and then staining them

E 20. A microscope that forms an image by focusing a beam of electrons on a specimen; its resolution is approximately 1,000 times better than that of the light microscope

Y 21. An electron microscope that creates an image from transmitted electrons that are not scattered when they pass through a thin section of a specimen

V 22. A staining process in which heavy metals are applied to specimens at an approximately 45° angle; this provides a three-dimensional image similar to shadowing with light

I 23. An electron microscope that creates an image from electrons emitted from the surface of a specimen that has been excited by a beam of focused electrons

K 24. A procedure in which frozen specimens are broken along lines of greatest weakness, usually down the middle of internal membranes; exposed surfaces are then shadowed for production of a better image

U 25. A scanning probe microscope that uses voltage flow between the tip of the probe and the electron clouds of the surface atoms of the specimen

A 26. A scanning probe microscope that maintains a constant distance between the probe tip and the specimen surface by exerting a small amount of force on the tip; it is useful for surfaces that do not conduct electricity well

MICROSCOPE IDENTIFICATION

Using the terms listed below, label the parts of the microscope indicated on the accompanying picture. In the space beside each term provide a *brief* description of its function. (Courtesty of Leica, Inc.)

1. Ocular (eyepiece) lens:

2. Arm:

3. Objective lens:

4. Coarse focusing adjustment knob:

5. Fine focusing adjustment knob:

6. Base:

7. Nosepiece:

8. Stage:

9. Substage condenser:

10. Diaphragm lever:

11. Mechanical stage:

12. Lamp:

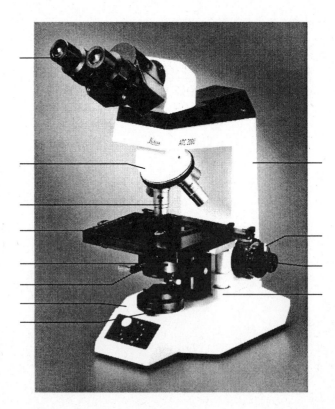

GRAM-STAINING PROCEDURE

Complete the table for the Gram-staining procedure by supplying the missing information.
    Color at Step Completion

Procedure Step	Reagent	Color at Step Completion	
		Gram Positive	Gram Negative
Primary stain	_____	_____	_____
_____	Iodine	_____	Purple
Decolorizer	_____	_____	Colorless
Counterstain	_____	_____	_____

LENSES AND MAGNIFICATION

Complete the table below by filling in the mixing information. Use line one as an example.

	Ocular Lens	Objective Lens	Magnification
1	10×	40×	400×
2	10×	_____	1000×
3	_____	20×	200×
4	15×	50×	_____
5	5×	100×	_____
6	15×	_____	600×
7	10×	4×	_____
8	10×	_____	800×
9	_____	100×	1500×
10	15×	_____	60×

FILL IN THE BLANK

1. The objective lens forms an enlarged image within the microscope called the _____ image. The eyepiece lens further magnifies this image to form the _____ image, which appears to lie just beyond the stage about 25 cm away.
2. Thin films of bacteria which have been air-dried onto a glass microscope slide are called _____.
3. A light microscope uses lenses made of _____ to focus light onto a specimen, while an electron microscope uses _____ lenses to focus beams of electrons onto a specimen.
4. Special dyes called _____ are used in fluorescence microscopy. These dyes are excited by light with a specific wavelength and emit light with a _____ wavelength thus having less energy than the light originally absorbed. In this way the dye gives up its trapped _____ and returns to a more stable state.
5. The presence of diffuse capsules surrounding many bacteria is commonly revealed by _____ staining, in which the background is stained dark. The cell can also be _____ for greater visibility, leaving the capsule colorless.
6. In scanning electron microscopy, when the electron beam strikes a _____ area, a large number of secondary electrons enter the detector; in contrast, fewer electrons escape a _____ in the surface to reach the detector.

MULTIPLE CHOICE

For each of the questions below select the *one best* answer.

1. Acid-fast organisms such as *Mycobacterium tuberculosis* resist decolorization by acid-alcohol solutions because of the high concentration of _____ in their cell walls.
   a. proteins
   b. carbohydrates
   c. lipids
   d. peptidoglycan
2. Smears are heat-fixed prior to staining in order to
   a. kill the organism.
   b. preserve the internal structures.
   c. attach the organism firmly to the slide.
   d. All of the above are correct.
3. Small morphological features within the cell interior are best visualized by using a(n)
   a. light microscope.
   b. transmission electron microscope.
   c. dark-field microscope.
   d. scanning electron microscope.
4. Transmission electron microscopy requires the use of thin slices of a microbial specimen. The thickness of the specimen should be approximately
   a. 20 to 100 mm.
   b. 100 to 200 nm.
   c. 20 to 100 nm.
   d. 0.2 to 10 nm.
5. Spreading a transmission electron microscopy specimen out in a thin film with uranyl acetate, which does not penetrate the specimen, is called
   a. negative staining.
   b. shadowing.
   c. freeze-etching.
   d. simple staining.
6. Surface features of an organism are best revealed using
   a. fluorescence microscopy.
   b. phase-contrast microscopy.
   c. scanning electron microscopy.
   d. transmission electron microscopy.

7. As the magnification of a series of objective lenses increase, the working distance
   a. increases.
   b. decreases.
   c. stays the same.
   d. cannot be predicted.
8. The Gram-staining procedure differentiates bacteria based on the chemical composition of the
   a. cytoplasmic membrane.
   b. cell wall.
   c. cytoplasm.
   d. chromosome.
9. The distance between the focal point of a lens and the center of the lens is called the
   a. working distance.
   b. numerical aperture.
   c. focal length.
   d. parallax distance.
10. A microscope that keeps objects in focus when the objective lens is changed is said to be
    a. equifocal.
    b. parfocal.
    c. optically constant.
    d. focally constant.
11. Denser regions of a specimen scatter _____ electrons and therefore appear _____ in the image projected onto the screen of a transmission electron microscope.
    a. fewer, darker
    b. fewer, lighter
    c. more, darker
    d. more, lighter

TRUE/FALSE

_____ 1. Light is refracted at the interface between two materials with different refractive indexes because the velocity of light is altered.

_____ 2. Resolution becomes greater as the wavelength of the illuminating light decreases.

_____ 3. Immersion oil used with the 100× objective lens increases the amount of light that comes to the objective lens because it has the same refractive index as glass.

_____ 4. Phase-contrast microscopy enhances density differences among internal cellular structures and therefore allows these structures to be visualized without stains or dyes.

_____ 5. Basic dyes are cationic (positively charged) and are commonly used to stain bacteria since the surfaces of these organisms are usually negatively charged.

_____ 6. The Gram-staining procedure is one of the most widely used differential stains because it divides bacterial species into two roughly equal groups.

_____ 7. Since transmission electron microscopy uses electrons rather than light, it is not necessary to stain biological specimens before observation.

_____ 8. Freeze-etching minimizes the production of artifacts since the cells are not subjected to chemical fixation, dehydration, or plastic embedding.

_____ 9. The resolution of a microscope is not related to its magnification. Therefore, although it is possible to build a microscope capable of 10,000× magnification, it would only be magnifying a blur.

_____ 10. Scanning tunneling electron microscopes can be used to visualize individual atoms.

CRITICAL THINKING

1. Explain why it is possible to increase the magnification of the light microscope above 1,500× and yet not be able to see any additional details.

2. Compare and contrast electron microscopy with light microscopy. Include in your answer the operation of the instruments, the degree of magnification and resolution possible, and the procedures used for preparation, fixation, and staining of specimens. What major advances in our knowledge of cell structure were made possible by the invention of the electron microscope that were not possible with the light microscope?

ANSWER KEY

Terms and Definitions

1. q, 2. r, 3. j, 4. i, 5. b, 6. o, 7. s, 8. l, 9. z, 10. c, 11. p, 12. h, 13. f, 14. w, 15. d, 16. m, 17. n, 18. x, 19. g, 20. e, 21. y, 22. v, 23. t, 24. k, 25. u, 26. a

Lenses and Magnification

2. 100×, 3. 10×, 4. 600×, 5. 500×, 6. 20×, 7. 40×, 8. 80×, 9. 15×, 10. 4×

Fill in the Blank

1. real; virtual 2. smears 3. glass; magnets 4. fluorochromes; longer; energy 5. negative; counterstained 6. raised; depression

Multiple Choice

1. c, 2. d, 3. b, 4. c, 5. a, 6. c, 7. b, 8. b, 9. c, 10. b, 11. c

True/False

1. T, 2. T, 3. T, 4. T, 5. T, 6. T, 7. F, 8. T, 9. T, 10. T

3 Procaryotic Cell Structure and Function

CHAPTER OVERVIEW

This chapter provides a description of the procaryotic cell, beginning with the general features of size, shape, and arrangement. Then the general features of biological membranes and the specific features of procaryotic membranes are given. Important internal structures of procaryotes, such as the cytoplasmic matrix, the ribosomes, the inclusion bodies, and the nucleoid are described, in addition to structures external to the cell, such as the cell wall, capsule, pili, and flagella. The differences between the cell walls of gram-positive organisms and gram-negative organisms are discussed and the mechanism of this differential staining reaction is explained. The chapter concludes with a discussion of bacterial chemotaxis and bacterial endospores.

CHAPTER OBJECTIVES

After reading this chapter you should be able to:

- describe the various sizes, shapes, and cellular arrangements exhibited by bacteria
- describe the bacterial plasma membrane and the limited internal membrane structures found in procaryotes
- describe the appearance, composition, and function of the various internal structures found in procaryotic organisms (such as inclusion bodies, ribosomes, and the nucleoid)
- define the composition of gram-positive and gram-negative cell walls and explain how these differences contribute to the differential reaction to the Gram-staining procedure
- describe external structures such as capsules, fimbriae and flagella
- diagram and describe the various arrangements of bacterial flagella
- describe how bacteria use their locomotive ability to swim toward chemical attractants and away from chemical repellents
- describe the production of the bacterial endospore and how it enables spore-forming bacteria to survive harsh environmental conditions and renew growth when the environment becomes conducive to growth

CHAPTER OUTLINE

I. An Overview of Procaryotic Cell Structure
   A. Size, shape, and arrangement
      1. Procaryotes come in a variety of shapes including spheres (cocci), rods (bacilli), ovals (coccobacilli), curved rods (vibrios), rigid helices (spirilla), and flexible helices (spirochetes)
      2. During the reproductive process, some cells remain attached to each other to form chains, clusters, square planar configurations (tetrads), or cubic configurations (sarcinae)
      3. A few bacteria are flat and some lack a single, characteristic form and are called pleiomorphic
   B. Procaryotic cells vary in size although they are generally smaller than most eucaryotic cells; recently, however, a large procaryote, *Epulopiscium fisheloni* was discovered that grows as large as 600μm $\times$ 80μm, a littler smaller than a printed hyphen
   C. Procaryotic cells contain a variety of internal structures

D. Not all structures are found in every genus, but procaryotes are consistent in their fundamental structure and most important components

E. Procaryotic cell organization—procaryotes are morphologically distinct from eucaryotic cells and have fewer internal structures.

II. Procaryotic Cell Membranes
    A. The plasma membrane
        1. The plasma membrane consists of a phospholipid bilayer with hydrophilic surfaces (interact with water) and a hydrophobic interior (insoluble in water); such asymmetric molecules are said to be amphipathic
        2. Archaeobacterial membranes have a monolayer instead of a bilayer structure
        3. Proteins are associated with the membrane and may be either peripheral (loosely associated and easily removed) or integral (embedded within the membrane and not easily removed)
        4. The membrane is highly organized, asymmetric, flexible, and dynamic
        5. The plasma membrane serves several functions for the cell
            a. It retains the cytoplasm and separates the cell from its environment
            b. It serves as a selectively permeable barrier, allowing some molecules to pass into or out of the cell while preventing passage of other molecules
            c. It is the location of a variety of crucial metabolic processes including respiration, photosynthesis, lipid synthesis, and cell wall synthesis
            d. It may contain special receptor molecules that enable bacterial detection of and response to chemicals in the surroundings

    B. Internal membrane systems
        1. Mesosomes are structures formed by invaginations of the plasma membrane that may play a role in cell wall formation during division, in chromosome replication and distribution, and in secretory processes; however, mesosomes may be artifacts generated during chemical fixation for electron microscopy
        2. Photosynthetic bacteria may have complex infoldings of the plasma membrane that increase the surface area available for photosynthesis
        3. Bacteria with high respiratory activity may also have extensive infoldings that provide a large surface area for greater metabolic activity
        4. These internal membranes may be aggregates of spherical vesicles, flattened vesicles, or tubular membranes

III. The Cytoplasmic Matrix
    A. The cytoplasmic matrix is the substance between the membrane and the nucleoid
    B. It is featureless in electron micrographs but is often packed with ribosomes and inclusion bodies
    C. Despite the homogenous appearance, the matrix is highly organized with respect to protein location
    D. Inclusion Bodies
        1. Inclusion bodies are granules of organic or inorganic material that are stockpiled by the cell for future use
            a. Some are not bounded by a membrane
            b. Others are enclosed by a single-layered membrane
        2. Gas vacuoles are a type of inclusion body found in cyanobacteria and some other aquatic forms; they provide buoyancy for these organisms and keep them at or near the surface of their aqueous habitat
    E. Ribosomes
        1. Ribosomes are complex structures consisting of protein and RNA
        2. They are responsible for the synthesis of cellular proteins
        3. Procaryotic ribosomes are similar in structure to, but smaller and less complex than, eucaryotic ribosomes
    F. Molecular Chaperones
        1. Helper proteins that aid the folding of nascent polypeptides during protein synthesis

2. Many molecular chaperones are heat-shock proteins that increase in concentration after cells are subjected to environmental stress; they promote proper folding of new proteins that are replacing heat-damaged existing proteins
3. Molecular chaperones also function to keep secretory proteins in an export-competent state until they are translocated across the plasma membrane

VI. The Nucleoid—an irregularly shaped region in which the single circular chromosome of the procaryote will be found; in most procaryotes it is not bounded by a membrane, but is sometimes found to be associated with the plasma membrane or with mesosomes; two genera of planctomycete bacteria have been shown to have membrane-bounded DNA-containing regions

A. The bacterial chromosome is an efficiently packed, closed circular DNA molecule that is looped and coiled extensively
B. In actively growing bacteria, the nucleoid has projections that extend into the cytoplasmic matrix; these projections probably contain DNA being actively transcribed
C. Plasmids are small, closed circular DNA molecules that can exist and replicate independently of the bacterial chromosome; they are not required for bacterial growth and reproduction, but they may carry genes that give the bacterium a selective advantage (e.g., drug resistance, enhanced metabolic activities, etc.)

VII. The Procaryotic Cell Wall—a rigid structure that results in the characteristic shapes of the various procaryotes and protects them from osmotic lysis

A. Peptidoglycan (murein) is a polysaccharide polymer found in procaryotic cell walls that consists of polysaccharide chains cross-linked by peptide bridges
B. Gram-positive cell walls consist of a thick layer of peptidoglycan and large amounts of teichoic acids
C. Gram-negative cell walls are more complex; they consist of a thin layer of peptidoglycan surrounded by an outer membrane composed of lipids, lipoproteins, and a large molecule known as lipopolysaccharide (LPS). There are no teichoic acids in gram-negative cell walls.
D. Archeaobacteria cell walls lack peptidoglycan and are composed of proteins, glycoproteins, or polysaccharides
E. The periplasmic space (or periplasm) is the gap between the plasma membrane and the cell wall (gram-positive organisms) or between the plasma membrane and the outer membrane (gram-negative organisms)
  1. Periplasmic enzymes are found in the periplasm of gram-negative bacteria; they generally participate in nutrient acquisition
  2. Exoenzymes are secreted by gram-positive bacteria and perform many of the same functions that periplasmic enzymes do for gram-negative bacteria
  3. Denitrifying and chemolithoautotrophic bacteria often have electron transport proteins in periplasm
  4. Enzymes involved in peptidoglycan synthesis and in the modification of toxic compounds may also be found in the periplasm
F. Adhesion sites are sites of direct contact or possibly true membrane fusions between the plasma membrane and the outer membrane of gram-negative bacteria; it has been proposed that substances can move into the cell through these sites rather than travel through the periplasm
G. The outer membrane is more permeable than the plasma membrane because of porin proteins that form channels through which small molecules (600-700 daltons) can pass
H. The mechanism of Gram staining involves constricting the thick peptidoglycan layer of gram-positive cells, thereby preventing the loss of the crystal violet stain during the brief decolorization step; the thinner, less cross-linked peptidoglycan layer of gram-negative organisms cannot retain the stain as well, and these bacteria are thus more readily decolorized when treated with alcohol
I. The cell wall and osmotic protection: The cell wall prevents swelling and lysis of bacteria in hypotonic solutions. However, in hypertonic habitats, the plasma membrane shrinks away from the cell wall in a process known as plasmolysis

19

VIII.   Components External to the Cell Wall
    A.   Capsules and slime layers are layers of polysaccharides lying outside the cell wall; they protect the bacteria from phagocytosis, viral infection, pH fluctuations, osmotic stress, hydrolytic enzymes, or the predacious bacterium *Bdellovibrio*
        1.   Capsules are well organized
        2.   Slime layers are diffuse and unorganized
    B.   A glycocalyx is a network of polysaccharides extending from the surface of bacteria and other cells
    C.   S layers are regularly structured layers of protein or glycoprotein
    D.   S layers are common among the archeaobacteria where it may be the only structure outside the plasma membrane
    E.   Pili and fimbriae are short, thin, hairlike appendages that mediate bacterial attachment to surfaces (fimbriae) or to other bacteria during sexual mating (pili)
    F.   Flagella and motility
        1.   Flagella are threadlike locomotor appendages extending outward from the plasma membrane and cell wall
        2.   Flagella may be arranged in various patterns:
            a.   Monotrichous—a single flagellum
            b.   Amphitrichous—a single flagellum at each pole
            c.   Lophotrichous—a cluster (tuft) of flagella at one or both ends
            d.   Peritrichous—a relatively even distribution of flagella over the entire surface of the bacterium
        3.   Flagellar ultrastructure: The flagellum consists of a hollow filament composed of a single protein known as flagellin. The hook is a short curved segment that links the filament to the basal body, a series of rings that drives flagellar rotation.
        4.   Flagellar synthesis involves many genes for the hook and basal body, as well as the gene for flagellin. New molecules of flagellin are transported through the hollow filament so that the growth of the flagellum is from the tip, not from the base.
        5.   The mechanism of flagellar movement appears to be rotation; the hook and helical structure of the flagellum causes the flagellum to act as a propeller, thus driving the bacterium through its watery environment
            a.   Counterclockwise rotation causes forward motion (called a run)
            b.   Clockwise rotation disrupts forward motion (resulting in a tumble)
        6.   Axial filaments cause flexing and spinning movements that allow spirochetes to move
    H.   Gliding motility is a mechanism used by some procaryotes by which they coast along solid surfaces; no visible structure is associated with this form of motility
IX.   Chemotaxis—directed movement of bacteria either towards a chemical attractant or away from a chemical repellent
    A.   The concentrations of these materials is detected by chemoreceptors in the surfaces of the bacteria
    B.   Directional travel toward a chemoattractant is caused by lowering the frequency of tumbles (twiddles), thereby lengthening the runs when traveling up the gradient, but allowing tumbling to occur at normal frequency when traveling down the gradient
    C.   Directional travel away from a chemorepellent involves similar but opposite responses
    D.   The mechanism of control of tumbles and runs is complex with several protein intermediates; nevertheless, it is fast with responses occurring in as little as 200 meters/second
    E.   Responses are triggered by methylation and/or phosphorylation of target proteins called methyl-accepting chemotaxis proteins (MCPs) to cycle them between active and inactive forms
X.   The Bacterial Endospore—a special, resistant, dormant structure formed by some bacteria, which enables them to resist harsh environmental conditions
    A.   Spore formation (sporulation) normally commences when growth ceases because of lack of nutrients, and is a complex, multistage process

B. Transformation of dormant spores into active vegetative cells is also a complex, multistage process that includes activation (preparation) of the spore, germination (breaking of the spores dormant state), and outgrowth (emergence of the new vegetative cell)

TERMS AND DEFINITIONS

Place the letter of each term in the space next to the definition or description that best matches it.

_D_ 1. Bacteria that are roughly spherical in shape
_G_ 2. Bacteria that are rod shaped
_H_ 3. Bacteria that are oval in shape
_JJ_ 4. Curved rod-shaped bacteria
_P_ 5. Long multinucleated filaments
_U_ 6. Branched network of hyphae
_FF_ 7. Long rod-shaped bacteria that are twisted into rigid helices
_GG_ 8. Long rod-shaped bacteria that are twisted into flexible helices
_BB_ 9. Bacteria that lack a single characteristic shape and that therefore vary in shape
_B_ 10. Structurally asymmetric lipids with polar and nonpolar ends
_N_ 11. Molecules or regions of molecules that readily interact with water
_O_ 12. Molecules or regions of molecules that are insoluble in water, or otherwise do not readily react with water
_X_ 13. Proteins that are loosely associated with a membrane and that can therefore be easily removed
_R_ 14. Proteins that are not easily extracted from membranes and that are insoluble in aqueous solutions when freed of lipids
_CC_ 15. Plasma membrane and everything contained within
_ii_ 16. A unit of measure of the sedimentation velocity in a centrifuge
_V_ 17. The nonmembrane-bound region of a procaryotic cell in which the DNA is located
_Z_ 18. A circular, double-stranded DNA molecule in procaryotes that can exist and replicate independently of the chromosome
_Y_ 19. A gap between the plasma membrane and the cell wall of bacteria
_J_ 20. Enzymes that are secreted out of the cell to aid in the acquisition and digestion of nutrients from the environment
_W_ 21. Movement of water across selectively permeable membranes from dilute solutions (higher water concentration) to more concentrated solutions (lower water concentration)
_S_ 22. Bursting of cells that occurs when cells are placed in hypotonic solutions so that water flows in the cell

a. adhesion sites
b. amphipathic
c. axial filament
d. bacilli
e. capsule
f. chemotaxis
g. cocci
h. coccobacilli
i. endospore
j. exoenzymes
k. fimbriae
l. germination
m. glycocalyx
n. hydrophilic
o. hydrophobic
p. hyphae
q. inclusion bodies
r. integral proteins
s. lysis
t. molecular chaperones
u. mycelia
v. nucleoid
w. osmosis
x. peripheral proteins
y. periplasmic space
z. plasmid
aa. plasmolysis
bb. pleomorphic
cc. protoplast
dd. structured (S) layer
ee. slime layer
ff. spirilla
gg. spirochete
hh. sporogenesis (sporulation)
ii. Svedberg
jj. vibrio

23. Shrinkage of the plasma membrane away from the cell wall that occurs when cells are placed in hypertonic solutions so that water flows out of the cell
24. Well-organized polysaccharide of material outside the cell wall that is not easily washed off
25. Diffuse, unorganized polysaccharides outside the cell wall that is easily removed
26. Protein or glycoprotein layer that exhibits a pattern not unlike floor tiles
27. Hairlike appendages that are thinner than flagella and are usually involved in attachment rather than motility
28. A flexible filament that enables motility for spirochetes
29. Movement towards chemical attractants or away from repellents
30. A structure formed by some species of bacteria that is resistant to some environmental stresses
31. The process of forming endospores within a vegetative cell
32. The process of forming a vegetative cell from an endospore
33. Granules of organic or inorganic material that are stockpiled by the cell for future use
34. Helper proteins that aid in the proper folding of nascent polypeptide chains
35. Site of direct contact or fusion between the outer membrane and the plasma membrane of gram-negative bacteria
36. A network of polysaccharides extending from the surface of bacteria and other cells

IDENTIFICATION OF PROCARYOTIC STRUCTURES

Label the appropriate structures indicated on the accompanying figure. The terms to be used are listed below. In the space beside each term provide a *brief* description of its structure and/or function.

1. Capsule:

2. Cell wall:

3. Flagellum:

4. Inclusion body:

5. Mesosome:

6. Nucleoid:

7. Periplasmic space:

8. Plasma membrane:

9. Ribosome:

10. Surface proteins:

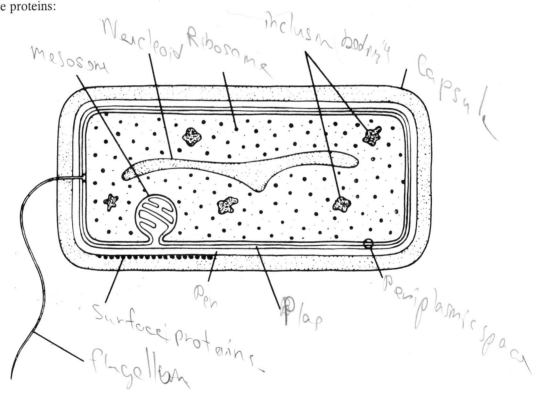

FLAGELLAR ARRANGEMENTS

Label each flagellar arrangement with the appropriate name.

1. _____

2. _____

3. _____

4. _____

BACTERIAL SHAPES

Label each bacterial shape.

1. _____

2. _____

3. _____

4. _____

5. _____

6. _____

Rigid

Flexible

BACTERIAL ARRANGEMENTS

Label each arrangement of spherical bacteria with the appropriate term.

1. _____

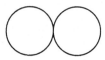

2. _____

3. _____

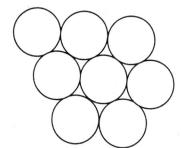

4. _____

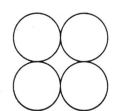

Planar configuration of 4 cells

5. _____

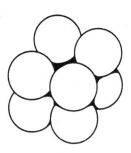

Cubic configuration of 8 cells

FILL IN THE BLANK

1. The _____ encompasses the cytoplasm of both procaryotic and eucaryotic cells. It is the chief point of contact with the cell's _____ and thus is responsible for much of its relationship with the outside world.

2. The amphipathic property of lipids enables them to form a _____ in membranes. Thus the outer surface is _____ while the _____ ends are buried in the interior away from the surrounding water.

3. The most widely accepted current structural model for membranes is called the _____ model.

4. The most common type of invagination of the plasma membrane is called a(n) _____. The function of this structure is not completely known but it is believed that it has a role in the formation of the _____ and/or the distribution of _____ during cell division. However, there is some concern that it may be a(n) _____ formed during fixation of bacteria for _____ _____ _____.

5. When placed in a(n) _____ environment, a cell will lose water so that the plasma membrane shrinks away from the cell wall. This is referred to as _____.

6. The space between the cell wall and the plasma membrane is called the _____. In gram-negative bacteria, this space contains many proteins that participate in _____ acquisition. In gram-positive bacteria, this same function is performed by _____ that are secreted out of the cell.

7. Solutes are less concentrated in _____ environments than in cells. When cells are placed in this type of environment, water will flow _____ causing the cell to swell and eventually burst if the plasma membrane is unprotected because the cell lacks a _____.

8. L-forms may arise by the complete (or partial) loss of the _____. If the loss is complete, the L-form will be stable because bacteria need a preexisting _____ in order to construct new _____.

9. The direction of flagellar rotation determines the nature of bacterial movement. For bacteria with peritrichous flagella, rotation in the _____ direction propels the bacterium forward, while rotation in the _____ direction disrupts the bundle of flagella and causes the bacterium to _____.

10. A bacterium travels in a straight line, a _____, for a few seconds; then it will stop and _____; this is then followed by a _____ in a different direction. When exposed to a chemical attractant gradient, it tumbles _____ frequently when traveling up the gradient, but tumbles at _____ frequency if moving down the gradient.

11. Molecular _____ increase in concentration after a cell is subjected to environmental stress. They promote proper folding of new proteins that are replacing _____-_____ proteins; for this reason, they are also known as _____-_____ proteins.

MULTIPLE CHOICE

For each of the questions below select the *one best* answer.

1. Which of the following is not a function of the plasma membrane?
   a. It retains the cytoplasm.
   b. It acts as a selectively permeable barrier, allowing some molecules to pass while preventing the movement of others.
   c. It maintains the various shapes of the bacteria.
   d. It provides the location for a variety of metabolic processes, including respiration and photosynthesis.

2. Which of the following is not true of bacterial plasmids?
   a. They can exist and replicate independently of the chromosome.
   b. They are required for host growth and/or reproduction.
   c. They may carry genes for drug resistance.
   d. They may carry genes that give the bacterium new useful metabolic activities.

3. Lipopolysaccharide (LPS), which is found in the outer membrane of gram-negative bacteria, is also known as
   a. teichoic acid.
   b. exotoxin.
   c. endotoxin.
   d. murein.

4. Gram-positive cells retain the primary stain while gram-negative cells do not because
   a. the stain is bound to the thicker peptidoglycan layer.
   b. the alcohol removes the lipids of the outer membrane, and thus the trapped stain, from gram-negative bacteria.
   c. the alcohol shrinks the pores of the thick peptidoglycan layer of gram-positive cells.
   d. Both (b) and (c) are correct.

5. Penicillin inhibits cell wall synthesis, but cells will continue to grow normally in the presence of penicillin if they are maintained in a(n) _____ environment.
   a. hypotonic
   b. isotonic
   c. hypertonic
   d. nonpolar

6. A network of polysaccharides extending from the surface of bacteria that aids in the attachment of bacteria to other surfaces is referred to as a(n)
   a. lipoteichoic acid.
   b. lipopolysaccharide.
   c. outer membrane.
   d. glycocalyx.

7. Which of the following is true of capsules?
   a. They help bacteria escape phagocytosis by host phagocytic cells.
   b. They retain water and protect bacteria from desiccation.
   c. They prevent the entry of bacterial viruses.
   d. All of the above are true of capsules.

8. In the presence of both attractants and repellents, bacteria will
   a. move towards the attractant.
   b. compare both signals and respond to the chemical with the most effective concentration.
   c. move away from the repellent.
   d. move in a random fashion.

9. Endospores are of great practical importance in industrial and medical microbiology because
   a. they are resistant to harsh environments and thus increase the survival of sporeforming bacteria as compared to that of bacteria that do not form spores.
   b. many sporeformers are dangerous pathogens.
   c. Both (a) and (b) are correct.
   d. Neither (a) nor (b) is correct.

10. Extensive invaginations of the plasma membrane are usually found in
    a. cyanobacteria and other photosynthetic bacteria.
    b. bacteria with high respiratory activity.
    c. Both (a) and (b) are correct.
    d. Neither (a) nor (b) is correct.

11. Which of the following is a function of molecular chaperones?
    a. ensure proper folding of nascent polypeptide chains
    b. maintain secretory proteins in an export-competent state
    c. Both (a) and (b) are correct.
    d. Neither (a) nor (b) is correct.

TRUE/FALSE

_____ 1. The cell membrane is a highly organized, rigid, and relatively static structure.

_____ 2. Gas vacuoles are membranous structures that regulate buoyancy in cyanobacteria and other photosynthetic bacteria. However, even though they are membranous, their vesicle walls contain no lipids.

_____ 3. Sedimentation coefficients (measured in Svedberg units) are directly proportional to the molecular weight of a particle and are unaffected by volume or shape.

_____ 4. Gram-positive cell walls have a thick layer of peptidoglycan but a rather simple overall structure, while gram-negative cell walls have a thinner peptidoglycan layer but a more complex overall structure.

_____ 5. Lipoteichoic acids are teichoic acids that are connected to lipids in the plasma membrane.

_____ 6. Porin proteins are found in the outer membrane and function in the transport of molecules into the cell.

_____ 7. Mycoplasmas must be maintained in an isotonic environment because they lack a cell wall.

_____ 8. In the construction of flagella, the protein flagellin is added at the base of the flagellum so that first the tip is constructed and then pushed outward by the addition of new material at the base.

_____ 9. Bacterial flagellar rotation appears to be powered by proton movement, not by ATP hydrolysis.

_____ 10. Procaryotic organisms are so remarkably uniform that nearly all genera contain all of the structures described in this chapter.

_____ 11. Pleiomorphic bacteria are uniformly club-shaped.

_____ 12. Chemotactic receptors are directly coupled to the basal body of the cells flagella.

_____ 13. Archaeobacterial membranes have a monolayer rather than a bilayer structure.

_____ 14. The cytoplasmic matrix is highly organized with respect to protein location.

_____ 15. Gliding motility is a means by which bacteria coast along solid surfaces; no visible structures have been associated with this type of movement.

CRITICAL THINKING

1. Discuss why the plasma membrane is considered the external boundary of the cell, even though other structures outside of the plasma membrane are considered part of the cellular anatomy.

2. Discuss the nature of flagellar-mediated movement of bacteria. In particular, discuss how the direction of rotation affects the direction of movement, and speculate how this movement is altered in the presence of chemical attractants and repellents for which the bacterium has the appropriate chemoreceptors. Give plausible mechanisms of the response to chemical attractants and/or repellents.

3. Discuss the evidence that suggests that mesosomes may be artifacts of preparation. Do you agree with this? Why or why not?

ANSWER KEY

Terms and Definitions

1. g, 2. d, 3. h, 4. jj, 5. p, 6. u, 7. ff, 8. gg, 9. bb, 10. b, 11. n, 12. o, 13. x, 14. r, 15. cc, 16. ii, 17. v, 18. z, 19. y, 20. j, 21. w, 22. s, 23. aa, 24. e, 25. ee, 26. dd, 27. k, 28. c, 29. f, 30. i, 31. hh, 32. l, 33. q, 34. t, 35. a, 36. m

Flagellar Arrangements

1. monotrichous, 2. amphitrichous, 3. lophotrichous, 4. peritrichous

Bacterial Shapes

1. coccus, 2. coccobacillus, 3. bacillus, 4. vibrio, 5. spirilla, 6. spirochete

Bacterial Arrangements

1. diplococcus, 2. streptococcus, 3. staphylococcus, 4. tetrad, 5. sarcinae

Fill in the Blank

1. plasma membrane; environment 2. bilayer; hydrophilic; hydrophobic 3. fluid mosaic 4. mesosomes; cell wall; chromosomes; artifacts; transmission electron microscopy 5. hypertonic; plasmolysis 6. periplasmic space; nutrient; exoenzymes 7. hypotonic; inward; cell wall 8. cell wall; cell wall; peptidoglycan 9. counterclockwise; clockwise; tumble (twiddle) 10. run; tumble (twiddle); run; less; normal 11. chaperones; heat-damaged; heat-shock

Multiple Choice

1. c, 2. b, 3. c, 4. d, 5. b, 6. d, 7. d, 8. b, 9. c, 10. c, 11. c

True/False

1. T, 2. T, 3. F, 4. T, 5. T, 6. T, 7. F, 8. F, 9. T, 10. F, 11. F, 12. F, 13. T, 14. T, 15. T

4 Eucaryotic Cell Structure and Function

CHAPTER OVERVIEW

This chapter focuses on eucaryotic cell structure and function. Although procaryotic organisms are immensely important in microbiology, eucaryotic microorganisms—such as fungi, algae, and protozoa—are also prominent members of many ecosystems, and some have medical significance as etiological agents of disease as well. The chapter concludes with a comparison of eucaryotic and procaryotic cells.

CHAPTER OBJECTIVES

After reading this chapter you should be able to:

- discuss the various elements of the cytoskeleton (microfilaments, intermediate filaments, microtubules) with regard to their structure and various functions within the cell
- discuss the composition, structure, and function of each of the internal organelles, such as the endoplasmic reticulum, Golgi apparatus, lysosomes, ribosomes, mitochondria, chloroplasts, nucleus, and nucleolus
- discuss the mechanism of endocytosis and the difference between phagocytosis and pinocytosis
- compare mitosis and meiosis
- compare and contrast procaryotes and eucaryotes

CHAPTER OUTLINE

I. An Overview of Eucaryotic Cell Structure
   A. Eucaryotic cells have membrane-delimited nuclei
   B. Eucaryotic cells have membrane-bound organelles that perform specific functions within the cells; this allows simultaneous independent control
   C. The large membrane surface area of eucaryotic cells allows greater respiratory and photosynthetic activity
II. The Cytoplasmic Matrix, Microfilaments, Intermediate Filaments, and Microtubules
   A. The cytoplasmic matrix, although superficially featureless, provides the complex environment required for many cellular activities
   B. Microfilaments (4 to 7 nm) may be scattered throughout the matrix or organized into networks and parallel arrays; they play a major role in cell motion and cell shape changes
   C. Microtubules are hollow cylinders (25 nm) that help maintain cell shape, that are involved (with microfilaments) in cellular movement, and that also participate in intracellular transport of substances
   D. Microtubules also form the mitotic spindle during cell division and are present in cilia and flagella
   E. Intermediate filaments (8 to 10 nm) are major components of the cytoskeleton, an intricate network of interconnected filaments that helps maintain cell shape and contributes to cellular movement

III. The Endoplasmic Reticulum (ER)—a complex set of internal membranes that may have ribosomes attached (rough or granular endoplasmic reticulum; RER or GER), or that may be devoid of ribosomes (smooth or agranular endoplasmic reticulum; SER or AER)
    A. The ER transports proteins, lipids, and other materials within the cell
    B. The ER is a major site of cell membrane synthesis
    C. Lipids and many proteins are synthesized by ER-associated enzymes and ribosomes
    D. New ER is produced through expansion of old ER

IV. The Golgi Apparatus—a set of membrane sacs (cisternae) that is involved in the modification, packaging, and secretion of materials; they exist in stacks called dictyosomes
    A. Materials move to the cis (forming) face of the Golgi apparatus from the ER
    B. These materials are then transported from the cis to the trans (maturing) cisternae by vesicles that bud off from the cisternal edges and move to the next sac
    C. As the materials move through the Golgi apparatus, they undergo further modification
    D. Finally, they bud off from the trans face for transport to their final destination

V. Lysosomes and Endocytosis
    A. Lysosomes are membrane-bound vesicles that contain enzymes needed for intracellular digestion of all types of macromolecules
        1. Primary lysosomes are newly formed lysosomes
        2. Secondary lysosomes (food vacuoles) result from the fusion of primary lysosomes with phagocytic vesicles (endosomes)
        3. Residual bodies are lysosomes that have accumulated large quantities of indigestible material
    B. Endocytosis is the process in which the cell takes up solutes or particles by enclosing them in vesicles (endosomes) pinched off from the plasma membrane
        1. Phagocytosis—endocytosis of large particles by engulfing them into a phagocytic vacuole
        2. Pinocytosis—endocytosis of small amounts of liquid with its solute molecules
    C. Lysosomes fuse with endosomes in order to digest the materials within the endosome
    D. Lysosomes join with phagosomes for defensive purposes as well as to acquire nutrients
    E. Autophagic vacuoles are lysosomes that selectively digest portions of the cell's own cytoplasm as part of the normal turnover of cellular components

VI. Eucaryotic Ribosomes—generally larger and more complex than procaryotic ribosomes; however, like their procaryotic counterparts, they are responsible for the synthesis of cellular proteins
    A. Ribosomes may be attached to the ER or they may be free
    B. ER-associated ribosomes synthesize integral membrane proteins or proteins that are secreted out of the cell
    C. Free ribosomes synthesize nonsecretory, nonmembrane proteins
    D. Molecular chaperones aid the proper folding of proteins after synthesis and also assist the transport of proteins into eucaryotic organelles such as mitochondria

VII. Mitochondria—the site of tricarboxylic acid cycle activity and the generation of ATP by electron transport and oxidative phosphorylation
    A. Mitochondria have both an inner membrane and an outer membrane enclosing a fluid matrix
    B. The inner and outer membrane have different lipids and enzymes
    C. The enzymes of the tricarboxylic acid cycle and the β-oxidation pathway for fatty acids are located within the matrix
    D. Electron transport and oxidative phosphorylation occur only on the inner mitochondrial membrane
    E. Mitochondria use their own DNA and their own ribosomes to synthesize some of their proteins

VIII. Chloroplasts—the site of both the light and the dark reactions of photosynthesis
    A. Chloroplasts have an outer membrane and an inner membrane system of flattened sacs called thylakoids that often form stacks known as grana; the fluid matrix compartment is called the stroma
    B. The formation of carbohydrate from carbon dioxide and water (dark reaction) occurs in the stroma

C. The trapping of light energy to generate ATP, NADPH, and oxygen (light reaction) occurs in the thylakoid membranes of the grana

IX. The Nucleus and Cell Division
    A. Nuclei are membrane-bound structures that house the chromatin (genetic material) of the cell
        1. Euchromatin is loosely organized and genetically active
        2. Heterochromatin is tightly coiled and contains dormant genes
    B. The nuclear envelope is a double-membrane structure penetrated by nuclear pores that allow materials to be transported into or out of the nucleus
    C. The nuclear lamina, a network of intermediate filaments, lies against the inner surface of the nuclear envelope and supports it
    D. The nucleolus is a highly active region of the chromatin involved in the synthesis of ribosomes
    E. Mitosis is a process of nuclear division in which the (duplicated) genetic material is distributed equally to two daughter nuclei so that each has a full set of chromosomes and genes
    F. Meiosis is a complex, two-stage process of nuclear division in which the number of chromosomes in the resulting daughter cells is reduced from the normal (diploid) number to one-half of that number (haploid)
    G. Cytokinesis is the process by which the cytoplasm and its components are distributed to the new daughter cells; it usually occurs in association with, but independently from, mitosis and/or meiosis

X. External Cell Coverings
    A. Some cells have a rigid cell wall
    B. Other cells, such as some protozoa, have a pellicle, which is a rigid layer of components just within the plasma membrane

XI. Cilia and Flagella—complex locomotor structures composed of a series of microtubules (the axoneme) located within the matrix of a membrane-bound hairlike appendage; they are very different from procaryotic flagella

XII. Comparison of Procaryotic and Eucaryotic Cells
    A. Eucaryotes have a membrane-delimited nucleus and many complex membrane-bound organelles, each of which perform a separate function for the cell
    B. Procaryotes lack a membrane-delimited nucleus and internal membrane-bound organelles; they are functionally simpler and do not undergo mitosis, meiosis, endocytosis, and other complex activities performed by many eucaryotes
    C. Despite the significant differences between procaryotes and eucaryotes, they have remarkable biochemical similarities: the same basic chemical composition, the same genetic code, and the same basic metabolic processes

TERMS AND DEFINITIONS

Place the letter of each term in the space next to the definition or description that best matches it.

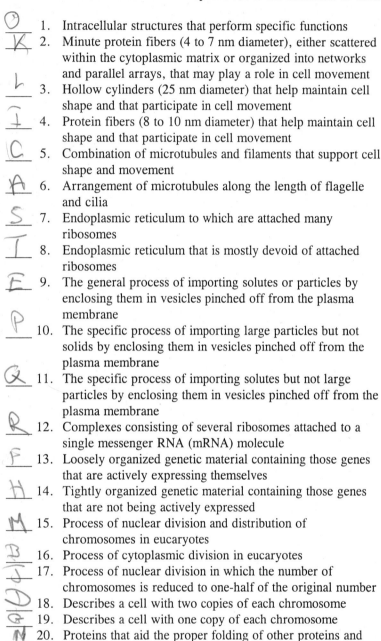

O   1. Intracellular structures that perform specific functions

K   2. Minute protein fibers (4 to 7 nm diameter), either scattered within the cytoplasmic matrix or organized into networks and parallel arrays, that may play a role in cell movement

L   3. Hollow cylinders (25 nm diameter) that help maintain cell shape and that participate in cell movement

I   4. Protein fibers (8 to 10 nm diameter) that help maintain cell shape and that participate in cell movement

C   5. Combination of microtubules and filaments that support cell shape and movement

A   6. Arrangement of microtubules along the length of flagelle and cilia

S   7. Endoplasmic reticulum to which are attached many ribosomes

T   8. Endoplasmic reticulum that is mostly devoid of attached ribosomes

E   9. The general process of importing solutes or particles by enclosing them in vesicles pinched off from the plasma membrane

P   10. The specific process of importing large particles but not solids by enclosing them in vesicles pinched off from the plasma membrane

Q   11. The specific process of importing solutes but not large particles by enclosing them in vesicles pinched off from the plasma membrane

R   12. Complexes consisting of several ribosomes attached to a single messenger RNA (mRNA) molecule

F   13. Loosely organized genetic material containing those genes that are actively expressing themselves

H   14. Tightly organized genetic material containing those genes that are not being actively expressed

M   15. Process of nuclear division and distribution of chromosomes in eucaryotes

B   16. Process of cytoplasmic division in eucaryotes

J   17. Process of nuclear division in which the number of chromosomes is reduced to one-half of the original number

D   18. Describes a cell with two copies of each chromosome

G   19. Describes a cell with one copy of each chromosome

N   20. Proteins that aid the proper folding of other proteins and also assist transport of proteins to target organelles

a. axoneme
b. cytokinesis
c. cytoskeleton
d. diploid
e. endocytosis
f. euchromatin
g. haploid
h. heterochromatin
i. intermediate filaments
j. meiosis
k. microfilaments
l. microtubules
m. mitosis
n. molecular chaperones
o. organelles
p. phagocytosis
q. pinocytosis
r. polysomes
s. rough endoplasmic reticulum
t. smooth endoplasmic reticulum

IDENTIFICATION OF EUCARYOTIC STRUCTURES

Label the appropriate structures indicated on the accompanying figure using the following terms. In the space beside each term provide a *brief* description of its structure and/or function.

1. Centrioles:

2. Chromatin:

3. Golgi apparatus:

4. Lysosome:

5. Mitochondrion:

6. Nucleolus:

7. Nucleus:

8. Plasma membrane:

9. Ribosomes:

10. RER (GER)

11. SER (AER)

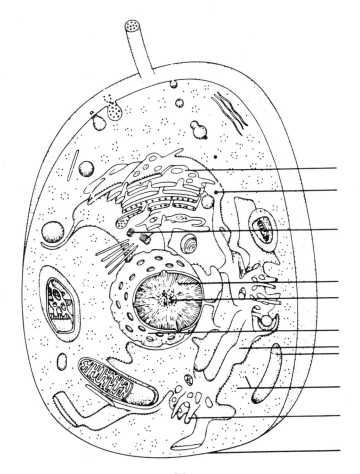

35

COMPARISON OF PROCARYOTIC AND EUCARYOTIC CELLS

Complete the following table by providing a *brief* description of the properties/structures listed for both procaryotes and eucaryotes.

Property/Structure	Procaryote	Eucaryote
Organization of Genetic Material		
1. Membrane-bound nucleus		
2. DNA complexed with histones		
3. Number of chromosomes		
4. Nucleolus		
Organelles		
5. Mitochondria		
6. Chloroplasts		
7. Plasma membrane with Sterols		
8. Flagella		
9. Endoplasmic reticulum		
10. Golgi apparatus		
11. Lysosomes		
12. Ribosomes		
Functions		
13. Photosynthesis		
14. Mitosis		
15. Meiosis		
16. Differentiation		

MATCHING

Match the following structures with the appropriate functions. Answers may be used once, more than once, or not at all.

_____ 1. Major route by which proteins, lipids, and other materials are transported through the cell

_____ 2. Involved in the modification, packaging, and secretion of materials

_____ 3. Contain the enzymes needed to digest all types of macromolecules

_____ 4. Structures responsible for the synthesis of proteins

_____ 5. Responsible for ATP synthesis by processing chemical nutrients through electron transport and oxidative phosphorylation

_____ 6. Major site of cell membrane synthesis

_____ 7. Responsible for the synthesis of ATP and carbohydrate using light as the energy source

_____ 8. Responsible for the production of ribosomes

_____ 9. Repository for the cell's genetic information

_____ 10. A complex structure or set of structures lying underneath the plasma membrane that gives some cells their characteristic shape

_____ 11. A rigid structure outside the plasma membrane that gives some cells their characteristic shape

_____ 12. Short fibers containing microtubules which are responsible for locomotion for some eucaryotes

_____ 13. Long fibers containing microtubules which are responsible for locomotion for some eucaryotes

a. cell wall
b. chloroplasts
c. cilia
d. endoplasmic reticulum
e. flagella
f. Golgi apparatus
g. lysosomes
h. mitochondria
i. nucleolus
j. nucleus
k. pellicle
l. ribosomes

FILL IN THE BLANK

1. The environment of the _____ and the location of many important biochemical processes is an apparently featureless, homogeneous substance called the _____ _____.

2. The Golgi apparatus consists of _____ in stacks, often called _____, which can be clustered in one region or scattered throughout the cell.

3. The primary functions of the Golgi apparatus is the _____ and _____ of materials.

4. Newly formed lysosomes, or _____ lysosomes, fuse with _____ vesicles to yield _____ lysosomes; _____ _____ are lysosomes that have accumulated large quantities of indigestible material.

5. Enzymes and electron carriers involved in electron transport and _____ _____ are located only in the _____ membrane of the mitochondrion, while the enzymes of the tricarboxylic acid cycle are located in the _____ of that organelle.

6. Photosynthetic reactions occur in separate compartments of the chloroplast. The formation of _____ from carbon dioxide and water, (the _____ reaction) takes place in the _____ or fluid matrix, while the trapping of light energy to form _____, _____, and _____ (the light reaction) takes place in the _____ membrane or _____ of the chloroplast.

7. The nucleus is the repository for the cell's _____ _____ and its control center.

8. The nucleus is bounded by the nuclear _____ that consists of an inner and outer membrane, and that is penetrated by many _____, each of which is surrounded by a ring of granular and fibrous material called the _____, which may facilitate the movement of material into and out of the nucleus.

9. The process of nuclear division and distribution of chromosomes in eucaryotic cells is known as _____, while the division of the cytoplasm is called _____. The period between mitotic divisions during which most cell growth takes place is known as _____.

10. Many protozoa and some algae have an external structure consisting of a relatively rigid layer just beneath the plasma membrane and including the plasma membrane. This structure is called the _____, and although it is not as rigid as a _____ _____, it does give a cell that possesses one its characteristic shape.

11. Flagella sometimes have lateral hairs called _____ _____ that change flagellar action so that a wave moving down the flagellum towards the tip _____ the cell along rather than _____ it. This type of flagellum is often called a _____ flagellum, while a naked flagellum is referred to as a _____ flagellum.

12. Cilia beats have two distinctive phases. In the _____ stroke, the cilium moves like an oar to propel the organism through the water. The cilium then bends along its length while it is pulled forward during the _____ stroke.

13. Material is transported from the ER to the _____ face of the Golgi apparatus. It moves through the Golgi apparatus by budding off the cisternal edges and moving to the next sac. After it has completed its journey through the Golgi apparatus and has been appropriately modified, it is secreted from the _____ face for transport to its final destination.

MULTIPLE CHOICE

For each of the questions below select the *one best* answer.

1. The flattened sacs of the endoplasmic reticulum are called
   a. thylakoids.
   b. cristae.
   c. cisternae.
   d. vacuomes.

2. One of the more important functions of the Golgi apparatus is the synthesis of
   a. ribosomes.
   b. lysosomes.
   c. nucleosomes.
   d. mesosomes.

3. Cells can selectively digest potions of their own cytoplasm in a type of lysosome called a(n)
   a. autophagic vacuole.
   b. turnover lysosome.
   c. suicide vacuole.
   d. recycling vacuole.

4. Which of the following is not a function of the mitochondrion?
   a. tricarboxylic acid cycle enzyme reactions
   b. electron transport
   c. oxidative phosphorylation
   d. All of the above are functions of the mitochondrion.

5. The nucleolar organizer is a part of a specific chromosome that directs the synthesis of
   a. transfer RNA (tRNA).
   b. ribosomal RNA (rRNA).
   c. messenger RNA (mRNA).
   d. heterogeneous nuclear RNA (hnRNA).

6. A cell with a diploid number of 8 chromosomes undergoes mitosis. The number of chromosomes in each of the daughter cells is
   a. 16.
   b. 4.
   c. 8.
   d. 12.

7. If the same cell as described in question #6 undergoes meiosis instead of mitosis, the resulting daughter cells will each have _____ chromosomes.
   a. 16
   b. 4
   c. 8
   d. 12

8. Construction of flagella and/or cilia is directed by the
   a. axoneme.
   b. tubulin.
   c. centriole.
   d. basal body.

9. A set of intermediate filaments underlying the nuclear envelope and providing support for that structure is called the
   a. nuclear annulus.
   b. nuclear lamina.
   c. nuclear lattice.
   d. nuclear pellicle.

10. Stacks of cisternae in the Golgi apparatus are referred to as
    a. stigmata.
    b. golgisomes.
    c. dictyosomes.
    d. lamellosomes.

TRUE/FALSE

____ 1. In cells that produce large quantities of lipid, the endoplasmic reticulum is mostly devoid of ribosomes and is referred to as smooth endoplasmic reticulum.

____ 2. The Golgi apparatus is the major site of cell membrane synthesis.

____ 3. Both eucaryotes and procaryotes have ribosomes, but the eucaryotic ribosome is generally larger and more complex than the procaryotic ribosome.

____ 4. The majority of mitochondrial proteins are manufactured under the direction of mitochondrial DNA by mitochondrial ribosomes.

____ 5. DNA is replicated during the S period of interphase.

____ 6. Coenocytic cells result when mitosis occurs in the absence of cytokinesis.

____ 7. In contrast with bacteria, many eucaryotes lack an external cell wall.

____ 8. Proteins to be secreted out of the cell are synthesized by free ribosomes (those that are not attached to the ER).

____ 9. Molecular chaperones in eucaryotic cells perform the same basic functions as they do in procaryotic cells.

____ 10. Lysosomes join with phagosomes for defensive purposes as well as for acquisition of nutrients.

CRITICAL THINKING

1. Describe how the "division of labor" associated with internal, membrane-delimited organelles enables eucaryotic cells to be more efficient and thereby allows them to grow much larger than procaryotes grow.

2. Both procaryotes and eucaryotes are capable of using large particles as nutrient sources. Eucaryotes bring them into the cell by endocytosis. Procaryotes cannot do this; they export digestive enzymes to digest the particles externally and then transport the resulting small nutrient molecules into the cell. Why is the procaryotic process much less efficient? Use diagrams to support your answer.

ANSWER KEY

Terms and Definitions

1. o, 2. k, 3. l, 4. i, 5. c, 6. a, 7. s, 8. t, 9. e, 10. p, 11. q, 12. r, 13. f, 14. h, 15. m, 16. b, 17. j, 18. d, 19. g, 20. n

Matching

1. d, 2. f, 3. g, 4. l, 5. h, 6. d, 7. b, 8. i, 9. j, 10. k 11. a 12. c 13. e

Fill in the Blank

1. organelles; cytoplasmic matrix 2. cisternae; dictyosomes 3. packaging; secretion 4. primary; phagocytic; secondary; residual bodies 5. oxidative phosphorylation; inner; matrix 6. carbohydrate; dark; stroma; ATP; NADPH; oxygen; thylakoid; grana 7. genetic information 8. envelope; pores; annulus 9. mitosis; cytokinesis; interphase 10. pellicle; cell wall 11. flimmer filaments; pulls; pushes; tinsel; whiplash 12. effective; recovery 13. cis (forming); trans (maturing)

Multiple Choice

1. c, 2. b, 3. a, 4. d, 5. b, 6. c, 7. b, 8. d, 9. b, 10. c

True/False

1. T, 2. F, 3. T, 4. F, 5. T, 6. T, 7. T, 8. F, 9. T, 10. T

Part II MICROBIAL NUTRITION, GROWTH, AND CONTROL

5 Microbial Nutrition

CHAPTER OVERVIEW

This chapter describes the basic nutritional requirements of microorganisms. Cells must have a supply of raw materials and energy in order to construct new cellular components. This chapter also describes the processes by which microorganisms acquire nutrients and provides information about the cultivation of microorganisms.

CHAPTER OBJECTIVES

After reading this chapter you should be able to:

- list the ten elements that microorganisms require in large amounts (macronutrients/macroelements) and the six elements that they require in trace amounts (micronutrients)
- list the major nutritional categories and give the source of carbon, energy, and hydrogen/electrons for each of the categories
- compare the various processes (passive diffusion, facilitated diffusion, active transport, group translocation) by which cells can obtain nutrients from the environment
- describe the various types of culture media for microorganisms (synthetic, defined, selective, differential) and tell how each is normally used in the study of microorganisms
- describe the techniques used to obtain pure cultures (spread plate, streak plate, pour plate)

CHAPTER OUTLINE

I. The Common Nutrient Requirements
- A. Macroelements or macronutrients (C, O, H, N, S, P, K, Ca, Mg, Fe) are required by microorganisms in relatively large amounts
- B. Trace elements or micronutrients (Mn, Zn, Co, Mo, Ni, Cu) are required in trace amounts by most cells and are often adequately supplied in the water used to prepare the media or in the regular media components
- C. Other elements may be needed by particular types of microorganisms

II. Requirements for Carbon, Hydrogen, and Oxygen—often satisfied together
- A. Autotrophs use carbon dioxide as their sole or principal carbon source
- B. Heterotrophs use reduced, preformed organic molecules (usually from other organisms) as carbon sources

41

C. Prototrophs are microorganisms requiring the same nutrients as most naturally occurring members of their species
D. Auxotrophs are mutated microorganisms that lack the ability to synthesize an essential nutrient and therefore must obtain it or a precursor from the surroundings

III. Nutritional Types of Microorganisms
   A. Energy
     1. Phototrophs use light as their energy source
     2. Chemotrophs obtain energy from the oxidation of organic or inorganic compounds
   B. Hydrogen/electrons
     1. Lithotrophs use reduced inorganic compounds as their electron source
     2. Organotrophs use reduced organic compounds as their electron source
   C. Nutritional types of microorganisms—most microorganisms can be categorized as belonging to one of four major nutritional types depending on their sources of carbon, energy, and electrons:
     1. Photolithotrophic autotrophs
     2. Photoorganotrophic heterotrophs
     3. Chemolithotrophic autotrophs
     4. Chemoorganotrophic heterotrophs
   D. Mixotrophic organisms combine autotrophic and heterotrophic metabolic processes, relying on inorganic energy sources and organic carbon sources
   E. Chemolithotrophs contribute greatly to the chemical transformation of elements that continually occur in the ecosystem
   F. Some organisms show great metabolic flexibility and alter their metabolic patterns in response to environmental changes

IV. Requirements for Nitrogen, Phosphorus, and Sulfur—can be met by either organic or inorganic sources; some organisms have specific requirements for sources of these elements while others are more general

V. Growth Factors—organic compounds required by the cell because they are essential cell components (or precursors of these components) that the cell cannot synthesize
   A. Amino acids—needed for protein synthesis
   B. Purines and pyrimidines—needed for nucleic acid synthesis
   C. Vitamins—function as enzyme cofactors
   D. Knowledge of specific growth factor requirements makes possible quantitative growth-response assays

VI. Uptake of Nutrients by the Cell
   A. Passive diffusion—a phenomenon in which molecules move from an area of high concentration to an area of low concentration because of random thermal agitation
     1. Requires a large concentration gradient for significant levels of uptake
     2. Limited to only a few small useful molecules (e.g., glycerol, H_2O, O_2, and CO_2)
   B. Facilitated diffusion—a process that involves a carrier molecule (permease) to increase the rate of diffusion; net effect is limited to movement from an area of higher concentration to an area of lower concentration
     1. Requires a smaller concentration gradient than passive diffusion
     2. The rate plateaus when the carrier becomes saturated (i.e., when it is binding and transporting molecules as rapidly as possible)
     3. Generally more important in eucaryotes rather than procaryotes
   C. Active transport—a process in which metabolic energy is used to move molecules to the cell interior where the solute concentration is already higher (i.e., it runs against the concentration gradient)
     1. Characteristics of active transport
       a. Saturable uptake rate
       b. Requires an expenditure of metabolic energy

                c.   Can concentrate molecules inside the cell even when the concentration inside the cell is already higher than that outside the cell

      2.   ATP hydrolysis or protonmotive forces are the usual sources of metabolic energy

      3.   Types of active transport

                a.   Symport is the linked transport of two substances in the same direction

                b.   Antiport is the linked transport of two substances in opposite directions

    D.   Group translocation—a process in which molecules are modified as they are transported across the membrane

    E.   Iron uptake—the organism secretes siderophores that complex with the very insoluble ferric ion, which is then transported into the cell

VII.   Culture Media

    A.   Synthetic (defined) media are media in which all components and their concentrations are known

    B.   Complex media are media that contain some ingredients of unknown composition and/or concentration; this type supplies amino acids, vitamins, growth factors, and other nutrients

      1.   Peptones—protein hydrolysates prepared by partial proteolytic digestion of various protein sources

      2.   Extracts—aqueous extracts, usually of beef or yeast

    C.   Agar is a sulfated polymer used to solidify liquid media

    D.   Types of Media

      1.   General purpose media will support the growth of many microorganisms

      2.   Enriched media are supplemented by blood or other special nutrients to encourage the growth of fastidious heterotrophs

      3.   Selective media favor the growth of particular microorganisms and inhibits the growth of others

      4.   Differential media distinguish between different groups of bacteria on the basis of their biological characteristics

    E.   Some media can exhibit characteristics of more than one type (e.g., blood agar is enriched and differential, and distinguishes between hemolytic and nonhemolytic bacteria)

VIII. Isolation of pure cultures (a population of cells arising from a single cell)—can be accomplished from mixtures by a variety of procedures, including spread plates, streak plates, and pour plates

    A.   Colonies are macroscopically visible growths or clusters of microorganisms on solid media

    B.   Colony growth is most rapid at the colony's edge because oxygen and nutrients are more available; growth is slowest at the colony's center

    C.   Colony morphology helps microbiologists identify bacteria because individual species often form colonies of characteristic size and appearance

TERMS AND DEFINITIONS

Place the letter of each term in the space next to the definition or description that best matches it.

_____ 1. Elements required by microorganisms in relatively large amounts

_____ 2. Elements required in trace amounts sufficiently supplied in water and regular media components

_____ 3. Organisms that can use carbon dioxide as their sole or principal source of carbon

_____ 4. Organisms that use reduced, preformed organic molecules as carbon sources

_____ 5. Organisms that require the same nutrients as the majority of naturally occurring members of its species

_____ 6. Organisms that have mutated so that they lack the ability to synthesize a molecule essential for growth and reproduction

_____ 7. Organisms that obtain energy from the oxidation of organic or inorganic compounds

_____ 8. Organisms that obtain electrons from the oxidation of inorganic compounds

_____ 9. Organisms that obtain electrons from the oxidation of organic compounds

_____ 10. Organisms that obtain energy from light

_____ 11. Organic compounds that are required because they are essential cell components (or precursors of such components) that cannot be synthesized by the organism

_____ 12. Small organic molecules that make up all or part of enzyme cofactors

_____ 13. The process in which molecules move from a region of higher concentration to one of lower concentration as a result of random thermal agitation

_____ 14. Carrier proteins embedded in the plasma membrane that increase the rate of diffusion of specific molecules across selectively permeable membranes

_____ 15. The diffusion process that is aided by action of a carrier protein

_____ 16. Transport of molecules to areas of higher concentration (i.e., against a concentration gradient) with the input of metabolic energy

_____ 17. Linked transport of two substances in the same direction

_____ 18. Linked transport of two substances in opposite directions

_____ 19. A process in which molecules are chemically modified and simultaneously transported into the cell

_____ 20. Low molecular weight molecules that complex with ferric ion and supply it to the cell

_____ 21. A growth medium in which all components and their specific concentrations are known

_____ 22. A growth medium that contains some ingredients of unknown composition and/or concentration

a. active transport
b. antiport
c. autotrophs
d. auxotrophs
e. chemotrophs
f. complex medium
g. differential medium
h. facilitated diffusion
i. group translocation
j. growth factors
k. heterotrophs
l. lithotrophs
m. macronutrients
n. micronutrients
o. organotrophs
p. passive diffusion
q. permeases
r. phototrophs
s. prototrophs
t. selective medium
u. siderophores
v. symport
w. synthetic (defined) medium
x. vitamins

44

_____ 23. A growth medium that favors the growth of some microorganisms and inhibits the growth of other microorganisms

_____ 24. A growth medium that distinguishes between different groups of bacteria on the basis of their biological characteristics

MACRONUTRIENTS/MICRONUTRIENTS

Indicate whether each of the following is a macronutrient or a micronutrient.

1. Magnesium (Mg) _____

2. Zinc (Zn) _____

3. Oxygen (O) _____

4. Sulfur (S) _____

5. Iron (Fe) _____

6. Manganese (Mn) _____

7. Nitrogen (N) _____

8. Cobalt (Co) _____

9. Phosphorus (P) _____

10. Carbon (C) _____

11. Hydrogen (H) _____

12. Potassium (K) _____

13. Molybdenum (Mo) _____

14. Nickel (Ni) _____

15. Calcium (Ca) _____

16. Copper (Cu) _____

NUTRITIONAL TYPES OF MICROORGANISMS

Complete the table listed below by supplying the missing information.

Nutritional Type	Energy	Electrons	Carbon
Photolithotrophic autotrophy		Inorganic compounds	
Photoorganotrophic heterotrophy			Organic compounds
Chemolithotrophic autotrophy	Inorganic compounds		
	Organic compounds		Organic compounds

SELECTIVE AND DIFFERENTIAL MEDIA

Mannitol salt agar is a culture medium that contains a high salt (NaCl) concentration, mannitol (a fermentable sugar), and a chemical pH indicator that is yellow at acidic conditions and red at alkaline conditions. (Acids are released when microorganisms ferment mannitol.) This medium also contains other carbohydrates that allow growth of nonfermenting, halophilic organisms (i.e., nonfermenting organisms that tolerate high salt concentrations). Nonhalophilic organisms will not grow on mannitol salt agar.

For each of the following situations, assume that the organisms described are the *only* organisms involved. Place the letter of the term that *best* describes the way the medium is being used. Then list the organism(s) that will grow, the organisms that will not grow, and the color of the pH indicator.

_____ 1. Onto mannitol salt agar you inoculate a halophilic mannitol fermenter, a halophilic mannitol nonfermenter, and a nonhalophilic mannitol fermenter.

_____ 2. Onto mannitol salt agar you inoculate a halophilic mannitol fermenter and a halophilic mannitol nonfermenter.

_____ 3. Onto mannitol salt agar you inoculate a halophilic mannitol nonfermenter that is pigmented yellow, and a halophilic mannitol nonfermenter that is pigmented red. These two organisms show the same pigmentation (yellow and red, respectively) on a general purpose medium such as nutrient agar.

_____ 4. Onto mannitol salt agar you inoculate a halophilic mannitol nonfermenter and a nonhalophilic mannitol fermenter.

a. selective medium
b. differential medium
c. both selective and differential medium
d. neither selective nor differential medium

FILL IN THE BLANK

1. Although a particular microbial species usually belongs in only one of the four major nutritional classes, some show great metabolic _____ and alter their metabolic patterns in response to _____ changes.
2. Bacteria use _____ force in the form of a _____ that is generated during _____ to drive active transport by membrane-bound transport systems.
3. Linked transport of two substances in the _____ direction is called symport. In procaryotes, energy stored in the form of a _____ gradient is used to drive _____ transport.
4. Iron uptake is made difficult by the great _____ of ferric ion. Many bacteria overcome this by secreting _____, which are compounds of low molecular weight that can complex with ferric ion and supply it to the cell.
5. Media that contain some ingredients of unknown chemical composition are called _____ media. This type of medium is often used because the nutritional requirements of a particular microbe are unknown and thus a _____ medium cannot be constructed.
6. If a solid medium is needed for surface cultivation of microorganisms, liquid media can be solidified by the addition of _____, a sulfated polymer extracted from _____.
7. A macroscopically visible growth or cluster of microorganisms on a solid medium is called a _____, and can be obtained by using _____, _____, and _____.
8. Colony development on agar surfaces aids the microbiologist in identifying bacteria because individual species often form colonies of characteristic _____ and _____.
9. The growth of colonies on agar has been frequently studied. Generally, the most rapid growth occurs at the colony _____, where _____ and _____ are plentiful.
10. Mixotrophic organisms combine _____ and _____ metabolic processes, relying on _____ energy sources and _____ carbon sources.
11. Microorganisms of the nutritional type _____ contribute greatly to the chemical _____ of elements that continually occur in the ecosystem.

MULTIPLE CHOICE

For each of the questions below select the *one best* answer.

1. Bacteria that can metabolize only methane, methanol, carbon monoxide, formic acid, and a few related one-carbon molecules are called _____ bacteria.
    a. carbotrophic
    b. methylotrophic
    c. monotrophic
    d. autotrophic
2. Which of the following is not a major class of growth factors?
    a. amino acids
    b. purines and pyrimidines
    c. vitamins
    d. All of the above are major classes of growth factors.
3. Membranes that allow some molecules to pass but not others are called _____ membranes.
    a. permeable
    b. partially permeable
    c. selectively permeable
    d. None of the above are correct.
4. Which of the following processes can be used to concentrate nutrients from dilute nutrient sources?
    a. active transport
    b. group translocation
    c. Both (a) and (b) are correct.
    d. None of the above are correct.

5. When there are several transport systems for the same substance, in what way do the systems differ?
    a. in the energy source they use
    b. in their affinity for the transported solute
    c. in the nature of their regulation
    d. All of the above are correct.
6. Which of the following is not a characteristic of active transport?
    a. saturable rate of uptake
    b. requires an expenditure of metabolic energy
    c. can transport materials against a concentration gradient
    d. All of the above are characteristics of active transport.
7. Which of the following is/are energy source(s) used by bacteria to drive active transport?
    a. ATP hydrolysis
    b. protonmotive force
    c. both (a) and (b)
    d. Neither (a) nor (b) is correct.
8. Which of the following is a good method for obtaining isolated pure cultures of a microorganism?
    a. spread plate
    b. pour plate
    c. streak plate
    d. All of the above are good methods for obtaining isolated pure cultures.

TRUE/FALSE

____ 1. Micronutrients are normally a part of enzymes and cofactors where they aid in the catalysis of reactions and the maintenance of protein structure.
____ 2. Most microorganisms require large amounts of sodium (Na).
____ 3. Transport of materials against a concentration gradient that requires expenditure of metabolic energy is called facilitated diffusion.
____ 4. Permease proteins resemble enzymes in their specificity for the substance to be transported; each carrier is selective and will transport only closely related solutes.
____ 5. Microorganisms usually have only one transport system for each nutrient.
____ 6. Agar is an excellent hardening agent because it is not usually degraded by microorganisms.
____ 7. Media can be selective or differential, but cannot be both selective and differential.
____ 8. Heterotrophs usually obtain preformed, partially reduced organic molecules from other organisms.
____ 9. Facilitated diffusion is generally more important in procaryotes than in eucaryotes.
____ 10. Quantitative growth-response assays use the growth of microorganisms as a way of measuring the amount of a specific, limited growth nutrient in a particular growth medium.

CRITICAL THINKING

1. In this figure, which line corresponds to the situation for passive diffusion? Which line corresponds to the situation for facilitated diffusion? Explain.

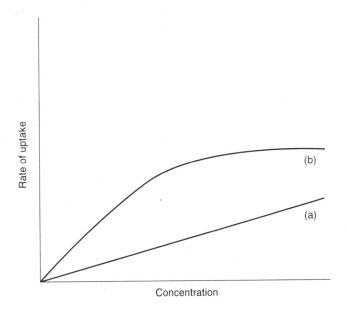

2. A great majority of microorganisms studied thus far are either photolithotrophic autotrophs or chemoorganotrophic heterotrophs. Why? Where would you expect to find photolithotrophic heterotrophs? Chemolithotrophic autotrophs?

3. Many bacteria have several different mechanisms for transport of a single substance. Compare the advantages of multiple mechanisms with those of a single transport mechanism for any given substance.

ANSWER KEY

Terms and Definitions

1. m, 2. n, 3. c, 4. k, 5. s, 6. d, 7. e, 8. l, 9. o, 10. r, 11. j, 12. x, 13. p, 14. q, 15. h, 16. a, 17. v, 18. b, 19. i, 20. u, 21. w, 22. f, 23. t, 24. g

Macronutrients/Micronutrients

1. macro, 2. micro, 3. macro, 4. macro, 5. macro, 6. micro, 7. macro, 8. micro, 9. macro, 10. macro, 11. macro, 12. macro, 13. micro, 14. micro, 15. macro, 16. micro

Selective and Differential Media

1. c, 2. b, 3. d, 4. a

Fill in the Blank

1. flexibility; environmental 2. protonmotive; proton gradient; electron transport 3. same; proton; solute 4. insolubility; siderophores 5. complex; defined 6. agar; red algae 7. colony; spread plates; streak plates; pour plates 8. size; appearance 9. edge; oxygen; nutrients 10. autotrophic; heterotrophic; inorganic; organic 11. chemolithotroph; transformation

Multiple Choice

1. b, 2. d, 3. c, 4. c, 5. d, 6. d, 7. c, 8. d

True/False

1. T, 2. F, 3. F, 4. T, 5. F, 6. T, 7. F, 8. T, 9. F, 10. T

6 Microbial Growth

CHAPTER OVERVIEW

This chapter describes the basic nature of microbial growth in the presence of an adequate nutrient supply. Several methods for the measurement of microbial growth are described and different systems used for microbial growth are also described. The chapter finishes with a discussion of the influence of various environmental factors on the growth of microorganisms.

CHAPTER OBJECTIVES

After reading this chapter you should be able to:

- name the various phases of growth that occur in closed culture systems and describe what is occurring in each phase
- determine from experimental data the various parameters (number of generations, specific growth rate constant, mean generation time) that describe microbial growth in mathematical terms
- explain the concept of growth yield and molar growth yield
- describe the various types of continuous culture systems and explain the differences in their function
- describe the influence of various environmental factors (water availability, pH, temperature, oxygen concentration, pressure, radiation) on the growth of microorganisms
- categorize microorganisms according to the environmental factors that are conducive to optimal growth of the organism

CHAPTER OUTLINE

I.   Growth—an increase in cellular constituents that may result in an increase in cell size, an increase in cell number, or both
II.   The Growth Curve—usually analyzed in a closed system called a batch culture; usually plotted as the logarithm of cell number versus the incubation time
   A.   Lag phase is the period of apparent inactivity in which the cells are adapting to a new environment and preparing for reproductive growth, usually by synthesizing new cell components; it varies considerably in length depending upon the condition of the microorganisms and the nature of the medium
   B.   Exponential (log) phase is the period in which the organisms are growing at the maximal rate possible given their genetic potential, the nature of the medium, and the conditions under which they are growing; the population is most uniform in terms of chemical and physical properties during this period
   C.   Stationary phase is the period in which the number of viable microorganisms remains constant either because metabolically active cells stop reproducing or because the reproductive rate is balanced by the rate of cell death; it may result from:
      1.   Nutrient limitation
      2.   Toxic waste accumulation

      D.   Death phase is the period in which the cells are dying at an exponential rate

      E.   The mathematics of growth—microbial growth can be described by certain mathematical terms:

          1.   Mean generation (doubling) time is the time required for the population to double

          2.   Mean growth rate constant is the number of generations per unit time, often expressed as generations per hour

          3.   Generation times vary markedly with the species of microorganism and environmental conditions; they can range from 10 minutes for a few bacteria to several days with some eucaryotic microorganisms

III.   Measurement of Microbial Growth

      A.   Measurement of cell numbers

          1.   Direct count methods do not distinguish between living and dead cells, and may be accomplished by direct microscopic observation on specially etched slides (such as Petroff-Hausser chambers or hemacytometers) or by using electronic counters (such as Coulter Counters, which count microorganisms as they flow through a small hole or orifice)

          2.   Viable cell counts involve plating diluted samples (using a pour plate or spread plate) onto suitable growth media and monitoring colony formation; this type of method counts only those cells that are reproductively active; because it is not possible to be certain that each colony arose from a single cell, results are usually expressed as colony forming units (CFU)

          3.   Microbial numbers are frequently determined from counts of colonies growing on membrane filters having pores small enough to trap bacteria

      B.   Measurement of cell mass may be used to approximate the number of microorganisms if a suitable parameter proportional to the number of microorganisms present is used (suitable parameters may be dry weight, light scattering in liquid solutions, or biochemical determinations of specific cellular constituents such as protein, DNA, or ATP)

IV.   Growth Yields and the Effects of a Limiting Nutrient

      A.   Growth yield is expressed as the mass of cells formed per gram of nutrient consumed

      B.   Molar growth yield is expressed as the mass of cells formed per mole of nutrient consumed

V.   The Continuous Culture of Microorganisms—used to maintain cells in the exponential growth phase at a constant biomass concentration for extended periods of time (these conditions are met by continual provision of nutrients and removal of wastes)

      A.   A chemostat is a continuous culture device that maintains a constant growth rate by supplying a medium containing a limited amount of an essential nutrient at a fixed rate and by removing medium that contains microorganisms at the same rate

      B.   A turbidostat is a continuous culture device that regulates the flow rate of media through the vessel in order to maintain a predetermined turbidity or cell density; there is no limiting nutrient

VI.   Balanced and Unbalanced Growth

      A.   Balanced (exponential) growth occurs when all cellular components are synthesized at constant rates relative to one another

      B.   Unbalanced growth occurs when the rates of synthesis of some components change relative to the rates of synthesis of other components. This usually occurs when the environmental conditions change

VII.   The Influence of Environmental Factors on Growth

      A.   Although most microorganisms only grow in fairly moderate environmental conditions, some, referred to as extremophiles, can grow under harsh conditions that would kill most other organisms

      B.   Solutes and water activity (a quantitative measurement of the availability of water) is inversely related to osmotic pressure and may have a profound effect on cell growth

          1.   Osmotolerant organisms can grow in solutions of both high and low water activity

          2.   Halophiles require environments of low water activity (high osmotic pressure) in order to grow

          3.   Microorganisms growing in a habitat with low water activity (a_w) usually maintain a high internal solute concentration in order to retain water

C. pH is the negative logarithm of the hydrogen ion concentration
   1. Acidophiles grow best between pH 0 and 5.5
   2. Neutrophiles grow best between pH 5.5 and 8.0
   3. Alkalophiles grow best between pH 8.5 and 11.5
   4. Extreme alkalophiles grow best at pH 10.0 or higher
   5. Despite wide variations in habitat pH, the internal pH of most microorganisms is maintained near neutrality either by proton/ion exchange or by internal buffering

D. Temperature
   1. Temperature has a profound effect on microorganism viability, primarily because enzyme-catalyzed reactions are sensitive to temperature
   2. At low temperatures, a temperature rise increases the growth rate by increasing the rate of enzyme reactions
   3. At high temperatures, microorganisms are damaged by enzyme denaturation, membrane disruption, and other phenomena
   4. Organisms exhibit distinct cardinal temperatures (minimal, maximal, and optimal growth temperatures)
      a. Psychrophiles can grow well at $0°C$, have optimal growth at $15°C$ or lower, and usually will not grow above $20°C$
      b. Psychrotrophs (facultative psychrophiles) can also grow at $0°C$, but have growth optima between $20°C$ and $30°C$, and growth maxima at about $35°C$
      c. Mesophiles have growth minima of 15 to $20°C$, optima of 20 to $45°C$, and maxima of about $45°C$ or lower
      d. Thermophiles have growth minima around $45°C$, and optima of 55 to $65°C$
      e. Hyperthermophiles have growth minima around $55°C$ and optima of 80 to $110°C$
   5. Stenothermal organisms have a narrow range of cardinal growth temperatures; eurythermal organisms have a wide range of cardinal growth temperatures

E. Oxygen concentration
   1. Obligate aerobes are completely dependent on atmospheric O_2 for growth
   2. Facultative anaerobes do not require O_2 for growth, but do grow better in its presence
   3. Aerotolerant anaerobes ignore O_2 and grow equally well whether it is present or not
   4. Obligate (strict) anaerobes do not tolerate O_2 and die in its presence
   5. Microaerophiles are damaged by the normal atmospheric level of O_2 (20%) but require lower levels (2 to 10%) for growth

F. Pressure
   1. Barotolerant organisms are adversely affected by increased pressure, but not as severely as are nontolerant organisms
   2. Barophilic organisms require, or grow more rapidly in the presence of, increased pressure

G. Radiation
   1. Ultraviolet radiation damages cells by causing the formation of thymine dimers in DNA
      a. Photoreactivation repairs thymine dimers by direct splitting when the cells are exposed to blue light
      b. Dark reactivation repairs thymine dimers by excision and replacement in the absence of light
   2. Ionizing radiation such as X rays or gamma rays are even more harmful to microorganisms than ultraviolet radiation
      a. How levels produce mutations and may indirectly result in death
      b. High levels are directly lethal by direct damage to cellular macromolecules or through the production of oxygen free radicals

TERMS AND DEFINITIONS

Place the letter of each term in the space next to the definition or description that best matches it.

_____ 1. An increase in cellular constituents that may or may not be accompanied by an increase in cell number

_____ 2. A culture in a closed vessel with a single batch of medium to which no fresh medium is added and from which no waste products are removed

_____ 3. The length of time it takes for a population of microorganisms to double in number

_____ 4. The number of generations per unit time, usually expressed as the number of generations per hour

_____ 5. The number of live or dead cells per unit volume

_____ 6. The number of cells per unit volume that are able to grow and reproduce

_____ 7. An approximation of the number of viable microorganisms per unit volume based on the number of colonies that form on solid media after plating dilute solutions such that each colony probably arises from a single viable microorganism

_____ 8. The amount of microbial mass produced from a nutrient; usually expressed as grams of cells formed per gram of substrate utilized

_____ 9. The amount of microbial mass produced from a nutrient; usually expressed as grams of cells formed per mole of substrate utilized

_____ 10. A culture system with constant environmental conditions maintained through continual nutrient provision and waste removal

_____ 11. An open system in which sterile medium containing an essential nutrient in limiting quantities is fed into the culture vessel at the same rate as medium containing microorganisms is removed

_____ 12. The rate at which medium flows through a chemostat

_____ 13. An open system in which the flow rate of media through the vessel is automatically regulated to maintain a predetermined turbidity or cell density

_____ 14. Organisms that can grow in habitats with low water activity (high osmotic pressure) by maintaining a high internal solute concentration to prevent water loss

_____ 15. Organisms that require high levels of sodium chloride in order to grow

_____ 16. Organisms that have their growth optimum between pH 0 and 5.5

_____ 17. Organisms that have their growth optimum between pH 5.5 and 8.0

_____ 18. Organisms that have their growth optimum between pH 8.5 and 11.5

_____ 19. Minimum, maximum, and optimum growth temperatures

_____ 20. Describes organisms that have a small range of growth temperatures

a. acidophiles
b. aerotolerant anaerobes
c. alkalophiles
d. barophiles
e. barotolerant
f. batch culture
g. cardinal temperatures
h. chemostat
i. colony forming units
j. continuous (open) culture system
k. dilution rate
l. eurythermal
m. extremophiles
n. facultative anaerobes
o. growth
p. growth yield (Y)
q. halophiles
r. hyperthermophiles
s. mean generation time (g)
t. mean growth rate constant (k)
u. mesophiles
v. microaerophiles
w. molar growth yield (Ym)
x. neutrophiles
y. obligate aerobes
z. obligate anaerobes
aa. osmotolerant
bb. psychrophiles
cc. psychrotrophs
dd. stenothermal
ee. thermophiles
ff. total cell count
gg. turbidostat
hh. viable cell count

_____ 21. Describes organisms that have a wide range of growth temperatures

_____ 22. Organisms that will grow well at 0°C and have optimum growth temperatures of 15°C or lower

_____ 23. Organisms that will grow well at 0°C and have optimum growth temperatures between 20°C and 30°C

_____ 24. Organisms with growth temperature minima of 15 to 20°C, optima around 20 to 45°C, and maxima at 45°C or lower

_____ 25. Organisms that have growth temperature minima around 45°C and optima between 55 to 65°C

_____ 26. Organisms that are completely dependent upon atmospheric oxygen for growth

_____ 27. Organisms that do not require oxygen for growth, but that do grow better in its presence

_____ 28. Organisms that ignore oxygen and grow equally well in its presence or in its absence

_____ 29. Organisms that do not tolerate oxygen at all and die in its presence

_____ 30. Organisms that are damaged by the normal atmospheric levels of oxygen (20%) but require oxygen at levels below the range of 2 to 10% for growth

_____ 31. Organisms that are not adversely affected by increased barometric pressure

_____ 32. Organisms that grow more rapidly at increased barometric pressure

_____ 33. Organisms that can grow in harsh environments that would kill most living organisms

_____ 34. Organisms with growth optima of 80 to 100°C

BACTERIAL GROWTH CURVE

Look at the accompanying bacterial growth curve. Label each of the four phases and describe what is occurring in each phase.

(a) _____
_____
_____
_____
(b) _____
_____
_____
_____
(c) _____
_____
_____
_____
(d) _____
_____
_____
_____

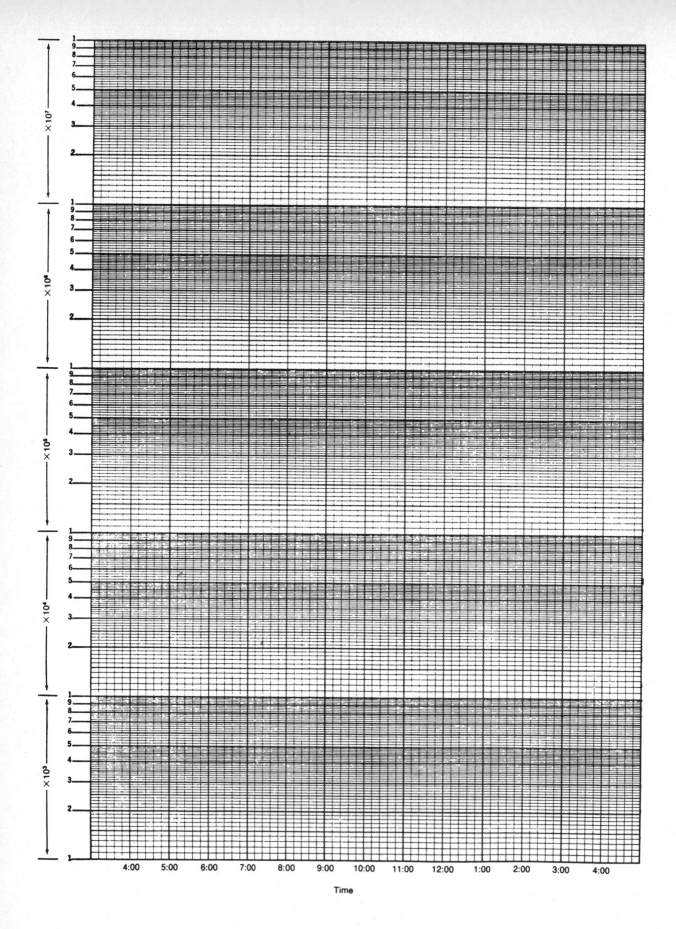

Time

56

MICROBIAL GROWTH PROBLEMS

1. At 4:00 P.M. you inoculate 1×10^3 cells into a closed flask of nutrient broth. The cells have a lag phase that lasts one hour. At 9:00 P.M. the culture enters the stationary phase. At that time there are 6.5×10^7 cells in the flask. Calculate the number of generations that have occurred (n), the mean generation time (g) and the mean growth rate constant (k).

2. Continuing with the same culture of question #1, you measure the number of viable cells at 10:00 P.M., at 10:30 P.M., and at 11:00 P.M. and obtain the following results:

    10:00 P.M. 6.5×10^7
    10:30 P.M. 3.3×10^7
    11:00 P.M. 1.6×10^7

    Construct a growth curve for this culture on the semilogarithmic graph paper provided. Be sure that you correctly label each axis. Explain what is occurring at each phase.

MICROBIAL GROWTH TEMPERATURES

The following is a list of organisms and their cardinal temperatures. For each organism, complete column A by deciding whether the organism is psychrophilic, psychrotrophic, mesophilic, thermophilic, or hyperthermophilic. Complete column B by deciding whether the organism is stenothermal or eurythermal.

Cardinal Temperatures (°C)

Organism	Min.	Opt.	Max.	A	B
1. *M. cryophilus*	-4	10	24	__	__
2. *S. aureus*	6.5	30 to 37	46	__	__
3. *S. acidocaldarius*	60	80	85	__	__
4. *T. acidophilum*	45	59	62	__	__
5. *E. coli*	10	37	45	__	__
6. *N. gonorrhoeae*	30	35 to 36	38	__	__
7. *F. sublinearis*	-2	5 to 6	8 to 9	__	__
8. *P. fluorescens*	4	25 to 30	40	__	__
9. *B. psychrophilus*	-10	23 to 24	28 to 30	__	__

FILL IN THE BLANK

1. An increase in cellular constituents is termed _____. If a microorganism is _____, an increase in cell size will result, but not an increase in the number of cells. If cells reproduce by processes such as _____ or _____, an increase in cell number could result.

2. Special slides with chambers of known depth and an etched grid that are designed for direct counts of bacteria are called _____ counting chambers. Similar slides called _____ are used for counting larger eucaryotic microorganisms.

3. Rapid sensitive techniques for quantifying microbial populations are based on the fact that microbial cells _____ light striking them. Since the microbial cells are of roughly constant size, the amount of _____ is proportional to the _____ of cells present. The instrument used to measure this response is called a _____.

4. In a _____ culture system, exponential growth lasts for only a few generations; in a _____ culture system, a microbial population can be maintained in the exponential growth phase for extended periods of time.

5. In a chemostat, both the _____ and the _____ are related to the dilution rate.

6. A _____ is a continuous culture system equipped with a photocell that measures the _____ of the culture and automatically regulates the flow rate of media through the vessel in order to maintain a predetermined cell _____.

7. When a microorganism is placed in a _____ solution, water will _____ the cell and cause it to burst unless something is done to prevent this.

8. The _____ of a solution is 1/100 the relative humidity of the solution and is equivalent to the ratio of the _____'s vapor pressure to that of _____.

9. Microorganisms that can grow in habitats of high and low water activity are said to be _____, while those that require low water activity are called _____.

10. The cell membranes of psychrophilic microorganisms have high levels of _____ fatty acids and remain _____ when cold.

11. In obligate aerobes, oxygen serves as the _____ during respiration.

12. For organisms that are _____ anaerobes, the efficiency of growth increases when the organism is shifted from an anaerobic environment to an aerobic environment, while _____ anaerobes show no change in growth efficiency when the same shift occurs.

13. Organisms that are not affected by increased pressure are _____, while those which grow more rapidly at increased pressure are _____.

14. Ultraviolet light damages organisms, primarily by inducing the formation of _____ in DNA. These flaws can be repaired by blue light, which directly splits them—a process called _____. They can also be repaired by excision and replacement in the absence of light—a process called _____.

15. Organisms that have growth optima of 55 to 65°C are called _____ while those with growth optima of 80 to 100°C are called _____.

MULTIPLE CHOICE

For each of the questions below select the *one best* answer.

1. Which of the following is not a reason for the occurrence of a lag phase in a bacterial growth curve?
   a. The cells may be old and depleted of ATP, essential cofactors, and ribosomes, which must be synthesized before growth can begin.
   b. The current medium may be different from the previous growth medium; therefore, the cells must synthesize new enzymes to utilize different nutrients.
   c. The organisms may have been injured and thereby may require time to recover.
   d. All of the above are potential reasons for the occurrence of a lag phase.

2. Which of the following does not affect the growth rate observed during the exponential (log) growth phase?
   a. the genetic potential of the organism
   b. the number of cells originally inoculated into the culture vessel
   c. the composition of the medium
   d. the conditions of the incubation

3. In the stationary phase, the total number of viable microorganisms remains constant. This may result from
   a. a balance between the rate of cell division and the rate of cell death.
   b. a cessation of cell division even though the cells remain metabolically active.
   c. Both (a) and (b) may occur.
   d. Neither (a) nor (b) is correct.

4. Which of the following may be the reason for cells entering the stationary phase?
   a. depletion of an essential nutrient
   b. lack of available oxygen
   c. accumulation of toxic waste products
   d. All of the above are correct.

5. Which of the following is not a reason that electronic counters are not useful for counting bacteria?
   a. Bacteria are too small to be detected.
   b. Small particles of debris interfere with accurate counts.
   c. The formation of filaments interferes with accurate counts.
   d. All of the above are reasons that electronic counters aren't useful for counting bacteria.

6. Which of the following is not true about the use of colony forming units as a measurement of bacterial population numbers?
   a. For best results, the samples should yield between 25 and 250 colonies.
   b. Because each colony arises from a single viable cell, the number of colonies is an accurate measure of the number of viable cells in the sample.
   c. Hot agar used in the pour plate technique may lead to an underestimate of the viable cell number compared to the streak plate technique.
   d. All of the above are true about the use of colony forming units as a measurement of bacterial population numbers.

7. Membrane filter techniques have been used for
   a. viable cell counts.
   b. direct cell counts.
   c. Both viable and direct cell counts.
   d. Neither viable nor direct cell counts.

8. An organism with a pH growth optimum of 10.5 is best described as a(n)
   a. acidophile.
   b. neutrophile.
   c. alkalophile.
   d. extreme alkalophile.

9. Compatible solutes are best described as
   a. solutes that allow otherwise incompatible microorganisms to grow in the same habitat.
   b. solutes that can coexist in the same cell without forming complexes that are toxic to the cell.
   c. solutes that are compatible with metabolism and growth when at high intracellular concentrations.
   d. None of the above correctly describe compatible solutes.

10. Which of the following is not a mechanism by which ionizing radiation kills cells.
   a. indirectly by accumulation of harmful mutations
   b. directly by damaging cellular macromolecules
   c. directly by the production of oxygen free radicals
   d. all of the above are mechanisms by which ionizing radiation kills cells

TRUE/FALSE

_____ 1. When a young, vigorously growing culture is transferred to fresh medium with the same composition as the original medium, the lag phase is usually short or absent.

_____ 2. When a young, vigorously growing culture is transferred to fresh medium with a different composition from the original medium, the lag phase is usually short or absent.

_____ 3. The rate of growth in the exponential (log) phase is constant and is independent of the composition of the medium.

_____ 4. Exponential (log) phase cultures are usually used in biochemical studies because populations in this phase are almost uniform in terms of their chemical and physiological properties.

_____ 5. The progress of the death phase can usually be followed experimentally by measuring total cell number, because cells lyse (disintegrate) after dying.

_____ 6. In a chemostat, when the dilution rate is increased while the nutrient concentration is kept constant, the microbial population density remains unchanged and the generation time shortens.

_____ 7. Continuous culture systems are very useful because they enable us to study microbial growth at very low nutrient levels that approximate nutrient levels of natural environments.

_____ 8. The water activity of a solution is directly proportional to the osmotic pressure.

_____ 9. Most microorganisms maintain their internal pH at approximately the same level as their optimum growth pH.

_____ 10. Psychrotrophic and psychrophilic bacteria can both grow at 0°C, but they differ in their other cardinal temperatures: the psychrotrophs' optimum and maximum growth temperatures are higher than those of the psychrophiles'.

_____ 11. Obligate anaerobes are usually poisoned by oxygen, but may grow in aerobic habitats if they are associated with facultative anaerobes that use up all of the available oxygen.

_____ 12. The mean generation time is the reciprocal of the mean growth rate constant.

CRITICAL THINKING

1. In this chapter, several methods used to determine microbial numbers were presented. These included total cell counts, viable cell counts, turbidity measurements, and biomass determinations, such as dry weight, protein, and DNA. What are the advantages and disadvantages of these various methods, and under what circumstances might one method be preferred to another?

2. Thermophilic bacteria that can grow in temperatures above 100°C have been found. However, this is above the boiling point of water. In what natural habitats might you find these conditions, and how is an aqueous environment maintained at that high a temperature? What types of adaptations would you envision for organisms living in those habitats?

ANSWER KEY

Terms and Definitions

1. o, 2. f, 3. s, 4. t, 5. ff, 6. hh, 7. i, 8. p, 9. w, 10. j, 11. h, 12. k, 13. gg, 14. aa, 15. q, 16. a, 17. x, 18. c, 19. g, 20. dd, 21. l, 22. bb, 23. cc, 24. u, 25. ee, 26. y, 27. n, 28. b, 29. z, 30. v, 31. e, 32. d, 33. m, 34. r

Microbial Growth Problems

1. $n = 16$ generations; $g = 15$ min (0.25 hr); $k = 4$ generations/hr

Microbial Growth Temperatures

1. psychrophilic, eurythermal 2. mesophilic, eurythermal 3. hyperthermophilic, eurythermal 4. thermophilic, stenothermal 5. mesophilic, eurythermal 6. mesophilic, stenothermal 7. psychrophilic, stenothermal 8. mesophilic, eurythermal 9. psychrotrophic, eurythermal

Fill in the Blank

1. growth; coenocytic; budding; binary fission 2. Petroff-Hausser; hemacytometer; 3. scatter; scattering; concentration; spectrophotometer 4. batch; continuous 5. microbial population level; generation time 6. turbidostat; turbidity; density 7. hypotonic; enter 8. water activity; solution; pure water 9. osmotolerant; halophiles 10. unsaturated; semi-fluid 11. terminal electron acceptor 12. facultative; aerotolerant 13. barotolerant; barophilic 14. thymine dimers; photoreactivation; dark reactivation 15. thermophiles; hyperthermophiles

Multiple Choice

1. d, 2. b, 3. c, 4. d, 5. a, 6. b, 7. c, 8. d, 9. c, 10. d

True/False

1. T, 2. F, 3. F, 4. T, 5. F, 6. T, 7. T, 8. F, 9. F, 10. T, 11. T, 12. T

7 Control of Microorganisms by Physical and Chemical Agents

CHAPTER OVERVIEW

This chapter focuses on the control and the destruction of microorganisms by physical and chemical agents. This is a topic of great importance, because microorganisms may have deleterious effects, such as food spoilage and disease. It is therefore essential to be able to kill or remove microorganisms from certain environments in order to minimize their harmful effects.

CHAPTER OBJECTIVES

After reading this chapter you should be able to:

- compare and contrast the processes of disinfection, sanitization, antisepsis, and sterilization
- compare the difficulties encountered when trying to kill endospores with those encountered when trying to kill vegetative cells
- discuss the exponential pattern of microbial death
- discuss the influence of environmental factors on the effectiveness of various agents used to control microbial populations
- discuss the uses and limitations of various physical and chemical agents used to control microbial populations
- describe the procedures used to evaluate the effectiveness of various antimicrobial agents

CHAPTER OUTLINE

I.   Definition of Frequently Used Terms
   A.   Sterilization—destruction or removal of all viable organisms from an object or from a particular environment
   B.   Disinfection—killing, inhibition, or removal of pathogenic microorganisms (usually on inanimate objects)
   C.   Antisepsis—prevention of infection of living tissue by microorganisms
   D.   Sanitization—reduction of the microbial population to a safe level as determined by public health standards
   E.   -cide—a suffix indicating that the agent will kill the kind of organism in question (e.g., viricide, fungicide)
   F.   -static—a suffix indicating that the agent will prevent the growth of the type of organism in question (e.g., bacteriostatic, fungistatic)
II.   The Pattern of Microbial Death—microorganisms are not killed instantly when exposed to a lethal agent; rather, the population decreases by a constant fraction at constant intervals (exponential killing); a microorganism is considered dead when it is unable to grow in conditions that would normally support its growth

III. Conditions Influencing the Effectiveness of Antimicrobial Agent Activity
    A. Population size—larger populations take longer to kill than smaller populations
    B. Population composition—microorganisms differ markedly in their sensitivity to various agents
    C. Concentration or intensity of the antimicrobial agent—higher concentrations or intensities are generally more efficient, but the relationship is not linear
    D. Duration of exposure—the longer the exposure, the greater the number of organisms killed
    E. Temperature—a higher temperature will usually (but not always) increase the effectiveness of killing
    F. Local environment—environmental factors, such as pH, viscosity, and concentration of organic matter can profoundly influence the effectiveness of a particular antimicrobial agent
IV. The Use of Physical Methods in Control
    A. Heat
        1. Moist heat
            a. Boiling water is effective against vegetative cells and eucaryotic spores
            b. Autoclaving (steam under pressure) is effective against vegetative cells and most bacterial endospores
            c. Pasteurization, a process involving brief exposure to temperatures below the boiling point of water, reduces the total microbial population and thereby increases the shelf life of the treated material; it is often used for heat-sensitive materials that cannot withstand prolonged exposure to high temperatures
                1) Low-temperature long-term (LTLT) pasteurization-63°C for 30 min
                2) High-temperature short-term (HTST) flash pasteurization-72°C for 15 sec
                3) Ultrahigh temperature (UHT) pasteurization-140 to 150°C for 1 to 3 sec
            d. Tyndallization (fractional steam sterilization), a process that kills spore-forming microorganisms, involves exposing the material to elevated temperatures (killing the vegetative cells), then incubation at 37°C (to allow spores to germinate to form new vegetative cells), and then exposure to elevated temperatures again (to kill the newly germinated vegetative cells); it is used to sterilize heat-sensitive materials or to kill particularly resistant endospores
        2. Dry heat can be used to sterilize moisture-sensitive materials such as powders, oils, and similar items; it is less efficient than moist heat because it usually requires higher temperatures (160 to 170°C) and longer exposure times (2 to 3 hrs)
        3. The thermal death time (TDT) is the shortest time necessary to kill all microorganisms in a suspension at a specific temperature and under defined conditions
        4. The decimal reduction time (D, or D value) is the time required to kill 90% of the microorganisms or spores in a sample at a specific temperature
        5. The Z value is the increase in temperature required to reduce D to 1/10 of its previous value
        6. The F value is the time in minutes at a specific temperature (usually 250°F or 121.1°C) necessary to kill a population of cells or spores
    B. Filtration sterilizes heat-sensitive liquids and gases by removing microorganisms rather than destroying them
        1. Depth filters are thick fibrous or granular filters that remove microorganisms by physical screening, entrapment, and/or adsorption
        2. Membrane filters are thin filters with defined pore sizes that remove microorganisms, primarily by physical screening
        3. High-efficiency particulate air (HEPA) filters are used in laminar flow biological safety cabinets to sterilize the air circulating in the enclosure
    C. Radiation
        1. Ultraviolet (UV) radiation is effective, but its use is limited to surface sterilization because UV radiation does not penetrate glass, dirt films, water, and other substances

2. Ionizing radiation (X rays, gamma rays, etc.) is effective and penetrates the material; the Food and Drug Administration and the World Health Organization have approved food irradiation and declared it safe; however, it is not widely used because of cost and concerns about the effects of the radiation on food

V. The Use of Chemical Agents in Control
   A. Phenolics—laboratory and hospital disinfectants; act by denaturing proteins
   B. Alcohols—widely used disinfectants and antiseptics; will not kill endospores; act by denaturing proteins and possibly by dissolving membrane lipids
   C. Halogens—widely used antiseptics and disinfectants; iodine acts by oxidizing cell constituents and iodinating cell proteins; chlorine acts primarily by oxidizing cell constituents
   D. Heavy metals—effective but usually toxic; act by combining with proteins and inactivating them
   E. Quaternary ammonium compounds—cationic detergents used as disinfectants for food utensils and small instruments, and because of low toxicity, as antiseptics for skin; act by disrupting biological membranes and possibly by denaturing proteins
   F. Aldehydes—reactive molecules that can be used as chemical sterilants; may irritate the skin; act by combining with proteins and inactivating them
   G. Sterilizing gases (e.g., ethylene oxide, betapropiolactone)—can be used to sterilize heat-sensitive materials such as plastic petri dishes and disposable syringes; act by combining with proteins and inactivating them
   H. Recently, vapor-phase hydrogen peroxide has been used to decontaminate biological safety cabinets

VI. Evaluation of Antimicrobial Agent Effectiveness—agents' effectiveness and safety must be tested under a variety of conditions, including conditions approximating those of normal use

TERMS AND DEFINITIONS

Place the letter of each term in the space next to the definition or description that best matches it.

_____ 1. Destruction or removal of all viable microorganisms from an object or a particular environment
_____ 2. A chemical that can be used to sterilize materials
_____ 3. The killing, inhibition, or removal of vegetative forms of pathogenic organisms or viruses
_____ 4. Agents used to carry out disinfection of inanimate objects
_____ 5. Reduction of the microbial population to safe levels as determined by public health standards
_____ 6. The prevention of infection
_____ 7. Chemical agents applied to living tissue to prevent infection, which act by killing or damaging pathogens
_____ 8. The time in minutes at a specific temperature needed to kill a population of cells or spores
_____ 9. The shortest period of time needed to kill all the organisms in a microbial suspension at a specific temperature and under defined conditions
_____ 10. The time required to kill 90% of the microorganisms or spores in a sample at a specific temperature
_____ 11. The increase in temperature required to reduce D to 1/10 its value

a. antisepsis
b. antiseptic
c. decimal reduction time (D)
d. detergent
e. disinfectant
f. disinfection
g. filtration
h. F value
i. pasteurization
j. phenol coefficient
k. sanitization
l. sterilant
m. sterilization
n. thermal death time (TDT)
o. Tyndallization (fractional steam sterilization)
p. Z value

_____ 12. The process in which a microbial population is reduced by raising the temperature of heat-sensitive materials to something less than the boiling point of water for relatively short periods of time

_____ 13. The process in which spore-forming organisms are killed by alternately heating to kill vegetative cells, and then incubating to allow germination of endospores into new vegetative cells, which are then killed by another cycle of heating

_____ 14. The process in which microorganisms are removed from liquids and gases by passing the material through a substance with pore sizes designed to prevent microorganism penetration

_____ 15. Organic molecules that serve as wetting agents and emulsifiers because they have both polar hydrophilic ends, and nonpolar hydrophobic ends; such molecules that are cationic are useful as disinfectants

_____ 16. A measure of disinfectant efficiency in which the disinfectant being evaluated is compared to phenol

FILL IN THE BLANK

1. A sterile object is _____ free of _____ microorganisms. Sterilization can be achieved by either _____ or _____ of contaminating microorganisms.

2. A disinfectant will not necessarily _____ an object because viable _____ may still remain.

3. Moist heat is thought to kill effectively by degrading _____, denaturing _____, and disrupting _____.

4. Pasteurization of milk involves heating it to 63°C for _____ minutes or to _____ for 15 seconds or to _____ for 1 to 3 seconds. This process does not _____ the milk but does kill _____ and reduce the total microbial population, thereby drastically slowing _____ and _____ shelf life.

5. In tyndallization, the first heating will destroy _____ cells but not _____. The latter will germinate during the subsequent 37°C incubation period and will be killed during the second heating period.

6. Gamma radiation has been used to _____ meats, fish, and other foods. However, this process has not been widely employed because of _____ and because of concerns about the long-term effects on the irradiated foods.

7. Phenolics are germicidal because they _____ and disrupt _____. However their use is limited because they have a disagreeable _____ and may cause irritation of the _____.

8. One of the most popular antiseptics has been _____ because it reduces skin bacteria for long periods owing to its persistence on the skin once applied. However, its use has been limited in recent years because it has been shown to cause _____.

9. The two most popular alcohol germicides are _____ and _____.

10. Iodine is an effective antiseptic, but there are some problems associated with its use. Recently, iodine has been complexed with an _____ carrier to form an _____. This preparation is water soluble, stable, and nonstaining, and releases _____ slowly, thus minimizing skin burns and irritation. This eliminates most of the problems associated with iodine use.

11. _____ is the disinfectant of choice for municipal water supplies and _____, and is also employed in the _____ and food industries. Campers frequently use this in the form of _____ tablets to purify small volumes of drinking water.

12. Heavy metals are no longer widely used as germicides. Two exceptions, however, are _____, which is applied to the eyes of infants to prevent ophthalmic gonorrhea, and _____, which is used as an algicide in lakes and swimming pools.
13. Quaternary ammonium compounds such as benzalkonium chloride will kill most bacteria, but will not kill _____ or _____. They have the advantage of being stable, nontoxic, and bland, but they are inactivated by _____ and _____.
14. The two commonly used aldehydes are _____ and _____. They are highly reactive and are _____, and can therefore be used as chemical sterilants.
15. Disinfectants are generally regulated by the _____, while antiseptics are generally regulated by the _____.

MULTIPLE CHOICE

For each of the questions below select the *one best* answer.

1. Which of the following is not a reason for studying methods of destroying microorganisms?
   a. It makes microbial research possible.
   b. preservation of food
   c. prevention of disease
   d. All of the above are reasons for studying methods of destroying microorganisms.
2. An agent that specifically kills fungi but not other kinds of microorganisms is best described as a
   a. germicide.
   b. fungicide.
   c. germistatic agent.
   d. fungistatic agent.
3. Which will require a longer time to kill?
   a. a large population of microorganisms
   b. a small population of microorganisms
   c. neither, killing will be equally rapid in either a large or a small microbial population
   d. There is no way to predict which will require a longer time to kill.
4. Which of the following may contribute to resistance to killing?
   a. formation/existence of endospores
   b. increased age of the culture
   c. inherent/genetic differences among organisms
   d. All of the above may contribute to resistance to killing.

5. Which of the following is not a method by which depth filters normally sterilize materials passing through them?
   a. exclusion of materials too large to penetrate the filter
   b. entrapment of material in the channels of the filter
   c. destruction of cells by interaction with the filter's harmful chemical composition
   d. adsorption of cells to the surface of the filter material
6. Chemical disinfectants and antiseptics are greatly influenced by
   a. temperature.
   b. concentration.
   c. duration of exposure.
   d. All of the above are correct.
7. Why is betapropiolactone not as useful as ethylene oxide as a sterilizing agent?
   a. It decomposes to an inactive form after several hours and, therefore, is more easily eliminated than ethylene oxide.
   b. It does not destroy microorganisms as readily as does ethylene oxide.
   c. It does not penetrate materials as well as ethylene oxide penetrates.
   d. All of the above are reasons why betapropiolactone is not as useful as ethylene oxide as a sterilizing agent.
8. Which of the following is normally used to remove microorganisms from air?
   a. depth filters
   b. membrane filters
   c. HEPA filters
   d. None of the above is correct.

9. Which of the following represents the best definition for microbial death?
   a. The organism will not grow on minimal medium.
   b. The organism will not grow on a medium that normally supports its growth.
   c. The organism no longer retains its original shape and structures.
   d. None of the above adequately describe microbial death.
10. The process of heating milk products to 72°C for 15 seconds in order to reduce the microbial population is called _____ pasteurization.
    a. flash
    b. fast
    c. ultrafast
    d. None of the above are correct.

11. Which of the following is(are) properties of an ideal disinfectant?
    a. effective against a wide variety of infectious agents
    b. nontoxic to people
    c. noncorrosive for common materials
    d. All of the above are properties of an ideal disinfectant.

TRUE/FALSE

____ 1. In sanitization, the inanimate object is usually cleaned as well as partially disinfected.
____ 2. Antiseptics are generally not as toxic as disinfectants.
____ 3. Although many agents will kill or control the growth of nonpathogenic organisms as well as pathogenic organisms, only their effects on pathogens are important in evaluating their usefulness.
____ 4. Although agents normally kill a constant fraction of the microbial population in constant period of time, the rate of killing may decrease when the population has been greatly reduced. One reason for this may be the survival of a resistant strain of the organism.
____ 5. Dry heat sterilization is generally faster than moist heat sterilization.
____ 6. Since filtration removes rather than destroys microorganisms, it does not truly sterilize the materials passing through the filter.
____ 7. Ultraviolet radiation is normally used to sterilize air and exposed surfaces, but it cannot be used to sterilize interior compartments because it does not penetrate into those compartments.
____ 8. Only anionic detergents are useful as disinfectants.
____ 9. Ethylene oxide is used to sterilize heat-sensitive materials such as plastic petri dishes, and is particularly effective because it can penetrate even the plastic wrap used to package these plates. Therefore, they can be sealed before sterilization, and the chance of contamination after treatment can be eliminated.
____ 10. The phenol coefficient is a direct indication of disinfectant potency during normal use.
____ 11. Death of microorganisms is defined as the loss of the ability to grow on medium that would normally support growth.

CRITICAL THINKING

1. Pasteurization can take place by either of two methods. In one a lower temperature is used for a longer period of time, while in the other a higher temperature is used for a shorter period of time. Which method is more advantageous and why?

2. Consider the various chemicals that can be used as disinfectants and antiseptics. What criteria would you use to select the right chemical for a particular task. Give at least two specific examples of situations. Then give your selection for each situation and defend your choice.

ANSWER KEY

Terms and Definitions

1. m, 2. l, 3. f, 4. e, 5. k, 6. a, 7. b, 8. h, 9. n, 10. c, 11. p, 12. i, 13. o, 14. g, 15. d, 16. j

Fill in the Blank

1. totally; viable; destruction; removal 2. sterilize; spores 3. nucleic acids; enzymes (proteins); cell membranes 4. 30; 72°C; 140 to 150°C; sterilize; pathogens; spoilage; increasing 5. vegetative; endospores 6. pasteurize; expense 7. denature proteins; cell membranes; odor; skin 8. hexachlorophene; brain damage 9. ethanol; isopropanol 10. organic; iodophore; iodine 11. Chlorine; swimming pools; dairy; halazone 12. silver nitrate; copper sulfate 13. *Mycobacterium tuberculosis;* spores; hard water; soap 14. formaldehyde; glutaraldehyde; sporicidal 15. Environmental Protection Agency; Food and Drug Administration

Multiple Choice

1. d, 2. b, 3. a, 4. d, 5. c, 6. d, 7. c, 8. c, 9. b, 10. a, 11. d

True/False

1. T, 2. T, 3. F, 4. T, 5. F, 6. F, 7. T, 8. F, 9. T, 10. F, 11. T

8 Metabolism: Energy and Enzymes

CHAPTER OVERVIEW

This chapter discusses energy and the laws of thermodynamics. The participation of energy in cellular metabolic processes and the role of adenosine-5'-triphosphate (ATP) as the energy currency of cells is examined. The chapter concludes with a discussion of enzymes as biological catalysts and the ways in which enzymes work and are affected by their environment.

CHAPTER OBJECTIVES

After reading this chapter you should be able to:

- discuss the first and second laws of thermodynamics and show how they apply to biological systems
- discuss enthalpy, entropy, and free energy and their application to biological reactions
- discuss the use of ATP as the energy currency of the cell and show how it is used to couple energy-yielding exergonic reactions with energy-requiring endergonic reactions
- discuss reduction potential and its relationship to exergonic and endergonic processes
- describe the role of enzymes in the catalysis of biological reactions, and discuss the ways in which enzymes are influenced by their environment

CHAPTER OUTLINE

I.   Energy and Work
  A.   Energy cycle
    1.   Cells must efficiently transfer energy from their energy-trapping systems to the systems that actually carry out work
    2.   Cells must also use various metabolic processes to replace the energy used in doing work
    3.   Energy currency is adenosine 5'-triphosphate-ATP
  B.   Cellular work
    1.   Chemical work—synthesis of complex molecules
    2.   Transport work—nutrient uptake, waste elimination, ion balance
    3.   Mechanical work—internal and external movement
  C.   The ultimate source of all biological energy is visible sunlight through photosynthesis
  D.   Complex molecules manufactured by photosynthetic organisms serve as a carbon source and energy source (through aerobic respiration) for chemoheterotrophs
  E.   The major energy currency in a living cell is adenosine-5'-triphosphate (ATP)

II. The Laws of Thermodynamics
   A. First law—energy can be neither created nor destroyed
      1. The total energy in the universe remains constant
      2. Energy may be redistributed either within a collection of matter called a system, or between the system and its surroundings
      3. Energy is measured in calories where 1 calorie is the amount of heat energy needed to raise 1 gram of water from 14.5 to 15.5°C
   B. Second law—physical and chemical processes proceed in such a way that the disorder of the universe increases to the maximum possible
III. Free Energy and Reactions
   A. Free energy change (ΔG) is the amount of energy in a system that is available to do work
      1. A negative ΔG indicates that the reaction is favorable and will proceed spontaneously (i.e., the reaction is exergonic)
      2. A positive ΔG indicates that the reaction is unfavorable and will only proceed if energy is supplied (i.e., the reaction is endergonic)
IV. The Role of ATP in Metabolism
   A. The reaction in which the terminal phosphate of ATP is removed goes to completion with a large negative standard free energy change (i.e., the reaction is strongly exergonic)
   B. Exergonic breakdown of ATP can be coupled with various endergonic reactions to facilitate their completion
   C. Metabolic-energy-trapping processes are used to catalyze the formation of ATP from ADP and P_i, and thus to restore the energy balance of the cell
V. Oxidation-Reduction Reactions and Electron Carriers
   A. Oxidation-reduction (redox) reactions involve the transfer of electrons from a donor (reducing agent or reductant) to an acceptor (oxidizing agent or oxidant)
   B. The equilibrium constant for the reaction is called the standard reduction potential (E_0) and is a measure of the tendency of the reducing agent to lose electrons
   C. Biological cells use electron carriers to transfer electrons from a reductant to an acceptor with a greater, more positive reduction potential, and they thereby allow the release of free energy, which is often used in the formation of ATP
   D. Biological cells have a variety of electron carriers, and each is used in particular types of redox reactions; the particular carrier used in any given reaction will depend on the nature and location of the reaction
VI. Enzymes
   A. Structure and classification of enzymes
      1. Enzymes are protein catalysts with great specificity for the reaction catalyzed and the molecules acted upon
      2. A catalyst is a substance that increases the rate of a reaction without being permanently altered itself
      3. The reacting molecules are called substrates and the substances formed are the products
      4. An enzyme may be a pure protein, or it may be a holoenzyme, which consists of a protein component (apoenzyme) and a nonprotein component (cofactor)
         a. Prosthetic group—a cofactor that is firmly attached to the apoenzyme
         b. Coenzyme—a cofactor that is loosely attached to the apoenzyme; it may dissociate from the apoenzyme and carry one or more of the products of the reaction to another enzyme
   B. The mechanism of enzyme reactions
      1. Enzymes increase the rate of a reaction, but do not alter its equilibrium constant (or its standard free energy change)
      2. Enzymes lower the activation energy required to bring the reacting molecules together correctly to form the transition-state complex, which resembles both the substrates and the products; once the transition state has been reached the reaction can proceed rapidly

3. Enzymes lower activation energy in several ways:
   a. Local concentrations of the substrates are increased at the active (catalytic) site of the enzyme
   b. Molecules at the active site are oriented properly for the reaction to take place
C. The effect of environment on enzyme activity
   1. The amount of substrate present affects the reaction rate, which increases as the substrate concentration increases until all available enzyme molecules are binding substrate and converting it to products as rapidly as possible; no further increase in rate occurs with subsequent increases in substrate concentration—the reaction is then said to be proceeding at maximal velocity (V_{max})
   2. The Michaelis constant (K_m) of an enzyme is the substrate concentration required for the reaction to reach half maximal velocity and is used as a measure for the apparent affinity of an enzyme for its substrate
   3. Enzyme activity is affected by alterations in pH and temperature; each enzyme has specific pH and temperature optima
   4. If the temperature rises too much above the optima, an enzyme's structure will be disrupted and its activity lost; this phenomena, known as denaturation, may be caused by extremes of pH or temperature extremes or by other factors
D. Enzyme inhibition
   1. Competitive inhibition occurs when the inhibitor binds at the active site and thereby competes with the substrate (if the inhibitor binds, then the substrate cannot, and no reaction occurs); this type of inhibition can be overcome by adding excess substrate
   2. Noncompetitive inhibition occurs when the inhibitor binds to the enzyme at some location other than the active site, and changes the enzyme's shape so that it is inactive or less active; this type of inhibition cannot be overcome by the addition of excess substrate

TERMS AND DEFINITIONS

Place the letter of each term in the space next to the definition or description that best matches it.

_____ 1. During this phenomenon, the breakdown of ATP to ADP and P_i releases energy to do work for the cell; other processes store energy by reforming ATP from ADP and P_i

_____ 2. The science that analyzes energy changes in a collection of matter

_____ 3. The postulate that energy can be neither created nor destroyed

_____ 4. The unit of measurement that describes the amount of heat needed to raise 1 gram of water from 14.5°C to 15.5°C

_____ 5. The unit of measurement that describes the amount of work capable of being done; 1 calorie = 4.184 J

_____ 6. The postulate that physical and chemical processes occur in such a way that randomness (disorder) increases to a maximum

_____ 7. Describes the randomness or disorder of a system

_____ 8. The total energy of a system

_____ 9. The energy of a reaction that is available to do useful work

_____ 10. A reaction in which the forward rate equals the reverse rate

_____ 11. A favorable reaction that releases energy (ΔG is negative)

_____ 12. An unfavorable reaction which requires an input of energy in order to proceed (ΔG is positive)

_____ 13. Reactions in which there is a transfer of electrons from a donor to an acceptor

_____ 14. The electron donor in a redox reaction

_____ 15. The electron acceptor in a redox reaction

_____ 16. Protein catalysts with great specificity for the reaction catalyzed and the molecules acted upon

_____ 17. A substance that increases the rate of a reaction without being permanently altered by the reaction

_____ 18. The reacting molecules in an enzyme-catalyzed reaction

_____ 19. The molecules formed by a chemical reaction

_____ 20. A complex formed during a reaction that is composed of the substrates; it resembles both the substrates and the products

_____ 21. The energy required to bring the reacting molecules together in the correct way to reach the transition state

_____ 22. A special place on the surface of an enzyme where the substrates are brought together in the proper orientation for a reaction to occur

_____ 23. Term that describes the velocity of a reaction when all available enzyme molecules are binding substrate and converting it to product as rapidly as possible

a. activation energy
b. active site
c. calorie
d. catalyst
e. competitive inhibitor
f. denaturation
g. endergonic reaction
h. energy cycle
i. enthalpy
j. entropy
k. enzyme
l. equilibrium
m. exergonic reaction
n. first law of thermodynamics
o. free energy change
p. joule
q. maximal velocity (V_{max})
r. Michaelis constant (K_m)
s. noncompetitive inhibitor
t. oxidation-reduction (redox) reactions
u. oxidizing agent (oxidant)
v. products
w. reducing agent (reductant)
x. second law of thermodynamics
y. substrates
z. thermodynamics
aa. transition state

____ 24. Constant that is equal to the substrate concentration at which an enzyme-catalyzed reaction reaches half maximal velocity

____ 25. A molecule that binds to an enzyme at the active site and thereby prevents the substrate from binding and reacting

____ 26. A molecule that binds to an enzyme at some location other than the active site and alters the enzyme's shape so that it is inactive or less active

____ 27. Disruption of an enzyme's structure with loss of activity caused by extremes of pH, temperature, or other factors

FILL IN THE BLANK

1. The flow of carbon and energy in an ecosystem are intimately related. Light energy is trapped by _____ and some of this energy is obtained by _____ when they use the former for nutrients. The _____ produced during respiration can be incorporated into complex organic molecules during _____.

2. The science of _____ analyzes energy changes in a collection of matter called a _____. All other matter in the universe is called the _____.

3. The first law of thermodynamics states that energy can be neither _____ nor _____. Thus the total energy in the universe remains constant, even though it may be _____.

4. For any reaction, when $\Delta G^{0'}$ is negative, the equilibrium constant is _____, the reaction is said to be _____, and the reaction goes to completion in the way it is written. However, in an _____ reaction, $G^{0'}$ is positive, the equilibrium constant is _____, and the reaction is unfavorable; therefore, little product will be formed at equilibrium under standard conditions.

5. ATP is ideally suited for its role as energy currency. It is formed in energy-trapping and energy-generating processes such as _____, _____, and _____. In the cells economy, _____ ATP breakdown can be coupled with various _____ reactions to facilitate their completion.

6. In a redox reaction, electrons are transferred from a donor, the _____, to an electron acceptor, the _____. The two molecules are referred to as a _____ _____. When an _____ accepts electrons, it becomes the _____ of the pair.

7. The equilibrium constant for a redox reaction is called the _____ and is a measure of the tendency of the reductant to _____ electrons.

8. A number of enzymes are pure proteins. However, some enzymes consist of a protein component, the _____, plus a nonprotein component called a _____. The two together constitute the _____. When the nonprotein component is firmly attached to the protein it is referred to as a _____; when it is loosely attached, it is referred to as a _____.

9. The _____, K_m, is equal to the substrate concentration at which an enzyme-catalyzed reaction reaches half maximal velocity. It is used as a measure of the apparent _____ of an enzyme for its substrate. The lower the value of K_m the _____ the substrate concentration at which the enzyme catalyzes the reaction.

10. An inhibitor that binds at the active site and thereby prevents the binding of the substrate is called a _____ inhibitor; while an inhibitor that binds at a location other than the active site, thus altering the enzyme's shape so that it is inactive or less active, is called a _____ inhibitor.

11. The _____ _____ involves efficient energy transfer from the cell's energy-trapping systems to the systems actually doing work and to the various _____ _____ that replace the energy used in doing that work.

12. The synthesis of complex molecules involves _____ work; nutrient uptake, waste elimination, and the maintenance of ion balances involves _____ work; and internal and external movement involves _____ work.

MULTIPLE CHOICE

For each of the questions below select the *one best* answer.

1. The amount of heat energy needed to raise 1.0 gram of water from 14.5°C to 15.5°C is called a(n)
   a.   joule.
   b.   calorie.
   c.   erg.
   d.   thermal unit.

2. For the reaction
   $$A + B \rightleftharpoons C + D$$
   the equilibrium constant (K_{eq}) is defined as
   a. $\dfrac{[A][B]}{[C][D]}$

   b. $\dfrac{[C][D]}{[A][B]}$

   c. $\dfrac{[A][D]}{[B][C]}$

   d. $\dfrac{[B][C]}{[A][D]}$

3. Living organisms use a variety of electron carriers to aid in the cycle of energy flow. Which of the following is used as an electron carrier?
   a. NAD^+
   b. $NADP^+$
   c.   Ubiquinone
   d.   All of the above are used as electron carriers.

4. Which of the following is not true about enzymes?
   a.   Enzymes are catalysts and therefore, they increase the rate of a reaction without being permanently altered by the reaction.
   b.   Enzymes are proteins that can be denatured by changes in pH or temperature.
   c.   Enzymes are highly specific for the substrates with which they react, but they catalyze several possible reactions with those substrates.
   d.   All of the above are true about enzymes.

5. The energy required to bring the reacting molecules together in the correct way to form the transition-state complex is called the
   a.   activation energy.
   b.   free energy.
   c.   entropy.
   d.   enthalpy.

6. Which of the following is not a way in which enzymes lower the activation energy required for a reaction?
   a.   bringing the substrates together at the active site; in effect, concentrating them
   b.   binding the substrates so that they are correctly oriented to form the transition-state complex
   c.   increasing molecular motion, thereby providing kinetic energy to drive the reaction
   d.   All of the above are ways in which enzymes lower the activation energy required for a reaction.

7. Which of the following is not a function of the transport work done by a cell?
   a.   uptake of nutrients
   b.   elimination of waste products
   c.   maintenance of internal/external ion balances
   d.   All of the above are functions of cellular transport work.

TRUE/FALSE

____ 1. A reaction will occur spontaneously if the free energy of the system decreases during the reaction (i.e., if ΔG is negative).

____ 2. Since ΔS (the change in entropy) is a measure of disorder, a decrease in ΔS will lead to a decrease in ΔG, and therefore, the reaction will proceed spontaneously.

____ 3. The value of $\Delta G^{0'}$ indicates how fast a reaction will reach equilibrium.

____ 4. Redox couples that have greater negative reduction potentials will donate electrons to couples that have higher positive potentials. This is the basis for the electron transport chain that functions during respiration.

____ 5. Ferredoxin is a nonheme iron protein that is active in photosynthetic electron transport.

____ 6. Enzymes increase the rate of a reaction but do not alter equilibrium constants.

____ 7. When the amount of enzyme present is held constant, the rate of a reaction will continue to increase as long as the substrate concentration increases.

____ 8. Enzyme activity can be greatly affected by the pH, and temperature of the environment in which the enzyme must function.

____ 9. The ultimate source of all biological energy is visible sunlight through the process of photosynthesis.

CRITICAL THINKING

1. Consider the following diagram of the energy flow for a particular reaction. Is the reaction exergonic or endergonic? What does the diagram indicate? How would the use of an enzyme catalyst affect the energy flow? Indicate this on the diagram, and also indicate the energy of activation and the free energy change of both the catalyzed and uncatalyzed reactions.

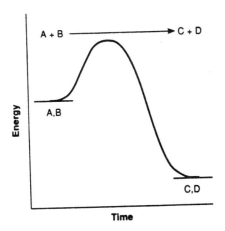

2. Each of the following two diagrams indicates the rate of a reaction as a function of substrate concentration. In each case an inhibitor is present (not the same one) and the substrate concentration is not saturating the enzyme. Explain the difference between the two situations.

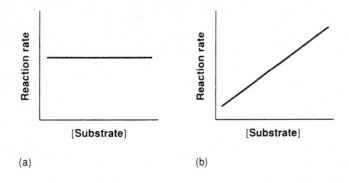

(a)   (b)

ANSWER KEY

Terms and Definitions

1. h, 2. z, 3. n, 4. c, 5. p, 6. x, 7. j, 8. i, 9. o, 10. l, 11. m, 12. g, 13. t, 14. w, 15. u, 16. k, 17. d, 18. y, 19. v, 20. aa, 21. a, 22. b, 23. q, 24. r, 25. e, 26. s, 27. f

Fill in the Blank

1. photoautotrophs; chemoheterotrophs; carbon dioxide; photosynthesis 2. thermodynamics; system; surroundings 3. created; destroyed; redistributed 4. greater than one; exergonic; endergonic; less than one 5. photosynthesis; fermentation; aerobic respiration; exergonic; endergonic 6. reductant; oxidant; redox couple; oxidant; reductant 7. standard reduction potential; lose 8. apoenzyme; cofactor; holoenzyme; prosthetic group; coenzyme 9. Michaelis constant; affinity; lower 10. competitive; noncompetitive 11. energy cycle; metabolic processes 12. chemical; transport; mechanical

Multiple Choice

1. b, 2. b, 3. d, 4. c, 5. a, 6. c, 7. d

True/False

1. T, 2. F, 3. F, 4. T, 5. T, 6. F, 7. F, 8. T, 9. T

9 Metabolism: The Generation of Energy

CHAPTER OVERVIEW

This chapter presents an overview of metabolism beginning with carbohydrate degradation and the aerobic generation of ATP through electron transport. Fermentation and anaerobic respiration are examined, followed by the catabolism of lipids, proteins, and amino acids. The chapter concludes with discussions of the function of inorganic molecules as electron acceptors and the trapping of energy by photosynthesis.

CHAPTER OBJECTIVES

After reading this chapter you should be able to:

- discuss the difference between catabolism and anabolism
- describe the various pathways for the catabolism of glucose to pyruvate
- discuss the tricarboxylic acid (TCA) cycle and its central role in aerobic metabolism
- describe the electron transport process, and compare and contrast the electron transport system of eucaryotes with that of procaryotes
- contrast the two major proposed mechanisms for oxidative phosphorylation (i.e., the chemiosmotic hypothesis and the conformational change hypothesis)
- discuss the different electron acceptors used during aerobic respiration, fermentation, and anaerobic respiration
- describe in general terms the catabolism of molecules other than carbohydrates
- discuss the photosynthetic light reactions, and compare and contrast the light reactions of eucaryotes (and cyanobacteria) with those of green (or purple) photosynthetic bacteria

CHAPTER OUTLINE

I. An Overview of Metabolism
   A. Catabolism—the breakdown of larger, more complex molecules into smaller, simpler ones, during which energy is released, trapped, and made available for work
   B. Anabolism—the synthesis of complex molecules from simpler ones during which energy is added as input
   C. Chemolithotrophy and photosynthesis are included as energy-yielding metabolic processes, even though they do not involve degradation of complex molecules
   D. Multi-stage process
      1. Stage 1—breakdown of large molecules (polysaccharides, lipids, proteins) into their component constituents with the release of little (if any) energy
      2. Stage 2—degradation of the products of stage 1 aerobically or anaerobically to even simpler molecules with the production of some ATP, NADH, and/or $FADH_2$
      3. Stage 3—complete aerobic oxidation of stage 2 products with the production of ATP, NADH, and $FADH_2$; the latter two molecules are processed by electron transport to yield much of the ATP produced

E.   Metabolic efficiency is maintained by the use of a few common catabolic pathways, each degrading many nutrients

F.   Microorganisms are catabolically diverse, but are anabolically quite uniform

G.   Amphibolic pathways function both catabolically and anabolically, and sometimes employ separate enzymes to catalyze the forward and reverse reactions; this separation enables independent regulation of the forward and reverse reactions

II.   The Breakdown of Glucose to Pyruvate

A.   The glycolytic (Embden-Meyerhof) pathway is the most common pathway and is divided into two parts:

1.   The 6-carbon sugar stage involves the phosphorylation of glucose twice to yield fructose 1,6-bisphosphate, and requires the expenditure of two molecules of ATP

2.   The 3-carbon sugar stage cleaves fructose 1,6-bisphosphate into two 3-carbon molecules, which are each processed to pyruvate; two molecules of ATP are produced by substrate-level phosphorylation from each of the 3-carbon molecules for a net yield of two molecules of ATP; 2 molecules of NADH are also produced per glucose molecule

B.   The pentose phosphate (hexose monophosphate) pathway uses a different set of reactions to produce a variety of 3-, 4-, 5-, 6-, and 7-carbon sugar phosphates

1.   These phosphates can be used to produce ATP and NADPH, as well as to provide the carbon skeletons for the synthesis of amino acids, nucleic acids, and other macromolecules

2.   The NADPH can be used to provide electrons for biosynthetic processes or can be converted to NADH to yield additional ATP through the electron transport chain

C.   The Entner-Doudoroff pathway can also be used to produce pyruvate with a lower yield of ATP, but is accompanied by the production of NADPH as well as NADH.

III.   The Tricarboxylic Acid Cycle—a series of reactions

A.   Acetyl-CoA (produced by decarboxylation of pyruvate) reacts with oxaloacetate to produce a 6-carbon molecule

B.   Subsequently, two molecules of carbon dioxide are released, regenerating the oxaloacetate

C.   ATP is produced by substrate-level phosphorylation

D.   Three molecules of NADH and one molecule of $FADH_2$ are produced per acetyl-CoA, and can be further processed to produce more ATP

E.   Even those organisms that lack the complete TCA cycle usually have most of the cycle enzymes because one of the TCA cycle's major functions is to provide carbon skeletons for use in biosynthesis

IV.   Electron Transport and Oxidative Phosphorylation

A.   The Electron Transport Chain

1.   Electrons from NADH and $FADH_2$ are transported in a series of redox reactions to a terminal electron acceptor

2.   Electron carriers are located within the inner mitochondrial membrane in eucaryotes, or within the plasma membrane in procaryotes

3.   The bacterial electron transport chain may be extensively branched with several terminal oxidases

4.   Bacterial electron transport chains may be shorter, and have lower P/O ratios than mitochondrial electron transport chains

5.   Procaryotic and eucaryotic electron transport chains, therefore, differ in the details of construction, although they operate according to the same fundamental principles

B.   Oxidative Phosphorylation

1.   Some of the energy liberated during electron transport is used to drive the synthesis of ATP in a process called oxidative phosphorylation

a.   The chemiosmotic hypothesis of oxidative phosphorylation postulates that the energy released during electron transport is used to establish a proton gradient, and that this protonmotive force is then used to drive ATP synthesis

      b.   The conformational change hypothesis of oxidative phosphorylation postulates that the energy released during electron transport causes conformational changes in the ATP-synthesizing enzyme, which drives ATP formation by increasing the substrates' binding affinities

      c.   There is evidence for conformational changes and molecular rotation in the ATP synthase complex during proton movement across the membrane and therefore the system exhibits features of both hypotheses

  2. Inhibitors of ATP synthesis fall into two main categories:

      a.   Blockers that inhibit the flow of electrons through the system

      b.   Uncouplers that allow electron flow, but disconnect it from oxidative phosphorylation

C.   The Yield of ATP in Glycolysis and Aerobic Respiration

  1. The yield of ATP in glycolysis and aerobic respiration varies with each organism, but has a theoretical maximum of 38 molecules of ATP per molecule of glucose catabolized

  2. Anaerobic organisms using glycolysis can only produce two molecules of ATP per molecule of glucose catabolized

  3. Aerobic respiration yields between 2 and 38 ATP molecules per glucose molecule, depending on the precise nature of the electron transport system

  4. The Pasteur effect is a regulatory phenomenon by which organisms lower their rate of sugar catabolism when conditions cause a shift from anaerobic to aerobic metabolism because the aerobic process is more efficient and generates greater energy per glucose molecule

V.   Fermentations

A.   In the absence of oxygen, NADH is not usually oxidized by the electron transport chain because no external electron acceptor is available

B.   However, NADH must still be oxidized to replenish the supply of NAD^+ for use in glycolysis

C.   Fermentations are reactions that regenerate NAD^+ from NADH in the absence of oxygen

  1. Fermentations involve pyruvate or pyruvate derivatives as electron acceptors

  2. Fermentations may or may not produce additional ATP for the cell

D.   Alcoholic fermentations produce ethanol and CO_2

E.   Lactic acid fermentations produce lactic acid (lactate)

  1. Homolactic fermenters reduce almost all pyruvate to lactate

  2. Heterolactic fermenters form substantial amounts of products other than lactate

F.   Formic acid fermentation produces either mixed acids or butanediol

VI.   Anaerobic Respiration

A.   Uses inorganic molecules other than oxygen as terminal electron acceptors; this produces additional ATP for the cell, but not usually as much as is produced by aerobic respiration

B.   The major electron acceptors are nitrate, sulfate, and CO_2 but some metals can also be reduced

VII. Catabolism of Carbohydrates and Intracellular Reserve Polymers—proceeds by either hydrolysis or phosphorolysis to produce molecules that can enter the common catabolic pathways already discussed

VIII. Lipid Catabolism—proceeds by the β-oxidation pathway, which produces acetyl-CoA, which can enter the TCA cycle

IX.   Protein and Amino Acid Catabolism

A.   Proteins are degraded by secreted proteases to their component amino acids, which are transported into the cell and catabolized

B.   The amino group is removed by deamination or transamination

C.   The resulting organic acids are converted to pyruvate, acetyl-CoA, or a TCA-cycle intermediate

X.   Oxidation of Inorganic Molecules

A.   A pathway used by a small number of microorganisms called chemolithotrophs

B.   Produces a significant but low yield of ATP

C.   The electron acceptor is usually O_2, but sulfate and nitrate are also used

D.   The most common electron donors are hydrogen, reduced nitrogen compounds, reduced sulfur compounds, and ferrous iron (Fe^{2+})

XI.   Photosynthesis—energy from light is trapped and used to produce ATP and NADPH (light reactions), and to reduce carbon dioxide to form carbohydrates (dark reactions)
   A.   Photosynthetic organisms serve as the base of most food chains in the biosphere
   B.   Photosynthesis is also responsible for replenishing our supply of O_2
   C.   Cyclic photophosphorylation is performed by photosystem I; an excited electron is recycled to the chlorophyll of origin with concomitant production of ATP
   D.   Noncyclic photophosphorylation is performed by photosystem II; excited electrons from chlorophyll molecules are used in the production of NADPH, with concomitant production of ATP; electrons are replaced from water and oxygen gas is produced as a waste product
   E.   Photosynthesis in green and purple photosynthetic bacteria differs from photosynthesis in eucaryotes and cyanobacteria: water is not used as an electron source and oxygen is not produced; thus, these organisms are said to be anoxygenic

TERMS AND DEFINITIONS

Place the letter of each term in the space next to the definition or description that best matches it.

____ 1. The total of all chemical reactions occurring in a cell	a. aerobic respiration
____ 2. The breakdown of larger, more complex molecules into smaller, simpler ones, which is accompanied by energy release and the trapping of some of that energy	b. amphibolic pathways
	c. anabolism
	d. anaerobic respiration
____ 3. The synthesis of complex molecules from simpler ones, accompanied by energy input	e. catabolism
	f. chemiosmotic hypothesis
____ 4. Metabolic pathways that function both catabolically and anabolically	g. conformational change hypothesis
____ 5. The most common pathway for degradation of glucose to pyruvate	h. cyclic photophosphorylation
	i. electron transport
____ 6. ADP phosphorylation that is coupled with the exergonic breakdown of a high-energy substrate molecule	j. Embden-Meyerhof (glycolytic) pathway
____ 7. An alternative to the glycolytic pathway that may operate simultaneously	k. Entner-Doudoroff pathway
	l. fermentation
____ 8. An alternative to glycolysis that is used by a few genera of bacteria	m. metabolism
	n. noncyclic photophosphorylation
____ 9. A cyclical set of reactions that is used by aerobic organisms to generate additional energy, as well as to provide carbon skeletons for use in biosynthesis	o. oxidative phosphorylation
	p. Pasteur effect
____ 10. The transfer of electrons in which a series of electron carriers is employed, and during which energy is liberated	q. pentose phosphate (hexose monophosphate) pathway
____ 11. Production of ATP using the energy liberated by the electron transport system	r. photosynthesis
	s. protonmotive force
____ 12. A gradient of protons and a membrane potential that are due to an unequal distribution of charges	t. substrate level phosphorylation
____ 13. The postulate that ATP synthesis is driven by proton diffusion back into the mitochondrion as a result of the protonmotive force established during electron transport	u. tricarboxylic (citric) acid cycle
	v. uncouplers
____ 14. The postulate that ATP synthesis is driven by change in the shape of the enzyme ATP synthase that is induced by electron transport	
____ 15. Inhibitors that stop ATP synthesis without inhibiting electron transport	

____ 16. The phenomenon in which the rate of sugar catabolism is drastically reduced when cells switch from anaerobic to aerobic conditions

____ 17. A process for oxidizing NADH to NAD^+ using organic molecules as electron donors and acceptors

____ 18. A process for oxidizing NADH to NAD^+ using molecular oxygen as the terminal electron acceptor

____ 19. A process for oxidizing NADH to NAD^+ using inorganic molecules other than oxygen as the terminal electron acceptors

____ 20. The process in which light energy is trapped and converted to chemical energy in the form of ATP and NADPH

____ 21. A process in which excited electrons from a chlorophyll molecule return to that molecule driving the synthesis of ATP

____ 22. A process in which excited electrons from a chlorophyll molecule are given to NADPH, are replaced with electrons from water, and drive the synthesis of ATP

METABOLIC ENERGY: NUTRIENT PROCESSING

Consider the following metabolic processes and answer the questions below. (Where numbers are asked for, give the number *per glucose molecule entering the pathway.*)

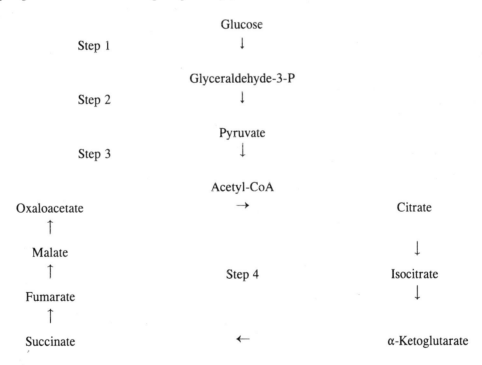

1. At which step(s) is there an investment of ATP? How many ATP molecules must be invested?

2. At which step(s) is there a production of ATP by substrate level phosphorylation? How many are produced?

3. At which step(s) is there a production of NADH? How many are produced?

4. At which step(s) is there a production of $FADH_2$? How many are produced?

5. Considering that NADH and FADH can be processed through the electron transport chain to produce additional ATP, what is the maximum net gain of ATP in eucaryotes?

6. Not all NADH molecules give the same net yield of ATP. Which ones are different and why?

7. At which step(s) are CO_2 molecules released? How many are released at each step?

ENERGY METABOLISM: PHOTOSYNTHESIS

Consider the process of photosynthesis and answer the following questions:

1. The goal of the light-dependent reactions of photosynthesis is to provide two molecules for the cell. What are they? What does the organism use them for?

2. Photosystem I (PS I), in the absence of photosystem II (PS II), can be used to provide one of these molecules. Which one?

3. The process by which PS I accomplishes this is diagrammed below. In this process the electrons given up by P700 are cycled back to P700. What is this process called?

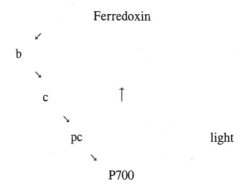

4. Why can't this process be used to provide the other molecule? (HINT: Consider the fate of the electrons given up by P700.)

5. How did the evolution of photosystem II (PS II) solve this problem? Diagram the overall process. (HINT: Use the diagram of PS I as your starting point.) On your diagram show where the two molecules referred to in question 1 are produced.

6. What is the fate of the electrons given up by P680 of PS II? How are they replaced?

ATP YIELD FROM THE AEROBIC OXIDATION OF GLUCOSE

Complete the table below by supplying the missing information. Answers should be expressed as molecules produced (yield) per molecule of glucose oxidized.

Metabolic Step	Number of Molecules Produced				Net Yield of ATP*
	ATP	GTP	NADH	FADH$_2$	
1. Glycolysis					
A. Glucose → Fructose 1,6-bisphosphate	——	——	——	——	——
B. Fructose 1,6-bisphosphate → Pyruvate	——	——	——	——	——
2. Pyruvate → Acetyl-CoA	——	——	——	——	——
3. Tricarboxylic Acid Cycle					
A. 6-Carbon Stage	——	——	——	——	——
B. 5-Carbon Stage	——	——	——	——	——
C. 4-Carbon Stage	——	——	——	——	——
Total Yield	——	——	——	——	——

*Assume maximum yield of ATP from NADH and FADH$_2$ processed through the electron transport chain

FILL IN THE BLANK

1. The breakdown of larger, more complex molecules into smaller, simpler ones with the release of energy is called _____ . Some of this energy is trapped and made available for work in _____ , which is the synthesis of complex molecules from simpler ones. The latter process takes energy to increase the _____ of a system.

2. The above definition does not encompass such energy-yielding processes as _____ and _____ , where complex molecules are not degraded but other energy sources are used.

3. Most metabolic reactions are freely reversible and can be used to both synthesize and degrade molecules. The few irreversible _____ steps are bypassed in biosynthesis with special enzymes that catalyze the reverse reaction, thereby allowing independent _____ of catabolic and anabolic functions of these _____ pathways.

4. In the tricarboxylic acid (TCA) cycle, two carbons in the form of _____ are added to oxaloacetate at the start of the cycle. Subsequently, two carbons in the form of _____ are removed, regenerating oxaloacetate. Thus, there is no net carbon loss or gain during the production of _____, _____, and _____, which are all used to store additional energy.

5. The mitochondrial electron transport system is arranged into four complexes of carriers, each capable of transporting electrons part of the way to oxygen. The complexes are connected to each other by _____ _____ and _____ _____.

6. The two most important hypotheses about the mechanism of oxidative phosphorylation are the _____ hypothesis and the _____ hypothesis.

7. The Pasteur effect is a drastic reduction in the rate of sugar catabolism that occurs in many microorganisms when they are changed from _____ to _____ conditions. This occurs because less sugar must be degraded to produce _____ amount of ATP when aerobic processes can be employed.

8. The reduction of pyruvate to lactate is called _____ fermentation. If organisms directly reduce almost all of their pyruvate to lactate, they are called _____ fermenters; organisms that form both lactate and substantial amounts of products other than lactate are called _____ fermenters.

9. Although electrons derived from sugars and other organic molecules are usually donated to organic electron acceptors during _____, or to molecular oxygen during _____, some bacteria have electron transport chains that can operate with inorganic electron acceptors such as _____, _____, and _____ in a process called _____.

10. Disaccharides and polysaccharides can be processed as nutrients after first being cleaved to monosaccharides by either _____ or _____.

11. Fatty acids are metabolized by the _____ pathway in which 2-carbon units are cleaved off to form _____, which is channeled into the TCA cycle.

12. Proteins are catabolized by hydrolytic cleavage to amino acids by the action of enzymes called _____. The amino acids are then processed by first removing the amino group through _____ or _____.

13. The energy of all living organisms is ultimately derived from the process of _____. It provides the organisms that actually undergo this process with the _____ and _____ necessary to manufacture the organic material required for their growth. These organisms then serve as the base for most of the _____ _____ in the biosphere, and in addition, these organisms are responsible for replenishing our supply of _____.

14. Photosynthesis as a whole is divided into two parts. In the _____ _____, light energy is trapped and converted to chemical energy, which is then used in the _____ _____ to reduce or fix _____ _____ and synthesize cellular constituents.

15. Green and purple photosynthetic bacteria do not use water as an electron source and do not produce _____. Therefore, they are said to be _____, in contrast with cyanobacteria and eucaryotic photosynthesizers, which are almost always _____.

16. In procaryotes the electron transport chain is located within the _____ _____, while in eucaryotes the electron transport chain is located within the _____ _____ _____.

17. Fermentation reactions oxidize _____ to _____ using _____ molecules as electron acceptors; in the course of this process, they may or may not generate additional _____.

MULTIPLE CHOICE

For each of the questions below select the *one best* answer.

1. Lehninger described three stages of metabolism. Which of the following is not one of those three stages?
   a. hydrolysis of large macromolecules (proteins, polysaccharides, and lipids)
   b. photosynthesis
   c. degradation of amino acids, fatty acids, glycerol, and monosaccharides to simpler molecules
   d. tricarboxylic acid cycle

2. Which of the following does not represent a major function of carbohydrates and other nutrients?
   a. use, without modification, as cellular constituents
   b. use, with modification, as cellular constituents
   c. oxidation to provide energy
   d. All of the above are major functions of carbohydrates and other nutrients.

3. Which of the following is the most common pathway for degradation of glucose to pyruvate?
   a. Entner-Doudoroff pathway
   b. pentose phosphate pathway
   c. phosphogluconate pathway
   d. Embden-Meyerhof pathway

4. The synthesis of ATP from ADP and P_i, when coupled with the exergonic enzymatic breakdown of a high-energy molecule, is called
   a. oxidative phosphorylation.
   b. chemiosmotic phosphorylation.
   c. substrate-level phosphorylation.
   d. conformational change phosphorylation.

5. Which of the following is not a useful outcome of the pentose phosphate pathway?
   a. production of ATP from the metabolism of glyceraldehyde 3-phosphate to pyruvate
   b. production of NADPH, which serves as a source of electrons during biosynthesis
   c. production of 4- and 5-carbon sugars for amino acid biosynthesis, nucleic acid synthesis, and photosynthesis
   d. All of the above are useful outcomes of the pentose phosphate pathway.

6. Which of the following statements is true about the similarities or differences between eucaryotic and procaryotic electron transport?
   a. They use the same electron carriers.
   b. In procaryotes electrons can enter at a number of points and leave through several terminal oxidases, while in eucaryotes electrons can enter at several points but can only leave at a single terminal oxidase.
   c. Procaryotic chains have the same or higher P/O ratios than eucaryotic mitochondrial electron transport chains.
   d. None of the above are true.

7. Which of the following is not true of fermentation reactions?
   a. NADH is oxidized to NAD^+.
   b. The electron acceptor is pyruvate or a pyruvate derivative.
   c. Some fermentations generate additional ATP for the organism.
   d. All of the above are true of fermentation reactions.

8. To reduce one molecule of carbon dioxide into carbohydrate it takes
   a. three molecules of ATP.
   b. two molecules of NADPH.
   c. 10 to 12 quanta of light.
   d. All of the above are correct.

9. Which of the following may serve as inorganic energy sources for chemolithotrophic organisms?
   a. hydrogen gas
   b. nitrogen compounds
   c. sulfur compounds
   d. All of the above are correct.

10. Which of the following best describes the evidence that currently exists about oxidative phosphorylation?
    a. The evidence currently supports the chemiosmotic hypothesis.
    b. The evidence currently supports the conformational change hypothesis.
    c. The evidence indicates that aspects of both hypotheses may be correct.
    d. The evidence favors neither of these hypotheses and suggests the existence of a different, as yet unknown, mechanism.

11. Which of the following is not true about fermentations?
    a. They always use pyruvate or pyruvate derivatives as electron acceptors
    b. They always generate additional ATP
    c. They always recycle NADH to NAD^+
    d. All of the above are true about fermentations

TRUE/FALSE

____ 1. Pathways are either catabolic or anabolic, but cannot be both.

____ 2. Glycolysis and the pentose phosphate pathway are mutually exclusive. An organism may use one or the other, but not both simultaneously.

____ 3. Organisms that lack some of the TCA cycle enzymes, and that therefore do not use the TCA cycle for energy production, usually have most of the TCA cycle enzymes in order to fulfill the other major function of the cycle, providing carbon skeletons for use in biosynthesis.

____ 4. The P/O ratio represents the number of ATP molecules generated when electrons are passed from a donor such as NADH or $FADH_2$ to oxygen.

____ 5. Most of the ATP that is generated aerobically comes from the oxidation of NADH and $FADH_2$ in the electron transport chain.

____ 6. There is considerable evidence for the generation of proton gradients across membranes, but the evidence that such a gradient is the direct driving force for oxidative phosphorylation is not yet conclusive.

____ 7. Uncouplers stop ATP synthesis without inhibiting electron transport and may even enhance the rate of electron flow.

____ 8. In the absence of oxygen, NADH is not usually oxidized by the electron transport chain because no external electron acceptor is available; therefore, NAD^+ is not regenerated.

____ 9. Unlike mitochondrial electron transport, photosynthetic electron transport takes place in either the cytoplasm or the fluid matrix (stroma) of chloroplasts rather than in a membrane.

____ 10. The electron transport chains of procaryotes and eucaryotes are fundamentally different processes that couple electron transport to ATP synthesis by very different mechanisms.

CRITICAL THINKING

1. The authors make the statement, "the use of a few common catabolic pathways, each degrading many nutrients, greatly increases metabolic efficiency by avoiding the need for a large number of less metabolically flexible pathways." Explain the significance of this statement.

2. The overall yield of ATP in eucaryotes has a theoretical maximum of 38 ATP molecules per molecule of glucose. However, usually only 36 ATP molecules are produced. Explain this apparent discrepancy. Procaryotes often have lower P/O ratios than eucaryotic systems. Explain the significance of this relative to the aerobic yield of ATP.

ANSWER KEY

Terms and Definitions

1. m, 2. e, 3. c, 4. b, 5. j, 6. t, 7. q, 8. k, 9. u, 10. i, 11. o, 12. s, 13. f, 14. g, 15. v, 16. p, 17. l, 18. a, 19. d, 20. r, 21. h, 22. n

Metabolic Energy: Nutrient Processing

1. Step 1, 2 molecules 2. Step 2, 4 molecules; Step 4, 2 molecules 3. Step 2, 2 molecules; Step 3, 2 molecules; Step 4, 6 molecules 4. Step 4, 2 molecules 5. 36 molecules 6. Those from step 1 must be transported into mitochondria at the expense of 1 ATP per NADH so that the net yield is only 2 ATP per NADH. 7. Step 3, 2 molecules; Step 4, 4 molecules

Metabolic Energy: Photosynthesis

1. ATP, NADPH 2. ATP 3. Cyclic photophosphorylation 4. The electrons given to NADPH cannot be recycled to P700, and the process would stop when P700 became electron deficient. 5. Electrons are replenished by P680. 6. They are replaced from water.

Fill in the Blank

1. catabolism; anabolism; order 2. chemolithotrophy; photosynthesis 3. catabolic; regulation; amphibolic 4. acetyl-CoA; carbon dioxide; GTP; NADH; FADH 5. coenzyme Q; cytochrome C 6. chemiosmotic; conformational change 7. anaerobic; aerobic; the same 8. lactic acid; homolactic; heterolactic 9. fermentation; aerobic respiration; nitrate; sulfate; carbon dioxide; anaerobic respiration 10. hydrolysis; phosphorolysis 11. β-oxidation; acetyl-CoA 12. proteases; deamination; transamination 13. photosynthesis; ATP; NADPH; food chains; oxygen 14. light reactions; dark reactions; carbon dioxide 15. oxygen; anoxygenic; oxygenic 16. plasma membrane; inner mitochondrial membrane 17. NADH; NAD; organic; ATP

Multiple Choice

1. b, 2. a, 3. d, 4. c, 5. d, 6. b, 7. d, 8. d, 9. d, 10. c, 11. b

True/False

1. F, 2. F, 3. T, 4. T, 5. T, 6. T, 7. T, 8. F, 9. F, 10. F

10 Metabolism: The Use of Energy in Biosynthesis

CHAPTER OVERVIEW

This chapter presents an overview of anabolism. It then focuses on the synthesis of carbohydrates, amino acids, purines and pyrimidines, and lipids. It also provides a description of the assimilation of carbon dioxide, phosphorus, sulfur, and nitrogen. The chapter then concludes with a discussion of the synthesis of peptidoglycan and bacterial cell walls.

CHAPTER OBJECTIVES

After reading this chapter you should be able to:

- discuss the use of energy to construct more complex molecules and structures from smaller, simpler precursors
- discuss the way that biosynthetic pathways are organized to conserve genetic storage space, biosynthetic raw materials, and energy
- discuss the way that autotrophs use ATP and NADPH to reduce carbon dioxide and incorporate it into organic material
- describe the assimilation of phosphorus, sulfur, and nitrogen
- discuss the use of the TCA cycle as an amphibolic pathway and the need for anaplerotic reactions to maintain adequate levels of TCA cycle intermediates
- discuss the synthesis of glucose (gluconeogenesis) as a reversal of glycolysis and the synthesis of fatty acids by a process quite different from the β-oxidation catabolic process
- describe in general terms the synthesis of peptidoglycan and the construction of new cell walls after the peptidoglycan repeat unit has been transported across the cell membrane

CHAPTER OUTLINE

I.   Introduction
    A.   Anabolism—the creation of order by the synthesis of complex molecules from simpler ones with the input of energy
    B.   Turnover—the continual degradation and resynthesis of cellular constituents
    C.   The rate of biosynthesis is approximately balanced by that of catabolism
II.   Principles Governing Biosynthesis
    A.   The synthesis of large complex molecules (macromolecules) from a limited number of simple structural units (monomers) saves much genetic storage capacity, biosynthetic raw material, and energy
    B.   The use of many of the same enzymes for both catabolism and anabolism saves additional materials and energy
    C.   Many enzymes participate in both catabolic and anabolic activities; however, some steps are catalyzed by two different enzymes: one catalyzes the reaction in the catabolic direction, while the other reverses the step, thus permitting independent regulation of catabolism and anabolism

D. Coupling some anabolic pathways with the breakdown of ATP (or other nucleoside triphosphates) drives the biosynthetic reaction to completion

E. In eucaryotic cells, anabolic and catabolic reactions involving the same constituents are frequently located in separate compartments for simultaneous but independent operation

F. Catabolic and anabolic pathways use different cofactors: catabolic oxidations produce NADH, which is a substrate for electron transport, while NADPH acts as a reductant for anabolic pathways

G. Large assemblies (e.g., ribosomes) form spontaneously from their macromolecular components by a process known as self-assembly

III. The Photosynthetic Fixation of Carbon Dioxide—Calvin Cycle

   A. Consists of three phases

      1. The carboxylation phase—the enzyme ribulose 1,5-bisphosphate carboxylase catalyzes the addition of carbon dioxide to ribulose 1,5-bisphosphate, forming two molecules of 3-phosphoglycerate

      2. The reduction phase—3-phosphoglycerate is reduced to glyceraldehyde 3-phosphate

      3. The regeneration phase—a series of reactions is used to regenerate ribulose 1,5-bisphosphate and to produce carbohydrates such as fructose and glucose

   B. Energy expenditure—each carbon dioxide takes three ATP molecules and two NADPH molecules, thus the formation of a single glucose molecule requires six turns through the cycle with an expenditure of 18 ATP molecules and 12 NADPH molecules

   C. Sugars formed in the Calvin cycle can then be used to synthesize other essential molecules

IV. Synthesis of Sugars and Polysaccharides

   A. Heterotrophs synthesize glucose from noncarbohydrate precursors in a process called gluconeogenesis

   B. Gluconeogenesis is a functional reversal of glycolysis

      1. Shares seven enzymes that also catalyze glycolytic reactions in the reverse direction

      2. Three steps cannot be directly reversed and therefore require separate enzymes or multi-enzyme systems

   C. Fructose is synthesized as part of the pathway

   D. Other sugars are manufactured from glucose or fructose, or phosphorylated derivatives of those sugars

   E. Polysaccharide production requires the use of nucleoside diphosphate sugars as precursors and are therefore not direct reversals of polysaccharide catabolism

V. The Assimilation of Inorganic Phosphorus, Sulfur, and Nitrogen

   A. Phosphorus assimilation

      1. Inorganic phosphates are incorporated through the formation of ATP by photophosphorylation, oxidative phosphorylation, and substrate-level phosphorylation

      2. Organic phosphates obtained from the surroundings are hydrolyzed to release inorganic phosphates by enzymes called phosphatases

   B. Sulfur assimilation

      1. Organic sulfur in the form of cysteine and methionine can be obtained from external sources

      2. Inorganic sulfate must first be reduced by a process called assimilatory sulfate reduction before it can be incorporated into cysteine

   C. Nitrogen assimilation

      1. Incorporation of ammonia can be easily and directly accomplished through reductive amination, glutamate formation, transamination, or through the actions of the enzymes glutamine synthetase and glutamate synthase

      2. Assimilatory nitrate reduction involves the reduction of nitrate to nitrite, then to hydroxylamine, and finally to ammonia, which can then be incorporated by the routes described

3. Nitrogen fixation involves the reduction of atmospheric nitrogen to ammonia; this is catalyzed by the enzyme nitrogenase, which is found in only a few species of bacteria; this requires a large energy expenditure consuming almost 20% of the ATP produced by the host plant

VI. The Synthesis of Amino Acids—involves the synthesis of carbon skeletons, often by complex routes, from acetyl-CoA, TCA cycle intermediates, glycolytic intermediates, and pentose phosphate pathway intermediates and then the addition of nitrogen through transaminase reactions

VII. Anaplerotic Reactions—replace TCA cycle intermediates that have been used to provide carbon skeletons for biosynthetic reactions
    A. Involve carbon dioxide fixation for most microorganisms
    B. Not limited to autotrophic organisms
    C. Heterotrophic organisms use certain dioxide fixation reactions only to replace TCA cycle intermediates and to maintain metabolic balance, not to supply the carbon atoms needed for cell growth

VIII. The Synthesis of Purines, Pyrimidines, and Nucleotides—these molecules are critical for all cells because they are used in the synthesis of ATP, several cofactors, RNA, and DNA
    A. Purine biosynthesis—very complex pathway in which seven different molecules contribute parts to the final purine skeleton; it involves the cofactor folic acid
    B. Pyrimidine biosynthesis—aspartic acid and carbamoyl phosphate form the initial pyrimidine product, which can then be converted to other pyrimidines
    C. Purine and pyrimidine bases are then joined with pentose sugars to form nucleosides
    D. Phosphorylation of nucleosides forms nucleotides

IX. Lipid Synthesis
    A. Fatty acids are synthesized using the substrates acetyl-CoA and malonyl-CoA, the reductant NADPH, and a small protein called an acyl carrier protein to carry the growing fatty acid
    B. Unsaturated fatty acids can be formed either aerobically or anaerobically
    C. Triacylglycerols are formed from the reduction of dihydroxyacetone phosphate (a glycolytic pathway intermediate) to glycerol 3-phosphate, which then undergoes esterification with two fatty acids to form phosphatidic acid; this can then be used to produce triacylglycerol
    D. Phospholipids are also produced from phosphatidic acid using a cytidine diphosphate (CDP) carrier

X. Peptidoglycan Synthesis—involves a complex 8-stage process that forms the peptidoglycan repeat unit, which is then attached to the growing peptidoglycan chain after being transported across the cytoplasmic membrane; crosslinks are then formed by transpeptidation

XI. Patterns of Cell Wall Formation
    A. Autolysins carry out limited digestion of peptidoglycan, and provide acceptor ends for the addition of new peptidoglycan units
    B. Some organisms, such as gram-positive cocci, have only one or a few growth zones, usually at the site of septum formation
    C. Rod-shaped organisms, however, usually have growth sites scattered along the cylindrical portion of the bacteria, as well as at the site of septum formation

TERMS AND DEFINITIONS

_____ 1. The process by which cellular molecules and constituents are continually being degraded and resynthesized

_____ 2. Very large molecules that are polymers of smaller units

_____ 3. The series of enzymatic metabolic steps that incorporates carbon dioxide into carbohydrate

_____ 4. The synthesis of glucose from nonglucose precursors

_____ 5. The reduction of atmospheric gaseous nitrogen to ammonia

_____ 6. Reactions that replenish TCA cycle intermediates that have been used in biosynthetic reactions

_____ 7. Enzymes that digest peptidoglycan just enough to provide acceptor ends for new synthesis

_____ 8. The localization of different metabolic processes in separate organelles in order to independently regulate those processes

a. anaplerotic reactions
b. autolysins
c. Calvin cycle
d. compartmentation
e. gluconeogenesis
f. macromolecules
g. nitrogen fixation
h. turnover

FILL IN THE BLANK

1. Cells are never the same from instant to instant. Despite the continual _____ of cell constituents, metabolism is carefully _____ so that the rate of _____ is approximately balanced by that of catabolism.

2. Proteins are made of only _____ common amino acids joined by _____ bonds; different proteins simply have different amino acid _____, but not new and dissimilar amino acids.

3. The cell saves energy and materials by using many of the same enzymes for both _____ and _____. However, some steps will be catalyzed by different enzymes to allow _____ of the two pathways.

4. After macromolecules have been constructed from simpler precursors, they are assembled into _____ or _____. Macromolecules often contain the information to do this spontaneously in a process known as _____.

5. The three phases of the Calvin cycle, or the _____ _____ _____ cycle, are the _____ phase, the _____ phase, and the _____ phase.

6. The photosynthetic production of one molecule of glucose requires _____ molecules of ATP and _____ molecules of NADPH, which are provided by the _____ reactions of photosynthesis.

7. The reduction of sulfate for use in the production of such compounds as the amino acid cysteine is called _____ sulfate reduction, while the reduction of sulfate as a terminal electron acceptor during anaerobic respiration is called _____ sulfate reduction.

8. Most microorganisms assimilate nitrogen as either _____ or _____; however, a few microorganisms have the ability to reduce atmospheric nitrogen to ammonia in a process known as _____.

9. _____ and _____ are cyclical nitrogenous bases with several double bonds and pronounced aromatic properties. Biosynthesis of these molecules is critical for all cells since these molecules are used in the synthesis of _____, _____, and _____, as well as several cofactors and other important cellular components.

10. A purine or pyrimidine base joined with a pentose sugar, either _____ or _____, is a _____. If one or more phosphate groups are attached to the sugar, it is called a _____.

MULTIPLE CHOICE

For each of the questions below select the *one best* answer.

1. Large molecules that are polymers of smaller units joined together are called
   a. monomers.
   b. micromolecules.
   c. multimers.
   d. macromolecules.
2. Which of the following represents ways in which cells independently regulate anabolic and catabolic pathways?
   a. use of separate enzymes for reversal of key steps
   b. compartmentation of anabolic and catabolic pathways
   c. use of different cofactors for anabolic and catabolic pathways
   d. All of the above are ways in which cells independently regulate anabolic and catabolic pathways.
3. Which of the following is not required to be assimilated or incorporated in large quantities into organic molecules?
   a. nitrogen
   b. sodium
   c. phosphorus
   d. sulfur
4. Which of the following is not a method of assimilation of inorganic phosphate?
   a. oxidative phosphorylation
   b. substrate-level phosphorylation
   c. periplasmic phosphorylation
   d. photophosphorylation
5. Which of the following is not a source of carbon skeletons that are used in the synthesis of amino acids?
   a. acetyl-CoA
   b. TCA cycle intermediates
   c. glycolysis intermediates
   d. All of the above are sources of carbon skeletons for the synthesis of amino acids.
6. Reactions that are used to replace TCA cycle intermediates so that the TCA cycle can continue to function when active biosynthesis is taking place are called
   a. anaplerotic reactions.
   b. amphibolic reactions.
   c. anabolic reactions.
   d. catabolic reactions.

7. Which of the following is not true about fatty acids?
   a. Fatty acids are monocarboxylic acids.
   b. Fatty acids have long alkyl chains.
   c. Fatty acids usually have an even number of carbons, and the average chain length is 18 carbons.
   d. Microbial fatty acids are always straight chained and are never branched.
8. In which of the following ways are the CO_2 fixation reactions of the anaplerotic reactions found in heterotrophs different from the CO_2 fixation reactions found in autotrophs?
   a. They usually use pyruvate as the acceptor molecule rather than ribulose 1,5-bisphosphate.
   b. They are not used as the primary source of carbon atoms for the increase in the biomass of the organism.
   c. Both (a) and (b) are correct.
   d. Neither (a) nor (b) is correct.
9. Cyanobacteria, some nitrifying bacteria, and thiobacilli have polyhedral inclusion bodies that contain the enzyme ribulose 1,5-bisphosphate carboxylase, and may be the site of CO_2 fixation in those organisms. These inclusion bodies are called
   a. fixosomes.
   b. carboxysomes.
   c. plastosomes.
   d. chloroplasts.
10. Which of the following is true about the synthesis of macromolecules from monomeric subunits?
    a. It saves genetic storage capacity.
    b. It saves biosynthetic raw materials.
    c. It saves energy.
    d. All of the above are true.

TRUE/FALSE

_____ 1. The process of linking a few monomers together with a single type of covalent bond makes the synthesis of macromolecules an inefficient process.

_____ 2. In eucaryotic microorganisms, biosynthetic pathways are frequently located in cellular compartments that are different from those in which their corresponding catabolic pathways are located, which makes it easier for simultaneous but independent operation.

_____ 3. Most microorganisms have the ability to incorporate or fix carbon dioxide, but only autotrophs can use carbon dioxide as their sole or principal source of carbon.

_____ 4. Nitrogen fixation does not require the expenditure of much energy.

_____ 5. The enzyme nitrogenase is not highly specific and can reduce a number of compounds containing triple bonds, such as acetylene, cyanide, and azide.

_____ 6. Both purines and pyrimidine nucleotides are synthesized by first synthesizing the nitrogenous base and then adding ribose 5-phosphate to form the nucleotide.

_____ 7. Unsaturated fatty acids are those containing one or more carbon-carbon double bonds.

_____ 8. Membrane phospholipids are constructed from the products of glycolysis, fatty acid biosynthesis, and amino acid biosynthesis.

_____ 9. Because peptidoglycan lies outside of the cytoplasmic membrane, all of the steps in the complex synthesis process for this molecule must take place outside the membrane at the cell wall.

_____ 10. There are two general patterns of peptidoglycan synthesis. In cocci there are only one or a few sites of growth, with the principal growth zone located at the site of septum formation. In bacilli, active growth also occurs at the site of septum formation, but there are also multiple growth sites scattered along the cylindrical portion of the rod so that growth is more diffusely distributed.

_____ 11. Nitrogen fixation can consume up to 20% of the ATP generated by the host plant

CRITICAL THINKING

1. What would be the consequences for a cell if *anaplerotic* reactions did not exist? Consider in your discussion the interrelationship between catabolic and anabolic reactions and the needs of an organism growing on limited nutritional sources.

2. There are a number of different ways available to various microorganisms to assimilate nitrogen. Only a relatively few species of bacteria, however, are capable of utilizing gaseous atmospheric nitrogen (nitrogen fixation). Why have organisms developed such a variety of assimilatory pathways and why are so few equipped to use gaseous atmospheric nitrogen, even though it is quite abundant?

3. Biosynthetic processes (e.g., gluconeogenesis) frequently are not direct reversals of the related catabolic processes (e.g., glycolysis). However, some of the steps involved in the overall pathway may be direct reversals. What is the advantage to the organism to have separate pathways for synthesis and degradation? Furthermore, what is the advantage to the organism for the substantial overlap (i.e., directly reversed steps) within these pathways?

ANSWER KEY

Terms and Definitions

1. h, 2. f, 3. c, 4. e, 5. g, 6. a, 7. b, 8. d

Fill in the Blank

1. turnover; regulation; biosynthesis 2. twenty; peptide; sequences 3. catabolism; anabolism; independent regulation 4. supramolecular complexes; organelles; self-assembly 5. photosynthetic carbon reduction; carboxylation; reduction; regeneration 6. 18; 12; light 7. assimilatory; dissimilatory 8. ammonia; nitrate; nitrogen fixation 9. Purines; pyrimidines; ATP; RNA; DNA 10. ribose; deoxyribose; nucleoside; nucleotide

Multiple Choice

1. d, 2. d, 3. b, 4. c, 5. d, 6. a, 7. d, 8. c, 9. b, 10. d

True/False

1. F, 2. T, 3. T, 4. F, 5. T, 6. F, 7. T, 8. T, 9. F, 10. T, 11. T

11   Metabolism: The Synthesis of Nucleic Acids and Proteins

CHAPTER OVERVIEW

This chapter presents an overview of the synthesis of two major classes of macromolecules: nucleic acids and proteins. Adequate knowledge of these processes is essential to the understanding of molecular biology and microbial genetics. The chapter begins with general information about nucleic acid structure to increase the comprehension of the processes of DNA replication, RNA transcription, and polypeptide (protein) translation.

CHAPTER OBJECTIVES

After reading this chapter you should be able to:

- discuss the structural and compositional differences between DNA and RNA
- discuss the association of proteins with DNA and describe the differences in the types of proteins associated with procaryotic and eucaryotic DNA
- discuss the flow of genetic information from DNA, to RNA, to protein, and discuss the relationship between the nucleotide sequences of DNA and RNA and the amino acid sequences of proteins
- describe the replication of DNA and the processes used to minimize errors and correct those errors which do occur
- discuss the transcription of RNA and describe the similarities and differences between eucaryotic and procaryotic RNA transcription
- discuss the translation of proteins and describe the role(s) of the various components required for this process

CHAPTER OUTLINE

    I.   Nucleic Acid Structure
        A.   DNA Structure
            1.   DNA is composed of purine and pyrimidine nucleosides that contain the sugar 2′-deoxyribose and are joined by phosphodiester bridges
            2.   DNA is usually a double helix consisting of two chains of DNA coiled around each other
            3.   The purine adenine (A) on one strand of DNA is always paired with the pyrimidine thymine (T) on the other strand, while the purine guanine (G) is always paired with the pyrimidine cytosine (C); thus, the two strands are said to be complementary
            4.   The two polynucleotide chains are antiparallel (i.e., their sugar-phosphate backbones are oriented in opposite directions)
            5.   The two strands are not positioned directly opposite one another; therefore, a major groove and a smaller minor groove are formed by the double helix backbone
        B.   RNA structure
            1.   RNA differs from DNA in that it is composed of the sugar ribose rather than 2′-deoxyribose
            2.   RNA differs from DNA in that it contains the pyrimidine uracil (U) instead of thymine

3. RNA differs from DNA in that it usually consists of a single strand that can coil back on itself, rather than two strands coiled around each other
4. Three different kinds of RNA exist—ribosomal (rRNA), transfer (tRNA), and messenger (mRNA)—and differ from one another in function, site of synthesis in eucaryotic cells, and structure

C. The organization of DNA in cells
1. In procaryotes, the DNA exists as a closed circular, supercoiled molecule associated with basic (histonelike) proteins
2. In eucaryotes, the DNA is more highly organized
   a. It is associated with basic (histone) proteins
   b. It is coiled into repeating units known as nucleosomes

II. DNA Replication
A. Pattern of DNA synthesis
1. DNA replication is semiconservative: each strand of DNA is conserved, but the two strands are separated from each other and serve as templates for the production of another strand (according to the base-pairing rules discussed earlier)
2. Replication forks are the areas of the DNA molecule where this strand separation occurs and the synthesis of new DNA takes place
3. A replicon consists of an origin of replication and all the DNA that is replicated as a unit from that origin
4. The bacterial chromosome is usually a single replicon
5. Closed circular DNA molecules replicate by means of a rolling-circle mechanism
6. The large linear DNA molecules of eucaryotes employ multiple simultaneous replicons to efficiently replicate the relatively large molecules within a reasonable time span

B. Mechanism of DNA replication
1. Helicases unwind the two strands of DNA
2. Single-stranded DNA binding proteins (SSBs) keep the single strands apart
3. Topoisomerases relieve the tension caused by the unwinding process; DNA gyrase is a topoisomerase that removes the supertwists produced during replication
4. Primases synthesize a small RNA molecule (approximately 10 nucleotides) that will act as a primer for DNA synthesis
5. DNA polymerase III synthesizes the complementary strand of DNA according to the base-pairing rules established; on one strand (the leading strand), synthesis is continuous, while on the other (the lagging strand), a series of fragments are generated by discontinuous synthesis; a multiprotein complex called a replisome organizes all of these processes
6. DNA polymerase I removes the primers and fills the gaps that result from the RNA deletion
7. DNA ligases join the discontinuous fragments to form a complete strand of DNA
8. DNA replication is extraordinarily complex; at least 30 proteins are required to replicate the *E. coli* chromosome
9. The rate of DNA synthesis is 750 to 1,000 base pairs per second in procaryotes, and 50 to 100 base pairs per second in eucaryotes

III. DNA Transcription or RNA Synthesis
A. Three types of RNA are produced by transcription
1. tRNA carries amino acids during protein synthesis
2. rRNA molecules are components of the ribosomes

3. mRNA carries the message that directs the synthesis of proteins; in addition to the translated regions of the mRNA, there are several untranslated regions that serve particular purposes
   a. Leader sequences consist of 25 to 150 bases at the 5′ end of the mRNA, and precede the initiation codon
   b. Spacer regions separate the segments that code for individual polypeptides in polygenic mRNAs (i.e., those RNAs that encode more than one polypeptide chain)
   c. Trailer regions are found at the 3′ end of the mRNA after the last termination codon
B. RNA polymerase (a large multi-subunit enzyme) is the enzyme responsible for the synthesis of RNA
C. A gene is a DNA segment or sequence that codes for a polypeptide, rRNA, or tRNA
D. In a given segment, only one strand (sense or template strand) is copied
E. A promoter is the region of the DNA to which RNA polymerase binds in order to initiate transcription
F. Terminators are regions of the DNA that, when transcribed, result in the termination of the transcription process
G. In eucaryotes, transcription yields large RNA precursors (heterogeneous nuclear RNA; hnRNA) that must be processed by posttranscriptional modification to produce mRNA
H. hnRNA is modified by the addition of adenylic acid to the 3′ end to produce a polyA sequence about 200 nucleotides long
I. hnRNA is also modified by the addition of 7-methylguanosine to the 5′ end by a tri-phosphate linkage
J. These two modifications are believed to protect the mRNA from exonuclease digestion
K. Eucaryotic genes are split or interrupted such that the expressed sequences (exons) are separated from one another by intervening sequences (introns); the introns are represented in the primary transcript but are subsequently removed by a process called RNA splicing; some splicing is self-catalyzed by RNA molecules called ribozymes
L. Interrupted genes have been found in cyanobacteria and archaeobacteria, but not in other procaryotes

IV. Protein Synthesis—translation
A. Ribosomes complex with mRNA and, using the information contained within the mRNA, combine the appropriate amino acids one at a time to form the complete polypeptide chain
B. The first stage of protein synthesis is the attachment of amino acids to tRNA molecules; this process is referred to as amino acid activation
C. Each tRNA can only carry a specific amino acid
D. Ribosomes are complex organelles constructed from several rRNA molecules and many polypeptides
E. Protein synthesis is divided into three phases:
   1. Initiation takes place at a special initiator codon (AUG)
      a. The two subunits of the ribosome complex with the mRNA
      b. Three protein initiation factors are also required in procaryotes
   2. Elongation involves the sequential addition of amino acids to the growing polypeptide chain
      a. Each amino acid is positioned when the anticodon region of its tRNA approaches its complementary codon on the mRNA molecule
      b. A ribosomal enzyme peptidyl transferase catalyzes the formation of the peptide bonds between adjacent amino acids; the 23S rRNA is a major component of this enzyme
      c. After each amino acid is added to the chain, translocation occurs and thereby moves the ribosome to position the next codon appropriately
      d. Several polypeptide elongation factors are required for this process
      e. Hydrolysis of ATP and GTP provide the energy needed for this process
   3. Termination takes place at any one of three special codons (UAA, UAG, or UGA)
      a. Three polypeptide release factors aid in the recognition of these codons
      b. No amino acids are incorporated into the growing polypeptide chain at these positions

      c.   The ribosome hydrolyzes the bond between the completed protein and the final tRNA, and the protein is released from the ribosome, which then dissociates into its two component subunits

      d.   As the protein leaves the ribosome it folds into its proper shape aided by molecular chaperones

F.   Eucaryotic protein synthesis is similar, but may require more protein factors to mediate various parts of the process; and usually is less dependent on molecular chaperones for proper folding of the newly synthesized protein

G.   Procaryotic proteins may undergo splicing after translation; such splicing removes intervening sequences (inteins) from the sequences (exteins) that remain in the final product

TERMS AND DEFINITIONS

_____ 1. Term that describes the way in which the two strands of DNA are oriented in opposite directions with respect to their sugar residues. In a given direction one strand is oriented 5′ to 3′, while the other is oriented 3′ to 5′

_____ 2. Term that describes the twisting of the double helix of a closed circular DNA molecule

_____ 3. Small basic proteins rich in the amino acids lysine and/or arginine, which help organize the structure of DNA in the eucaryotic ribosome

_____ 4. DNA that is found coiled around the surface of an ellipsoid composed of histones

_____ 5. The process by which DNA is very precisely copied

_____ 6. The process by which the base sequence of all (or a portion) of the DNA molecule is used to direct the synthesis of an RNA molecule

_____ 7. The process by which the base sequence of an RNA molecule is used to direct the synthesis of a protein

_____ 8. The Y-shaped part of the DNA molecule where the actual replication process takes place

_____ 9. The site on the DNA molecule where replication begins and from which two replication forks move outward in opposite directions

_____ 10. Small fragments (100 to 1,000 bases long) formed by the discontinuous replication of the lagging strand during the replication process

_____ 11. A nontranslated sequence of 25 to 150 bases that precedes the initiation codon on mRNA molecules

_____ 12. mRNA molecules that direct the synthesis of more than one polypeptide

_____ 13. Nontranslated sequences that separate the coding sequences on polygenic mRNA molecules

_____ 14. A nontranslated sequence that follows the last termination codon of a mRNA molecule

_____ 15. A segment or sequence of a DNA molecule that codes for a polypeptide, rRNA, or tRNA

a.   amino acid activation
b.   antiparallel
c.   exons
d.   exteins
e.   gene
f.   heterogeneous nuclear RNA (hnRNA)
g.   histones
h.   inteins
i.   introns
j.   leader sequence
k.   nucleosome
l.   Okazaki fragments
m.   origin of replication
n.   polygenic (polycistronic) mRNA
o.   polyribosome (polysome)
p.   promoter
q.   replication
r.   replication fork
s.   replicon
t.   ribozymes
u.   RNA splicing
v.   sense (template) strand
w.   spacer regions
x.   supercoiled DNA
y.   trailer region
z.   transcription
aa. translation
bb. transpeptidation reaction

_____ 16. The strand of DNA for a particular gene that is copied by RNA polymerase during transcription to form a mRNA molecule

_____ 17. The region of DNA to which RNA polymerase binds, which is required for the initiation of transcription but is not transcribed itself

_____ 18. The large RNA molecule produced by transcription that must be processed to produce the smaller mRNA molecules

_____ 19. The expressed (coding) sequences of split (interrupted) genes

_____ 20. The intervening (noncoding) sequences of split genes

_____ 21. The process by which the intervening sequences are removed from the RNA transcript to produce mRNA molecules

_____ 22. A complex between a single mRNA molecule and several ribosomes that organisms use to increase the efficiency of protein synthesis

_____ 23. The phenomenon in which amino acids attach to the appropriate tRNA molecules

_____ 24. The reaction by which the amino acids are added sequentially to form the growing polypeptide chain

_____ 25. The origin of replication and all the DNA that is replicated as a unit from that origin

_____ 26. RNA molecules that self-catalyze the removal of their own introns

_____ 27. Portions of a protein that remain after post-translational proteolytic cleavage

_____ 28. Portions of a protein that are removed by post-translational proteolytic cleavage

REPLICATION/TRANSCRIPTION/TRANSLATION

Complete the table by providing a brief description of the functions of each of the components listed below:

Component	Description of Function
1. Replication	
A. DNA polymerase III	
B. Helicase	
C. Single-strand binding proteins	
D. Topoisomerase	
E. Primase	

F.   DNA polymerase I

G.   DNA ligase

2. Transcription

   A.   RNA polymerase

   B.   Gene

   C.   Sense strand

   D.   Sigma factor

   E.   Promoter

   F.   Rho factor

   G.   Exons

   H.   Introns

3. Translation

   A.   Ribosome

   B.   tRNA

   C.   Codon

   D.   Anticodon

   E.   Polyribosome

   F.   Aminoacyl-tRNA synthetase

   G.   Initiator codon (AUG)

   H.   Initiation factors (IF-1, IF-2, IF-3)

   I.   Elongation factors (EF-Tu, EF-Ts, EF-G)

   J.   Nonsense codons (UAA, UAG, UGA)

   K.   Release factors (RF-1, RF-2, RF-3)

FILL IN THE BLANK

1. The two strands making up the double helix of DNA are said to be _____ (i.e., the purine adenine on one strand always pairs with the pyrimidine _____ on the other strand by two hydrogen bonds, while the purine _____ always pairs with the pyrimidine cytosine by _____ hydrogen bonds).

2. The two polynucleotide strands are not positioned directly opposite one another in the helical cylinder. Therefore, when the strands are twisted about one another, a wide _____ and a narrower _____ are formed by the backbone.

3. The two backbones are _____ with respect to the orientation of their sugars. One end of each strand has an exposed 5'-hydroxyl, often with _____ attached, while the other end has a free 3'-hydroxyl group. If the end of a double helix is examined, the 5' end of one strand and the 3' end of the other strand will be seen.

4. In procaryotes, DNA exists in a closed circular form that is further twisted to form _____. It is associated with basic proteins different from the _____ found associated with eucaryotic DNA.

5. The double helical structure of DNA was proposed by _____ and _____ in 1953. Subsequently they proposed the process by which the DNA could be replicated: The two strands would unwind from each other and separate. Free nucleotides now line up along the two parental strands according to the _____ rules already established. When these nucleotides are covalently linked by one or more enzymes, two replicas result, each containing a _____ DNA strand and a newly synthesized strand. This process is referred to as _____ replication because the parental molecule is not conserved intact, but the individual strands are conserved.

6. The two strands of a replicating DNA molecule do not completely dissociate from one another; instead, replicating DNA molecules have a Y shape and the actual replication process takes place at the junction of the arms of the Y. This junction is called the _____. Replication begins at a single point called the _____ and moves outward in both directions until the whole _____ (that portion of the DNA molecule containing a replication origin that is replicated as a single unit) has been replicated.

7. There is evidence that in procaryotes the replication fork and associated enzymes are attached to the _____. If this is the case, the _____ moves through the replication apparatus rather than the replication fork moving around the DNA.

8. The noncoding region at the 5' end of a mRNA molecule (before the initiator codon) is called the _____, while the noncoding region at the 3' end after the terminator codon is called the _____. In addition, polygenic mRNAs have _____ separating the segments that code for individual polypeptides.

9. Eucaryotic mRNA arises from _____ modification of large RNA precursors called _____ (hnRNA). After transcription, _____ are added to the 3' end, and a cap consisting of _____ is attached to the 5' end. In addition, RNA sequences corresponding to the _____ of split genes are removed by _____.

10. In _____, ribosomes can attach to the mRNA and begin translation even though transcription has not been completed. This is called _____ transcription and translation. This cannot happen in _____ because the _____ separates the translation machinery from the transcription machinery.

11. Amino acids are activated for protein synthesis through a reaction that is catalyzed by _____. There are at least _____ of these enzymes, each specific for a single amino acid and all the _____ tRNAs to which it may properly be attached.

12. The ribosome has two sites for binding aminoacyl-tRNA and peptidyl-tRNA: the _____ (or donor site) and the _____ (or acceptor site). If the fourth codon is in the donor site, then the _____ codon is in the acceptor site. In the transpeptidation reaction, the growing peptide chain is transferred to the tRNA in the _____ site, lengthening the chain by one amino acid. Then this tRNA carrying the growing chain is moved into the _____ site and the sixth codon is moved into the _____ site. The movement is referred to as _____.

MULTIPLE CHOICE

For each of the questions below select the *one best* answer.

1. Which of the following nitrogenous bases is found in RNA but not in DNA?
   a. adenine
   b. thymine
   c. uracil
   d. guanine

2. Which of the following is not true about the structure of DNA?
   a. Purine and pyrimidine bases are attached to the $1'$-carbon of the deoxyribose sugars.
   b. The bases extend towards the middle of the cylinder formed by the two chains.
   c. The bases are stacked on top of each other in the center with the rings forming parallel planes.
   d. All of the above are true about the structure of DNA.

3. In which of the following ways do mRNA, rRNA, and tRNA differ from one another?
   a. function
   b. site of synthesis in eucaryotes
   c. structure
   d. All of the above are correct.

4. Procaryotic DNA replication is very rapid and occurs at a rate of
   a. 750 to 1,000 base pairs per second.
   b. 750 to 1,000 base pairs per minute.
   c. 50 to 100 base pairs per second.
   d. 50 to 100 base pairs per minute.

5. RNA molecules that carry amino acids to the ribosome during translation are called
   a. mRNA.
   b. tRNA.
   c. rRNA.
   d. hnRNA.

6. In which of the following types of procaryotes have split genes not been found?
   a. eubacteria
   b. cyanobacteria
   c. archaeobacteria
   d. Split genes have been found in all of the above.

7. Which of the following is true when comparing initiation of translation in procaryotes and initiation of translation in eucaryotes?
   a. They are identical processes.
   b. They are similar, but procaryotes require more initiation factors.
   c. They are similar, but eucaryotes require more initiation factors.
   d. They are very different.

8. Translation is an energy-expensive process. The number of high-energy bonds required to add a single amino acid into a growing polypeptide chain is
   a. two.
   b. three.
   c. four.
   d. six.

9. Which of the following types of noncoding sequences is found in polygenic mRNAs but not monogenic mRNAs?
   a. leader sequences
   b. spacer regions
   c. trailer regions
   d. None of the above are correct.

10. The multiprotein complex responsible for the organization and coordination of the complex processes associated with DNA replication is called
    a. primosome.
    b. replicon.
    c. replication complex.
    d. replisome.

11. Which of the following statements is true about protein folding?
    a. Protein folding in procaryotes generally occurs after translation is complete.
    b. Protein folding in eucaryotes generally occurs while the protein is still being translated.
    c. Both (a) and (b) are true.
    d. Neither (a) nor (b) is true.

12. Post-translational processing of some procaryotic protein removes which of the following?
    a. exteins
    b. inteins
    c. exons
    d. introns

TRUE/FALSE

_____ 1. The basic differences between RNA and DNA reside in their sugar and pyrimidine bases: RNA has ribose and uracil, while DNA has deoxyribose and thymine.

_____ 2. In addition to differing chemically from DNA, RNA is usually single stranded rather than double stranded as are most DNA molecules.

_____ 3. Because of the single-stranded nature of RNA, complementary base pairing and helical structure are not observed in RNA molecules.

_____ 4. Many procaryotes have only one origin of replication per DNA molecule, while most eucaryotes use multiple origins in order to efficiently copy the long DNA molecules found in those organisms.

_____ 5. The DNA replication process is very complex, which presumably contributes to the associated high frequency of error.

_____ 6. The rate of translation in eucaryotes is the same as that in procaryotes (300 to 450 amino acids per minute).

_____ 7. Ribosomes are composed of two separate subunits that come together as part of the initiation process and then dissociate immediately after termination.

_____ 8. Bacteria start protein synthesis with formylmethionine. However, the formyl group is removed after translation has been completed. Therefore, the first amino acid in all bacterial proteins is methionine.

_____ 9. DNA gyrase is a topoisomerase that is responsible for restoring superhelical twists to the DNA after DNA replication has taken place.

_____ 10. The poly-A sequence at the 3′ end of eucaryotic mRNA is believed to protect the mRNA from degradation by nucleases.

_____ 11. Molecular chaperones are more important in the folding of eucaryotic proteins than procaryotic proteins.

CRITICAL THINKING

1. Diagram and discuss the process of semiconservative replication. Starting with a single bacterium having only one chromosome, trace the parental DNA through three replicative cycles (resulting in eight cells). In how many of these cells will you find DNA that was in the original parent? If the replication process were fully conservative rather than semiconservative, how would this affect your results?

2. Compare and contrast the transcription processes in procaryotes and eucaryotes. Be sure to include in your discussion the role(s) of promoters, various RNA polymerase enzymes, posttranscriptional modification, coupled transcription/translation, and split (interrupted) genes, where applicable.

ANSWER KEY

Terms and Definitions

1. b, 2. x, 3. g, 4. k, 5. q, 6. z, 7. aa, 8. r, 9. m, 10. l, 11. j, 12. n, 13. w, 14. y, 15. e, 16. v, 17. p, 18. f, 19. c, 20. i, 21. u, 22. o, 23. a, 24. bb, 25. s, 26. t, 27. d, 28. h

Fill in the Blank

1. complementary; thymine; guanine; three 2. major groove; minor groove 3. antiparallel; phosphates 4. supercoiled DNA; histones 5. James Watson; Francis Crick; complementary base pairing; parental; semiconservative 6. replication fork; origin; replicon 7. plasma membrane; DNA 8. leader sequence; trailer sequence; spacer regions 9. posttranscriptional; heterogeneous nuclear RNA; poly A sequences; 7-methylguanosine; introns; RNA splicing 10. procaryotes; coupled; eucaryotes; nuclear envelope 11. aminoacyl-tRNA synthetases; 20; cognate 12. peptidyl; aminoacyl; fifth; acceptor (aminoacyl); donor (peptidyl); acceptor (aminoacyl); translocation

Multiple Choice

1. c, 2. d, 3. d, 4. a, 5. b, 6. a, 7. c, 8. c, 9. b, 10. d, 11. c, 12. b

True/False

1. T, 2. T, 3. F, 4. T, 5. F, 6. F, 7. T, 8. F, 9. F, 10. T, 11. F

12   Metabolism: The Regulation of Enzyme Activity and Synthesis

CHAPTER OVERVIEW

This chapter introduces the principles of metabolic regulation to demonstrate how the various pathways discussed in previous chapters are coordinated. The chapter begins with a discussion of the need for regulation and then describes three mechanisms by which organisms regulate metabolic activities: metabolic channeling, regulation of enzyme activity, and regulation of enzyme synthesis.

CHAPTER OBJECTIVES

After reading this chapter you should be able to:

- discuss the need for metabolic regulation to maintain cell components at the proper levels and to conserve materials and energy
- discuss regulation by metabolic channeling (localization) of enzymes and metabolites
- discuss regulation of enzyme activity by reversible binding of effector molecules or by covalent modification of the enzyme
- discuss regulation of enzyme activity by feedback (end product) inhibition
- discuss regulation of enzyme synthesis by induction and repression of the genes coding for a set of related enzymes
- discuss regulation of gene expression by sigma factors
- discuss the coordination of DNA replication and cell division

CHAPTER OUTLINE

    I.   Introduction
        A.   Metabolic regulation is necessary to maintain cell components at appropriate levels
        B.   Metabolic regulation is necessary to conserve materials and energy
            1.   If a particular energy source is not available, the enzymes required for its utilization are unnecessary, and synthesis of these enzymes wastes carbon, nitrogen, and energy that could be better used for other necessary activities
            2.   If a particular end product were already abundantly present in the environment, it would be wasteful to continue producing enzymes that only manufacture more of that end product
        C.   Mechanisms of metabolic regulation
            1.   Metabolic channeling
            2.   Adjustment of enzyme activity
            3.   Regulation of gene expression (mRNA synthesis)

II. Metabolic Channeling
   A. Compartmentation
      1. The phenomenon in which various processes are allocated to separate cell structures or organelles
      2. Enables simultaneous but separate operation and regulation of similar pathways
      3. Particularly important in eucaryotic cells, which have many separate membrane-bounded organelles
   B. Channeling also occurs within any given compartment because enzymes and their metabolites have finite diffusion rates
   C. Substrate concentrations are often below saturation levels and, therefore, have a profound effect on metabolic activity
III. Control of Enzyme Activity
   A. Allosteric regulation
      1. Effector (modulator) molecules bind reversibly and noncovalently to regulatory sites that are separate from the catalytic site
      2. Positive effectors increase enzyme activity, while negative effectors decrease activity
      3. In some cases, regulatory sites may even reside on polypeptide chains separate from the catalytic site(s)
   B. Covalent modification of enzymes
      1. Some enzymes can be regulated by reversible covalent attachment of a particular chemical group
      2. In this way, enzymes can alternate between two forms, one with the chemical group attached and one without the chemical group attached
      3. One form usually exhibits a much higher activity than the other form
   C. Feedback (end product) inhibition
      1. The first committed step in a metabolic pathway is often catalyzed by a pacemaker enzyme that is regulated by the end product of the pathway
      2. This insures balanced production of a pathway end product
         a. If the end product becomes too concentrated, it inhibits the regulatory enzyme and slows its own synthesis
         b. As the end product concentration decreases, metabolic activity again increases and more product is formed
      3. In branched pathways, the end products of each branch should only inhibit the activity of that particular branch; abundance of the end products of all the branches that act in concert with each other should also inhibit the flow of carbon into the whole set of pathways
      4. The regulation of multiply branched pathways often involves isoenzymes to catalyze the pacemaker step; these isoenzymes are different enzymes that catalyze the same reaction and each is under separate and independent control; in this situation, an excess of a single end product reduces but does not completely block pathway activity because some isoenzymes are still active
IV. Regulation of mRNA Synthesis
   A. Regulation at the level of mRNA (and thereby enzyme) synthesis provides a long-term regulatory mechanism that can respond to major changes in environmental conditions
   B. This type of regulation is even more conservative of materials and energy than those mechanisms already discussed, but the response to changing conditions is not as rapid
   C. Regulation by Sigma Factors
      1. Sigma factors enable the RNA polymerase to recognize and bind to promoters
      2. Different sigma factors recognize different sets of promoters
      3. Substitution of the sigma factors immediately changes gene expression
      4. This has been demonstrated in a number of systems including heat shock response in *E. coli* and sporulation in *B. subtilis*

D. Induction and repression
   1. Enzymes involved in catabolic pathways are inducible, and the initial substrate of the pathway (or some derivative of it) is usually the inducer
   2. Enzymes involved in some anabolic pathways are repressible and the end product of the pathway usually acts as a corepressor
   3. Enzymes involved in other anabolic pathways may be regulated by a process known as attenuation
E. The mechanism of induction and repression
   1. The rate of mRNA synthesis is controlled by repressor proteins
   2. Repressor proteins bind to specific sites on the DNA called operators
   3. When bound to the operator, the repressor protein overlaps the promoter region (adjacent to the operator), and thereby prevents RNA polymerase from attaching to the promoter, stopping transcription of the genes downstream of that promoter
   4. Repressors must exist in active and inactive forms
      a. In inducible systems, the repressor protein is active until bound to the inducer, which renders it inactive
      b. In repressible systems, the repressor is inactive until bound to the corepressor, which renders it active
   5. The set of structural genes controlled by a particular operator together with the associated operator and promoter is called an operon
F. Positive operon control and catabolite repression
   1. Positive control occurs when the operon can function only in the presence of a controlling factor (e.g., the lactose operon is dependent on the presence of cyclic AMP)
   2. Catabolite repression occurs when the operon is under control of some catabolite other than the initial substrate of the metabolic pathway
      a. Allows organisms to use one source of carbon preferentially over another if both are present in the environment
      b. Initial growth occurs using one carbon source until it is gone (e.g., glucose); then after a short lag period, growth resumes using the second carbon source (e.g., lactose) if both are present in the initial environment; this is called diauxic growth
G. Attenuation—in systems where transcription and translation are tightly coupled, ribosome behavior in the leader region of the mRNA can control transcription of operons involved in the biosynthesis of some amino acids
   1. If ribosomes actively translate the leader region (attenuator), which contains several codons for the amino acid product of the operon genes, a terminator forms and transcription will not continue
   2. If ribosomes stall during translation of the leader region because the appropriate charged aminoacyl-tRNA is absent, the terminator does not form and transcription will continue to produce the appropriate amino acid

V. Gene Regulation by Antisense RNA
A. Antisense RNA is complementary to some RNA component necessary for gene expression and, therefore, will form hydrogen bonds with it, thereby preventing the utilization of that component and blocking the subsequent gene expression
B. Examples include binding with RNA primers to prevent DNA replication, and binding with mRNA to prevent ribosome binding and subsequent translation
C. Antisense RNA regulation has not yet been demonstrated in eucaryotic cells, although there is evidence for its existence

VI. Control of the Cell Cycle
A. The complete sequence of events extending from the formation of a new cell through the next division is called the cell cycle
B. Since each daughter cell receives at least one copy of the genetic material, DNA replication and cell division must be tightly coordinated

C. The precise mechanisms(s) for control of the cell cycle are not known, although several cell-division genes have been identified
D. These appear to be two separate controls for the cell cycle, one sensitive to cell mass and the other responding to cell length
E. A number of proteins have been shown to participate in cell division regulation

TERMS AND DEFINITIONS

Place the letter of each term in the space next to the definition or description that best matches it.

_____ 1. The phenomenon in which metabolic pathways are regulated by controlling the intracellular location of the metabolites and enzymes involved in the pathway

_____ 2. Differential distribution of enzymes and metabolites between separate cell structures or organelles

_____ 3. Enzymes whose control can affect the overall activity of multienzyme metabolic pathways

_____ 4. Enzymes whose activity is altered by a small molecule

_____ 5. Small molecules that alter the activity of allosteric enzymes

_____ 6. The enzyme that catalyzes the slowest or rate-limiting reaction in a pathway

_____ 7. The process by which the end product of a metabolic pathway inhibits the first enzyme in the pathway

_____ 8. Different enzymes that catalyze the same reaction, but that may be regulated independently of one another

_____ 9. Small molecules that inactivate repressor proteins and thereby increase the synthesis of certain enzymes

_____ 10. Small molecules that activate repressor proteins and thereby decrease the synthesis of certain enzymes

_____ 11. The site on the DNA to which a repressor binds

_____ 12. The sequence of bases on the DNA that code for one or more polypeptides, together with the associated promoter and the operator that control their expression

_____ 13. A rho-independent termination site found in the leader region of certain operons, which, under the influence of ribosome behavior, controls the continued transcription of that operon

_____ 14. The complete sequence of events extending from the formation of a new cell through the next division

_____ 15. Genes whose expression is not regulated (i.e., genes that are always expressed)

_____ 16. RNA molecules that are complementary to some RNA component necessary for gene expression and that will hydrogen bond with that component, thereby preventing the utilization of that component

a. allosteric enzymes
b. antisense
c. attenuator
d. cell cycle
e. compartmentation
f. constitutive
g. corepressors
h. effector (modulator)
i. feedback (end product) inhibition
j. inducers
k. isoenzymes
l. metabolic channeling
m. operator
n. operon
o. pacemaker enzyme
p. regulatory enzymes

FILL IN THE BLANK

1. The localization of _____ and _____ in different parts of a cell, a phenomenon known as _____, influences the activity of metabolic pathways.

2. Compartmentation is the differential distribution of _____ and _____ between separate cell structures or organelles.

3. Enzyme activity can be regulated by small molecules known as _____ or _____. Such enzymes are called _____ enzymes. The small molecules usually bind by noncovalent forces to a _____ site that is different from the catalytic site.

4. Every metabolic pathway has at least one _____ enzyme that catalyzes the slowest or _____ reaction in the pathway. Since other reactions proceed more _____ than this reaction, changes in the activity of this enzyme directly alter the speed with which a pathway operates.

5. In feedback inhibition, when the end product becomes _____ it inhibits the _____ enzyme and slows its own synthesis. As the end product concentration _____, pathway activity once again _____ and _____ product is formed.

6. If *E. coli* is grown in a medium that contains both glucose and lactose, it uses _____ preferentially until this sugar is exhausted. Then after a short lag, growth resumes using _____ as a carbon source. This biphasic growth pattern is called _____ growth.

7. For operons containing genes for a catabolic pathway, the initial substrate of that pathway will usually act as a(n) _____, while for operons containing genes for an anabolic pathway, the end product of the pathway will usually act as a(n) _____.

8. Attenuation is only possible when the processes of _____ and _____ are tightly coupled. Therefore, this type of regulation of gene expression is usually seen in _____ organisms but not _____ organisms.

9. The regulation of multiply branched pathways often involves _____ to catalyze the _____ step. In this situation, excess of a single end product _____ but does not completely block pathway activity because some _____ are still active.

10. Sigma factors enable the RNA polymerase to recognize and bind to _____. Different sigma factors recognize _____ sets of _____. Therefore, substitution of sigma factors immediately changes _____ _____.

MULTIPLE CHOICE

For each of the questions below select the *one best* answer.

1. Which of the following is a reason for metabolic regulation?
   a. conservation of material
   b. conservation of energy
   c. maintaining metabolic balance
   d. All of the above are reasons for metabolic regulation.

2. A small molecule that binds to an allosteric enzyme and thereby increases the activity of the enzyme is called a(n)
   a. positive effector.
   b. negative effector.
   c. inducer.
   d. corepressor.

3. Which of the following is not true about the regulation of branched metabolic pathways?
   a. There are usually separate regulatory enzymes for each branch, as well as a regulatory enzyme that controls the flow of carbon into the entire set of possible branches.
   b. An excess of one end product will usually completely inhibit the activity of the branch responsible for the synthesis of that particular end product.
   c. An excess of one end product will usually partially inhibit the flow of carbon into the entire set of branched pathways.
   d. All of the above are true about the regulation of branched metabolic pathways.

4. Which of the following is not a regulatory mechanism used to control the lactose operon in *E. coli?*
   a. induction
   b. catabolite repression
   c. attenuation

5. Which of the following is not a regulatory mechanism used to control the tryptophan operon in *E. coli?*
   a. repression
   b. catabolite repression
   c. attenuation
   d. All of the above are regulatory mechanisms used to control the tryptophan operon in *E. coli.*

6. Enzymes that are not regulated (i.e., they are produced all the time) are said to be
   a. inducible.
   b. repressible.
   c. constitutive.
   d. attenuated.

7. In the active state, the repressor binds to the operator and thereby
   a. inhibits the initiation of transcription.
   b. reduces the rate of transcription.
   c. terminates transcription that has already begun.
   d. does all three of the above.

8. Which of the following is least likely to respond rapidly to changes in environmental conditions?
   a. metabolic channeling
   b. adjustment of enzyme activity
   c. regulation of gene expression
   d. All of the above are equally likely to respond to rapid environmental changes.

9. Which of the following will most likely conserve the greatest amount of energy for the cell?
   a. metabolic channeling
   b. adjustment of enzyme activity
   c. regulation of gene expression
   d. All of the above conserve nearly equal amounts of energy.

10. Regulation of the cell cycle responds to
    a. cell mass.
    b. cell length.
    c. Both (a) and (b) are correct.
    d. Neither (a) nor (b) is correct.

11. Antisense RNA has been demonstrated to regulate gene expression in which of the following?
    a. procaryotes but not eucaryotes
    b. eucaryotes but not procaryotes
    c. both procaryotes and eucaryotes
    d. neither procaryotes nor eucaryotes

TRUE/FALSE

_____ 1. If a particular energy source is unavailable, the enzymes required for its utilization are needed and energy must be expended to produce them.

_____ 2. For allosteric enzymes, the regulatory site is always on a different polypeptide chain (subunit) from that of the catalytic site.

_____ 3. Covalent modification represents a reversible way of controlling enzyme activity because the modified form has an altered activity (either higher or lower) than the unmodified form.

_____ 4. Usually the last step in a pathway is a pacemaker reaction and is catalyzed by a regulatory enzyme.

_____ 5. When a biosynthetic pathway branches to form more than one end product, an excess of one of the end products will only inhibit the branch of the pathway involved in the synthesis of that particular product, while an excess of all the end products will usually inhibit the flow of carbon into the entire pathway.

_____ 6. In the presence of both glucose and lactose, the lactose repressor is not bound to the operator; however, the genes of the lactose operon are still not expressed: catabolite repression affects the binding of RNA polymerase to the promoter, but does not involve the operator.

_____ 7. Although DNA replication and cell division are separate processes, they are tightly coordinated. Therefore, if DNA synthesis is inhibited by a drug or a gene mutation, cell division is also blocked.

_____ 8. Even at high rates of cell division (doubling time less than 60 minutes), DNA replication for the next doubling is not initiated until the previous round of cell division has been completed.

CRITICAL THINKING

1. Look at the following two diagrams. Compare and contrast the regulatory mechanisms depicted in (a) with those in (b). Which is more likely to regulate the enzymes of a catabolic pathway and which is more likely to regulate an anabolic pathway? Explain.

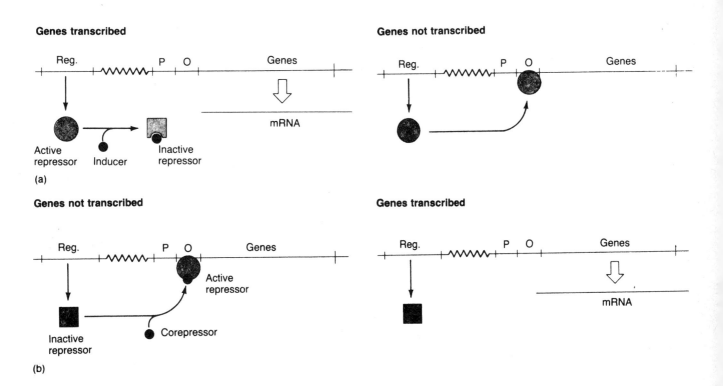

2. After inoculating a flask of minimal broth containing glucose and lactose with *E. coli* and following the growth kinetics, you obtain the following growth curve:

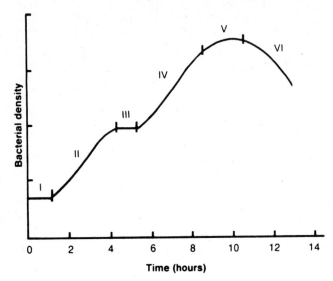

A. Describe each stage of the growth curve (I–VI) in terms of growth rates, growth substrates, and general metabolic activity.

B. Diagram the lactose operon. What differences are there in the state of the operon during phase I, phase II, phase III, and phase IV of the given growth curve? Be sure to discuss the promoter and operator regions as well as the roles of the lac repressor and the CAP protein in each of the first four phases.

ANSWER KEY

Terms and Definitions

1. l, 2. e, 3. p, 4. a, 5. h, 6. o, 7. i, 8. k, 9. j, 10. g, 11. m, 12. n, 13. c, 14. d, 15. f, 16. b

Fill in the Blank

1. metabolites; enzymes; metabolic channeling 2. enzymes; metabolites 3. effectors; modulators; allosteric; regulatory 4. pacemaker; rate-limiting; rapidly 5. too concentrated; regulatory (pacemaker); decreases; increases; more 6. glucose; lactose; diauxic 7. inducer; corepressor 8. transcription (translation); translation (transcription); procaryotic; eucaryotic 9. isoenzymes; pacemaker; reduces; isoenzymes 10. promoters; different; promoters; gene; expression

Multiple Choice

1. d, 2. a, 3. d, 4. c, 5. b, 6. c, 7. a, 8. c, 9. c, 10. c, 11. a

True/False

1. F, 2. F, 3. T, 4. F, 5. T, 6. T, 7. T, 8. F

13 Microbial Genetics: General Principles

CHAPTER OVERVIEW

This chapter presents the basic concepts of molecular genetics: storage and organization of genetic information in the DNA molecule, mutagenesis, and repair. The role of microorganisms in screening procedures for mutagenic agents is also described. Primary emphasis is given to the genetics of bacteria.

CHAPTER OBJECTIVES

After reading this chapter you should be able to:

- discuss the nature of the genetic code
- define a gene and discuss controlling elements, such as promoters and operators
- discuss the four parts (promoter, leader, coding region, trailer) associated with a bacterial gene
- discuss the nature and causes of mutations
- discuss the various genetic repair mechanisms and their limitations

CHAPTER OUTLINE

I. Introduction
    A. Clone—a population of cells that are genetically identical
    B. Genome—all the genes present in a cell or virus
    C. Genotype—the specific set of genes an organism possesses
    D. Phenotype—the collection of characteristics of an organism that an investigator can observe
II. DNA as the Genetic Material
    A. Griffith (1928) demonstrated the phenomenon of transformation: nonvirulent bacteria could become virulent when live, nonvirulent bacteria were mixed with dead, virulent bacteria
    B. Avery, MacLeod, and McCarty (1944) demonstrated that the transforming principle was DNA
    C. Hershey and Chase (1952) showed that for the T2 bacteriophage, only the DNA was needed for infectivity; therefore, they proved that DNA was the genetic material
III. The Genetic Code—the manner in which genetic instructions to direct polypeptide production are stored within the genome
    A. Colinearity—the DNA base sequence corresponds in linear fashion to the amino acid sequence of the protein whose synthesis it directs
    B. Establishment of the genetic code—each codon that specifies a particular amino acid must be three bases long for each of the 20 amino acids to have at least one codon

C. Organization of the code
　　1. Degeneracy—there are up to six different codons for a single amino acid
　　2. Sense codons—61 codons that specify amino acids
　　3. Stop (nonsense) codons—three codons (UGA, UAG, UAA) that do not specify an amino acid, and that are used as termination signals
　　4. Wobble—describes amino acid specificity leniency: the third position of an mRNA codon is less important than the first or second position in determining the encoded amino acid
　　　　a. Codons differing only in the third position frequently specify the same amino acid
　　　　b. Wobble eliminates the need for a unique tRNA for each codon because the first two positions are sufficient to establish hydrogen bonding between the mRNA and the aminoacyl-tRNAs

IV. Gene Structure
A. Gene—a linear sequence of nucleotides that is within the genomic nucleic acid molecule, and that has a fixed start point and end point
　　1. Encodes a polypeptide, a tRNA, or an rRNA
　　2. Controlling elements (e.g., promoters) that regulate expression of a gene may be considered as part of the gene itself, or they may be considered as separate regulatory sequences
　　3. With some exceptions, genes are not overlapping
　　4. The segment that encodes a single polypeptide is also called a cistron
B. Procaryotic vs. eucaryotic genes
　　1. Procaryote—coding information is normally continuous although some bacterial genes are interrupted
　　2. Eucaryote—most genes have coding sequences (exons) that are interrupted by noncoding sequences (introns); one exception are the genes that code for histones, which lack introns
C. Genes that code for proteins
　　1. Sense strand—the one strand which contains coding information
　　2. Antisense strand—the strand complementary to the sense strand
　　3. Promoter—a sequence of bases that is usually situated upstream from the coding region which serves as a recognition/binding site for RNA polymerase
　　　　a. Recognition site—site of initial association with RNA polymerase (35 bases upstream of transcription initiation site)
　　　　b. Binding site (Pribnow box)—sequence that favors DNA unwinding before transcription begins (approximately 10 bases upstream of transcription initiation site)
　　　　c. Consensus sequences—idealized base sequences at these and other positions found most often at those positions when comparing the sequences of different bacteria
　　4. Leader sequence—a transcribed sequence that is not translated but which contains a consensus sequence known as the Shine-Dalgarno sequence which serves as the recognition site for the ribosome
　　5. Coding region—the sequence that begins immediately downstream of the leader sequence; starts with the sense sequence 3′TAC5′, which gives rise to mRNA codon 5′AUG3′, the first translated codon (specifies N-formylmethionine)
　　6. Trailer region—nontranslated region located immediately downstream of the translation terminator sequence and before the transcription terminator
　　7. Regulatory sites—the operators and CAP binding sites previously discussed that are found associated with some genes
D. Genes that code for tRNA and rRNA
　　1. Promoter, leader, and trailer regions are also found for these genes; they are removed after transcription
　　2. More than one tRNA gene may be made from a single transcript; they are separated by a noncoding spacer region which is removed after transcription

V. Mutations and their Chemical Basis
   A. Mutation—a stable, heritable change in the genomic nucleotide sequence
      1. Conditional mutations—are expressed only under certain environmental conditions
      2. Morphological mutations—result in changes in colony or cell morphology
      3. Biochemical mutations—result in changes in the metabolic capabilities of a cell
         a. Auxotrophs—cannot grow on minimal media; require supplements
         b. Prototrophs—can grow on minimal media
      4. Resistance mutations—result in acquired resistance to some pathogen, chemical, or antibiotic
      5. Apparently directed- or adaptive-mutations may be the result of hypermutation followed by selection of favorable mutations
      6. Hypermutation involves activation of special mutator genes
   B. Spontaneous mutations—arise occasionally in all cells in the absence of any added agent
      1. Causes of spontaneous mutations
         a. Errors in DNA replication
         b. Damage to DNA from background gamma radiation or heat
         c. Insertion of transposons
      2. Types of mutations
         a. Transition—substitution of one purine for another, or of one pyrimidine for another
         b. Transversion—substitution of a purine for a pyrimidine or vice versa
         c. Frameshift—deletion of DNA segment resulting in an altered codon reading frame
   C. Induced mutations are caused by mutagens that damage DNA or alter its chemistry
      1. Base analogs are incorporated into DNA during replication and exhibit base-pairing properties different from the bases they replace
      2. Specific mispairing occurs when a mutagen changes a base's structure and thereby alters its pairing characteristics
      3. Intercalating agents, which become inserted between the stacked bases of the helix, distort the DNA and thus induce single nucleotide pair insertions or deletions that can lead to frameshifts
      4. Many mutagens can severely damage DNA so that it cannot act as a replication template; this would be lethal without the repair mechanisms to restore the DNA; however, the repair mechanisms are error prone, which also leads to mutations
   D. The expression of mutations
      1. Forward mutation—a conversion from the most prevalent gene form (wild type) to a mutant form
      2. Back mutation—a conversion of the mutant nucleotide sequence back to the wild type arrangement
      3. Suppressor mutation—a reestablishment of the wild type phenotype by a second mutation in the same or a different gene that overcomes the effect of the first mutation
      4. Point mutations—affect only one base pair and are more common than large deletions or insertions
         a. Silent mutations are alterations of the base sequence that do not alter the amino acid sequence of the protein because of code degeneracy
         b. Missense mutations are alterations of the base sequence that result in the incorporation of a different amino acid in the protein; at the level of protein function, the effect may range from complete loss of activity, to no change in activity at all
         c. Nonsense mutations are alterations that produce a translation termination codon, which results in premature termination of the protein during synthesis; location of the mutation within the protein will determine the extent of change in function
         d. Frameshift mutations are insertions or deletions of one or two base pairs that thereby alter the reading frame
      5. Regulatory mutations are changes in the promoter or operator, they affect the expression of the downstream genes

VI.   Detection and Isolation of Mutants
  A.   Detection
      1.   Visual observation of changes in colony characteristics
      2.   Auxotrophic mutants (i.e., those which have lost the ability to synthesize a particular end product and which therefore require its presence in the growth medium) can be detected by replica plating on media with and without the growth factor; mutants are those growing with the factor but not without it
  B.   Selection of mutations is achieved by finding the environmental condition in which the mutant will grow but the wild type will not (useful for auxotrophic revertants resulting from back mutations, and for resistance mutants)
  C.   Carcinogenicity testing
      1.   Most cancer-causing agents (carcinogens) are also mutagens
      2.   Tests for mutagenicity are used as a screen for carcinogenic potential
      3.   The Ames test is a widely used mutational reversion test for histidine-requiring auxotrophs of special strains of *Salmonella typhimurium*
VII.   DNA Repair—needed to correct errors in DNA sequences
  A.   Excision repair
      1.   Corrects damage that causes distortions of DNA
      2.   The damaged area is excised, producing a single-stranded gap
      3.   The gap is filled in by DNA polymerase I and DNA ligase
  B.   Removal of lesions (such as photoreactivation to remove thymine dimers) reverses damage without removing and replacing bases; although this process is error free, it is limited to the repair of certain kinds of damage
  C.   Postreplication repair
      1.   In this process, newly synthesized DNA is proofread for mismatched base pairs
      2.   Mismatched base pairs are removed and replaced by the action of DNA polymerase I and DNA ligase
  D.   Recombination repair restores DNA that has damage in both strands by recombination with an undamaged molecule if available (this frequently occurs in rapidly dividing cells where there is another copy of the chromosome not yet parceled out to a daughter cell)
  E.   SOS repair is used to repair excessive damage that halts replication, but this error-prone process results in many mutations

TERMS AND DEFINITIONS

____ 1. The concept that the DNA nucleotide sequence corresponds linearly to the amino acid sequence in the protein encoded by that gene

____ 2. The sequence of three nucleotides that specifies a particular amino acid

____ 3. The concept that there is more than one codon that specifies a particular amino acid

____ 4. Those sequences of three nucleotides that actually do specify amino acids

____ 5. Those sequences of three nucleotides that do not specify any amino acid; instead they cause termination of translation

____ 6. A sequence of nucleotides that codes for a polypeptide, a tRNA, or an rRNA

____ 7. The strand of DNA that contains coding information for mRNA

____ 8. The strand of DNA complementary to the one containing coding information for mRNA

____ 9. An idealized sequence composed of the bases most often found at each position of a regulatory sequence when such sequences from different bacteria are compared

____ 10. The region at which RNA polymerase binds before initiating transcription of the downstream sequences

____ 11. The transcribed, but not translated, sequence that is immediately upstream of the region that encodes the functional product

____ 12. The transcribed and translated region of a gene

____ 13. The transcribed, but not translated, sequence that is immediately downstream of the translation termination signal

____ 14. Alterations in the base sequence of the genomic nucleic acid

____ 15. Mutations that are only expressed under certain environmental conditions

____ 16. Mutations that arise in the absence of any added agent

____ 17. Mutations that arise as a result of exposure to some physical or chemical agent

____ 18. Physical or chemical agents that can cause mutation

____ 19. Mutations that result in purine-purine or pyrimidine-pyrimidine substitutions

____ 20. Mutations that result in purine-pyrimidine or pyrimidine-purine substitutions

____ 21. Mutations that result in a change in the reading frame

____ 22. The most prevalent gene form in a population

____ 23. A mutation from the most prevalent gene form in the population

____ 24. A mutation to the most prevalent gene form in the population

a. antisense strand
b. back mutation
c. cistron
d. clone
e. coding region
f. codon
g. colinearity
h. conditional mutations
i. consensus sequence
j. degeneracy
k. forward mutation
l. frameshift mutations
m. gene
n. induced mutations
o. leader sequence
p. missense mutation
q. mutagen
r. mutation
s. nonsense codons
t. nonsense mutation
u. point mutation
v. promoter
w. sense codons
x. sense strand
y. silent mutation
z. spontaneous mutations
aa. suppressor mutation
bb. trailer sequence
cc. transition mutations
dd. transversion mutations
ee. wild type

_____ 25. A second mutation that overcomes the effect of the first mutation
_____ 26. A mutation that involves only one base pair
_____ 27. A mutation that does not alter the amino acid sequence of the resulting protein
_____ 28. A mutation that changes the amino acid sequence of the resulting protein by substitution
_____ 29. A mutation that causes premature termination of the synthesis of the protein product
_____ 30. A population of cells that are genetically identical
_____ 31. A segment of DNA that encodes a single polypeptide

FILL IN THE BLANK

1. Avery, MacLeod, and McCarty provided the first evidence that _____ carried the genetic information when they exposed extracts from virulent strains of pneumococcus to enzymes that hydrolyze RNA, DNA, or protein. Only extracts exposed to enzymes that hydrolyzed _____ prevented the extract-treated nonvirulent strains from becoming virulent. This acquisition of virulence was called _____.

2. Hershey and Chase demonstrated that when the bacteriophage T2 infected its host cell, the _____ was injected into the host while the _____ remained outside. Since progeny virus was produced, it was clear that the _____ was carrying the genetic information.

3. The genetic code must be contained in some sequence of the four nucleotides found in the linear DNA sequence. If taken individually, only four amino acids could be specified. If taken two at a time only _____ amino acids could be specified. Therefore, a code word or _____ must involve at least three nucleotides. However, this would allow _____ possible combinations, which is more than the minimum of _____ needed to specify all the amino acids usually found in proteins.

4. There are 61 codons that specify amino acids. These are called _____ codons. The three codons, _____, _____, and _____, which do not specify amino acids, are called _____ codons and result in the termination of _____.

5. Although DNA is double stranded, only one of the strands contains coded information and directs RNA synthesis. This strand is called the _____ strand, and the complementing strand is called the _____ strand.

6. In addition to the coding region, each gene also has three noncoding regions. These are the _____, the _____, and the _____.

7. Mutations can alter the phenotype in several different ways. _____ mutations change the cellular or colonial characteristics. _____ mutations, when expressed, result in the death of the organism. These are usually only recovered if they are _____ in diploid organisms or _____ in haploid organisms.

8. Mutations that inactivate a metabolic pathway are called _____ mutations. A microorganism with this kind of mutation is often unable to grow on minimal medium, and for growth it requires an adequate supply of the pathway's end product. Such mutants are called _____, while microbial strains that can grow on minimal medium are called _____.

9. Frameshift mutations are usually very _____ and yield mutant phenotypes resulting from the synthesis of nonfunctional proteins. A second frameshift shortly _____ from the first may restore the reading frame, and thereby minimize the phenotypic effect. This second mutation is a good example of a(n) _____ suppressor mutation.

10. Thymine dimers are often directly repaired. They are split apart into separate thymines with the help of _____ light in a photochemical reaction catalyzed by the enzyme photolyase. This is called _____, and since it does not remove and replace nucleotides, it is relatively free of _____.

11. The _____ refers to all of the genes present in a cell or virus, while the _____ refers to the specific set of genes an organism possesses. The collection of characteristics for an organism that can be observed by an investigator is called the _____ .

12. The leader sequence of an mRNA is transcribed but not _____ . It contains a _____ sequence known as the _____ sequence which serves as a _____ binding site.

13. Mutations that appear to be _____ or _____ may be the result of _____ , which involves activation of special _____ genes, followed by selection of favorable mutants.

MULTIPLE CHOICE

For each of the questions below select the *one best* answer.

1. More than one codon will specify a particular amino acid. Therefore, the code is said to be
   a. ambiguous.
   b. degenerate.
   c. multiplicative.
   d. repetitious.

2. A gene is a sequence of bases in the genomic nucleic acid that encodes
   a. a polypeptide chain.
   b. a tRNA.
   c. an rRNA.
   d. any of the above.

3. Although key sequences within the promoter may vary somewhat, they are fairly constant and may be represented by idealized sequences composed of the bases most often found at each position when the sequences from different organisms are compared. These idealized sequences are called
   a. convergence sequences.
   b. idealized sequences.
   c. consensus sequences.
   d. common sequences.

4. Which of the following is not a function of the leader sequence?
   a. regulation of transcription by catabolite repression
   b. regulation of transcription by attenuation
   c. proper orientation of the mRNA on the ribosome by the interaction of a part of the leader with the 16S rRNA
   d. All of the above are functions of the leader sequence.

5. A particular mutation resulted in the substitution of cytosine for thymine in one strand of the DNA. Upon subsequent DNA replication, one of the daughter cells received a GC pair in this position instead of an AT pair. This type of mutation is called a(n)
   a. transversion.
   b. transition.
   c. frameshift.
   d. insertion.

6. Transitions can be caused by which of the following?
   a. incorporation of a base analog that exhibits different base-pairing properties from those of the base it replaces
   b. chemical modification of an existing base in the DNA so that during the next round of replication it will base pair differently from the unmodified base
   c. Both of the above are correct.
   d. None of the above are correct.

7. Back mutations that restore the wild type phenotype can occur by which of the following mechanisms?
   a. true reversion back to the wild type base sequence
   b. mutation that results in a different sequence from that of the wild type, but that restores the amino acid sequence in the protein to the wild type sequence
   c. a second mutation that overcomes the effect of the first mutation; the first mutation is not changed, but the function of the protein is restored
   d. All of the above can restore the wild type phenotype.

8. Which of the following is a reason for using mutation detection and selection methods?
   a. understanding more about the nature of genes
   b. understanding the biochemistry of a particular microorganism
   c. testing carcinogenic potential by determining mutagenic capability
   d. All of the above are reasons for using mutation detection and selection methods.

9. Which of the following make the tester strains of *Salmonella typhimurium* particularly useful in the Ames mutational reversion assay?
   a. Mutational alterations make them highly permeable to test substances.
   b. Their excision repair mechanisms are defective; therefore they cannot readily repair damage done by the test substances.
   c. They have plasmid genes that enhance error-prone repair, which increases mutagenic sensitivity even though some damage repair occurs.
   d. All of the above make these strains useful for mutagenicity testing.

10. Resistance mutations can confer on the organism resistance to which of the following?
    a. pathogens
    b. chemicals
    c. antibiotics
    d. All of the above are correct.

TRUE/FALSE

____ 1. Since a nucleotide sequence can be read in any of three different reading frames, each nucleotide sequence usually encodes three different polypeptide products in an overlapping fashion.

____ 2. In both procaryotes and eucaryotes the coding information within a gene is normally continuous (i.e., it is not interrupted by noncoding sequences).

____ 3. For any given gene, coding information is found in only one strand. However, different genes may use different strands for their information.

____ 4. The translation stop codon also functions as the site for termination of transcription.

____ 5. Several tRNA molecules may be produced from a single transcript by posttranscriptional processing. The segments coding for the tRNAs are separated from each other by short spacer sequences that are removed during the processing.

____ 6. All of the rRNAs are transcribed as a single, large precursor molecule that is cleaved to yield the final rRNA products. The spacer regions between the rRNA sequences usually contain tRNAs that are produced as well.

____ 7. Missense mutations may play an important role in providing new variability to drive evolution because they are often not lethal, and therefore, they remain in the gene pool.

____ 8. Nonsense mutations that cause premature termination of translation always severely affect the phenotypic expression of the gene by resulting in the production of a nonfunctional gene product.

____ 9. Controlling sequences (such as promoters) that regulate expression of a gene are not considered part of the gene.

____ 10. The concept of wobble in the codon-anticodon interaction eliminates the necessity for having a tRNA with an anticodon region specific for each of the 61 sense codons.

_____ 11. Point mutations are more common than large deletions or insertions.

_____ 12. Eucaryotic genes are usually interrupted. One interesting exception is the genes for histones, which contain not introns.

CRITICAL THINKING

1. A strain of bacteria requires the amino acid leucine (i.e., it is a leucine auxotroph) and is sensitive to the antibiotic streptomycin. You expose this strain to a mutagen. How would you isolate mutants that no longer require leucine (i.e., strains that have reverted to prototrophy)? How would you isolate mutants that are resistant to streptomycin? How would you isolate mutants that no longer require leucine and that are resistant to streptomycin?

2. There are a variety of suppressor mutations. These include intragenic suppressors, nonsense suppressors, and physiological suppressors. Describe each of these and discuss how they differ from true or equivalent reversions.

ANSWER KEY

Terms and Definitions

1. g, 2. f, 3. j, 4. w, 5. s, 6. m, 7. x, 8. a, 9. i, 10. v, 11. o, 12. e, 13. bb, 14. r, 15. h, 16. z, 17. n, 18. q, 19. cc, 20. dd, 21. l, 22. ee, 23. k, 24. b, 25. aa, 26. u, 27. y, 28. p, 29. t, 30. d, 31. c

Fill in the Blank

1. DNA; DNA; transformation 2. DNA; protein; DNA 3. 16; codon; 64; 20 4. sense; UAA; UAG; UGA; nonsense; translation 5. sense; antisense 6. promoter; leader; trailer 7. Morphological; Lethal; recessive; conditional 8. biochemical; auxotrophs; prototrophs 9. deleterious; downstream; intragenic 10. visible; photoreactivation; error; 11. genome; genotype; phenotype 12. translated; consensus; Shine-Dalgarno; ribosome 13. directed; adaptive; hypermutation; mutator

Multiple Choice

1. b, 2. d, 3. c, 4. a, 5. b, 6. c, 7. d, 8. d, 9. d, 10. d

True/False

1. F, 2. F, 3. T, 4. F, 5. T, 6. T, 7. T, 8. F, 9. F, 10. T, 11. T, 12. T

14   Microbial Genetics: Recombination and Plasmids

CHAPTER OVERVIEW

This chapter begins with a general discussion of bacterial recombination, plasmids, and transposable elements, and then examines the acquisition of genetic information by conjugation, transformation, and transduction. The way these recombination procedures are used to map the bacterial genome is explained. Finally, viral recombination and genome mapping are discussed.

CHAPTER OBJECTIVES

After reading this chapter you should be able to:

- discuss the nature of procaryotic recombination
- discuss the three ways (conjugation, transformation, and transduction) that bacteria acquire new genetic material
- discuss how plasmids and transposable elements can move genetic material between bacterial chromosomes and within a chromosome to cause changes in the genome and the phenotype of the organism
- discuss the use of these gene transfer procedures to map the bacterial genome
- discuss the sequencing of microbial genomes
- discuss the recombination that occurs when two viruses simultaneously infect the same host

CHAPTER OUTLINE

I.   Bacterial Recombination: General Principles
   - A.   Recombination is the process by which a new chromosome (with a genotype different from either parent) is formed when genetic material from two parent organisms combine
      1. General recombination usually involves a reciprocal exchange in which a pair of homologous sequences break and rejoin in a crossover
      2. Nonreciprocal recombination involves the incorporation of a single strand into the chromosome to form a stretch of heteroduplex DNA
      3. Site-specific recombination is the nonhomologous insertion of DNA into a chromosome
         - a.   Often occurs during viral genome integration into the host
         - b.   The enzymes responsible are specific for the virus and its host
      4. Replicative recombination accompanies replication and is used by genetic elements that move about the genome
   - B.   Terminology
      1. Horizontal gene transfer—transfer of genes from one mature, independent organism to another
      2. Vertical gene transfer—transmission of genes from parents to offspring
      3. Exogenote—donor DNA that enters the bacterium by one of several mechanisms
      4. Endogenote—the genome of the recipient

5. Merozygote—a recipient cell that is temporarily diploid for a portion of the genome during the replacement process
    C. Types of horizontal exogenote transfer
        1. Conjugation is direct transfer from another bacterium
        2. Transformation is transfer of a naked DNA molecule
        3. Transduction is transfer from a bacteriophage
    D. Intracellular fates of exogenote
        1. Integration into the host chromosome
        2. Independent functioning and replication of the exogenote without integration (a partial diploid clone develops)
        3. Survival without replication (only the one cell is a partial diploid)
        4. Degradation by host nucleases (host restriction)
II. Bacterial Plasmids—small, circular DNA molecules that are not part of the bacterium's chromosome
    A. Plasmids have their own replication origins; they replicate autonomously and are stably inherited
    B. Curing is the elimination of a plasmid; it can occur either spontaneously or as a result of treatments that inhibit plasmid replication but do not affect host cell reproduction
    C. Episomes are plasmids that can either exist independent of the host chromosome or be integrated into it
    D. Conjugative plasmids have genes for pili and can transfer copies of themselves to other bacteria during conjugation
    E. Types of plasmids
        1. Fertility plasmids (e.g., the F factor) are episomes that can direct the formation of sex pili and can transfer copies of themselves during conjugation
        2. Resistance factors (R plasmids) have genes for resistance to various antibiotics; some are conjugative; however, they are not episomal (they do not integrate into the host chromosome)
        3. Col plasmids carry genes for the synthesis of colicins that destroy *Escherichia coli;* other similar plasmids carry genes for bacteriocins that are directed against other bacterial species; some are conjugative and may also carry resistance genes
        4. Virulence plasmids make the bacterium more pathogenic by conferring resistance to host defense mechanisms or by carrying a code for the production of a toxin
        5. Metabolic plasmids carry genes for enzymes that utilize certain substances as nutrients (aromatic compounds, pesticides, etc.)
III. Transposable Elements—transposons
    A. Segments of DNA that can move about chromosomes within a single organism or between different organisms
    B. Differ from bacteriophages in that they lack an infectious viral life cycle
    C. Differ from plasmids in that they are unable to reproduce independently
    D. Types of transposable elements
        1. Insertion sequences (IS elements) contain genes only for those enzymes required for transposition; they are bound on both ends by inverted terminal repeat sequences
        2. Composite transposons carry other genes in addition to those needed for transposition (e.g., for antibiotic resistance, toxin production, etc.)
    E. Movement is typically by replicative transposition, during which a replicated copy of the transposon inserts at the target site on the DNA, while the original copy remains at the parental site
    F. Effects of transposable elements
        1. Insertional mutagenesis, including deletion of genetic material at or near the target site
        2. Arrest of translation or transcription due to stop codons or termination sequences located on the inserted material
        3. Activation of genes near the point of insertion due to promoters located on the inserted material

IV.   Bacterial Conjugation—the transfer of genetic information via direct cell-cell contact; this process is mediated by fertility plasmids (F plasmids)
    A. $F^+ \times F^-$ mating
        1.   Nonreciprocal exchange between donor (F^+) and recipient (F^-)
            a.   Recipients usually become F^+ (plasmid is transferred)
            b.   Donors remain F^+ (plasmid is retained)
            c.   The plasmid DNA replicates by the rolling-circle mechanism, and the displaced strand is transferred and then copied to produce double-stranded DNA; the other strand and its complement are retained by the donor
        2.   Chromosomal genes are not transferred
    B.   Hfr conjugation
        1.   F plasmid integration into the host chromosome results in an Hfr strain of bacteria
        2.   The mechanics of conjugation of Hfr strains are similar to those of F^+ strains
        3.   The initial break for rolling-circle replication is at the integrated plasmid's origin
            a.   Part of the plasmid is transferred first
            b.   Chromosomal genes are transferred next
            c.   The rest of the plasmid is transferred last
        4.   Complete transfer of the chromosome takes approximately 100 minutes, but the conjugation bridge does not usually last that long; therefore, the entire F factor is not transferred, and the recipient remains F^-
    C. F' conjugation (sexduction)
        1.   An integrated F plasmid leaving the chromosome incorrectly may take with it some chromosomal genes from one side of the integration site; this is called an F' plasmid
        2.   The F' cell retains all of the plasmid genes, although some of them are in the host chromosome and some are on the plasmid; in conjugation, it behaves as an F^+ cell, mating only with F^- cells
        3.   The chromosomal genes included in the plasmid are transferred with the rest of the plasmid, but other chromosomal genes are not
        4.   The recipient becomes an F' cell, and a partially diploid merozygote
V.   DNA Transformation
    A.   A naked DNA molecule from the environment is taken up by the cell and incorporated into its chromosome in some heritable form
    B.   A competent cell is one that is capable of acting as a recipient
    C.   This process naturally occurs in a limited number of species, but can be induced in other species in the laboratory
    D.   The mechanics of the process differ from species to species
    E.   Species that are not normally competent (such as *E. coli*) can be made competent by calcium chloride treatment, which makes the cells more permeable to DNA
VI.   Transduction—transfer of bacterial genes by viruses
    A.   Generalized transduction—any part of the bacterial genome can be transferred during lytic infection
        1.   The phage degrades host chromosome into randomly sized fragments
        2.   During assembly, fragments of host DNA of the appropriate size can be mistakenly packaged into the phage head
        3.   When the next host is infected, the bacterial genes are injected and a merozygote is formed
            a.   Preservation of the transferred genes requires their integration into the host chromosome
            b.   Much of the transferred DNA does not integrate into the host chromosome, but is often able to survive and be expressed; the host is called an abortive transductant
        4.   Because fragmentation is random, any bacterial gene can be transferred; therefore, this is called *generalized* transduction

     B.   Specialized (restricted) transduction—only temperate phages that have established lysogeny (their DNA has been integrated into the host chromosome and replicates along with it) are capable of specialized transduction
         1.   A prophage (an integrated, nondestructive viral genome) is sometimes excised incorrectly and contains portions of the bacterial DNA that was adjacent to the phage's integration site on the chromosome
         2.   The excised phage genome is defective because it lacks some of its own genes and carries some of the bacterial genes
         3.   When the next host is infected, the bacterial genes lead to the formation of a merozygote, and the phage cannot reproduce
         4.   Because only those genes adjacent to the integration site can be carried, and because integration sites are at specific locations, the number of transferable genes is limited; therefore, this is called *specialized* transduction

VII.   Mapping and Sequencing the Genome
     A.   Hfr mapping involves the use of an interrupted mating experiment
         1.   Chromosome transfer occurs at a constant rate
         2.   The frequency of a particular recombinant indicates the position of that gene relative to the plasmid integration site
            a.   High-frequency recombinants indicate that the gene is close to the integration site
            b.   Low-frequency recombinants indicate that the gene is farther from the integration site
         3.   The instability of the conjugation bridge makes it nearly impossible to map genes that are very distant
         4.   The use of several Hfr strains with different integration sites can generate overlapping maps, which can then be pieced together to form the entire genome map
     B.   Transformation mapping—the frequency with which two genes simultaneously transform recipient cells indicates the distance between the genes; overlapping maps can then be pieced together
     C.   Generalized transduction maps—as with transformation mapping, the frequency of cotransduction indicates the distance of two genes from each other
     D.   Specialized transduction maps provide distances from integration sites, which themselves must be mapped by conjugation mapping techniques
     E.   Sequencing a genome by the whole-genome shotgun approach is a multi-step process
         1.   Small fragments are generated and sequenced
         2.   Sequenced fragments are clustered into longer contiguous groupings (contigs) by analysis of sequence overlaps
         3.   Gaps between contigs are sequenced until the entire genome sequence has been determined
     F.   Annotation involves identifying open reading frames (ORFs), determining potential amino acid sequences and comparison to known proteins
     G.   Comparison among sequenced organisms allows estimates of functional similarity and determination of the minimum set of genes necessary to sustain life

VIII.   Recombination and Genome Mapping in Viruses
     A.   Recombination maps are generated from crossover frequency data obtained when cells are infected with two or more phage particles simultaneously
     B.   Denaturation maps—the effect of mild denaturing conditions is greater on AT-rich regions than on GC-rich regions; bubbles are thus formed, and can be seen in electron micrographs; maps are generated by comparing mutants for differences in bubbling
     C.   Heteroduplex mapping—wild type and mutant chromosomes are denatured and allowed to reanneal together; homologous regions pair normally, but mutant regions form bubbles that can be seen in electron micrographs
     D.   Restriction and endonuclease mapping—locates deletions and other mutations by examining the electrophoretic mobility (size) of the fragments generated
     E.   Sequence mapping—small phage genomes can be directly sequenced to map mutations

TERMS AND DEFINITIONS

Place the letter of each term in the space next to the definition or description that best matches it.

_____ 1. The process that occurs when genetic material from two organisms is combined, forming a genotype that differs from that of either parent

_____ 2. The piece of donor DNA in a recombination event

_____ 3. The piece of recipient DNA in a recombination event

_____ 4. A cell that is temporarily diploid for a portion of the genome during genetic transfer processes

_____ 5. Insertion of the donor DNA into the host chromosome mediated by sequence homologies

_____ 6. Small, circular DNA molecules that can exist independently of the host chromosome, that have their own replication origins, that are autonomously replicated, and that are stably inherited

_____ 7. A plasmid that can exist independent of the host chromosome or be integrated into it

_____ 8. A plasmid that has genes for pili and that can transfer copies of itself to other bacteria

_____ 9. A plasmid that carries genes that encode resistance to antibiotics

_____ 10. A plasmid that carries genes for a toxin, and thus renders the host bacterium pathogenic

_____ 11. A plasmid that carries genes for enzymes that degrade environmental substances such as aromatic compounds and pesticides

_____ 12. A piece of DNA that can move between chromosomes or within a single chromosome

_____ 13. Transfer of genetic information via direct cell-cell contact

_____ 14. Transfer of genetic information by uptake of a naked DNA molecule from the environment

_____ 15. Transfer of genetic information via viruses

_____ 16. The relationship between a phage and its host in which the phage genome is integrated into the host chromosome and is replicated along with the host chromosome without destroying the cell

_____ 17. The latent form of the virus genome that remains within the host without destroying it

_____ 18. Transfer of genes between independent, mature organisms

_____ 19. Transfer of genes from parent to offspring

_____ 20. A long sequence that has been determined by analysis of overlaps between shorter sequenced fragments

_____ 21. Identification of gene function by analysis of open reading frames and their associated amino acid sequences

a. annotation
b. conjugation
c. conjugative plasmid
d. contig
e. endogenote
f. episome
g. exogenote
h. horizontal gene transfer
i. integration
j. lysogeny
k. merozygote
l. metabolic plasmid
m. plasmids
n. prophage
o. recombination
p. resistance plasmid
q. transduction
r. transformation
s. transposable element
t. vertical gene transfer
u. virulence plasmid

FILL IN THE BLANK

1. The most common form of recombination is _____ recombination which involves a _____ exchange between a pair of _____ DNA sequences.
2. Integration of viral genomes into bacterial chromosomes involves another type of recombination known as _____ recombination, in which the genetic material is _____ with the host DNA.
3. Bacterial recombination normally takes place when a piece of donor DNA, the _____, enters the cell and become a stable part of the recipient's genome, its _____. During replacement of host genetic material, the recipient becomes a diploid for a portion of the genome, and is referred to as a _____.
4. The process of transposition involves a series of events, including _____ and _____ processes. Typically in bacteria, the original transposon remains at the parental site on the chromosome, while a _____ inserts at the target DNA site. This process is called _____ _____.
5. In 1952, William Hayes demonstrated that the gene transfer observed by Lederberg and Tatum was polar—that is, there were definite donor strains called _____ and recipient strains called _____—and that gene transfer was not _____ (i.e., it occurred in only one direction).
6. The F (fertility) factor carries genes for both _____ and _____. During mating of donor and recipient strains, a process called _____, the F factor replicates by the rolling-circle mechanism and one copy of it moves to the recipient.
7. Transfer of genetic information by uptake of a naked DNA molecule from the environment is called _____. In order to take up a naked DNA molecule, a cell must be _____, which may only occur at certain stages in the life cycle of the organism. Some organisms are not normally _____ and must be treated with _____ _____ in order to increase their permeability to DNA.
8. The transfer of bacterial genes by viruses is called transduction. When a virulent or temperate phage cleaves the host DNA into randomly sized fragments, some of these may be packaged into virions during the _____ stage of the viral life cycle. Upon subsequent infection of a new host, these genes may recombine with the host chromosome. Since these phages can transfer any part of the bacterial chromosome, this is called unrestricted, or _____, transduction. Some temperate phages can incorrectly excise from the host chromosome when switching from lysogeny to the lytic cycle, and these may carry genes that were adjacent to the integration site. Since the genes which may be carried are restricted to those located near specific integration sites, this is called restricted, or _____, transduction.
9. Conjugation involving _____ strains is frequently used to map the relative locations of bacterial genes. The technique involves disruption of the conjugation bridge in what is called a(n) _____ _____ experiment.

MULTIPLE CHOICE

1. Populations can be cured of their plasmids by treatments that inhibit plasmid replication but that do not affect host DNA replication. Which of the following have been used as curing agents?
    a. UV and ionizing radiation
    b. acridine mutagens
    c. thymine starvation
    d. All of the above have been used as curing agents.

2. Proteins produced by bacteria that destroy other bacteria are called
    a. colicins.
    b. bacteriocins.
    c. bacteriolysins.
    d. plasmolysins.

3. Which of the following is not encoded within a virulence plasmid?
    a. proteins that destroy other bacteria
    b. genes that render the bacterium less susceptible to host defense mechanisms
    c. genes for the production of toxigenic substances
    d. All of the above may be encoded within virulence plasmids.

4. Transposable elements that contain genes other than those required for transposition are called
    a. insertion sequences.
    b. plasmids.
    c. composite transposons.
    d. None of the above are correct.

5. In which of the following ways do transposable elements differ from temperate bacteriophages or from plasmids?
    a. Transposable elements lack a viral life cycle.
    b. Transposable elements are unable to reproduce autonomously.
    c. Transposable elements are unable to exist apart from the chromosome.
    d. All of the above are ways that transposable elements differ from bacteriophages or plasmids.

6. Which of the following is(are) possible effects of transposable elements?
    a. insertional mutagenesis by deletion
    b. insertion of stop codons or termination sequences that may block translation or transcription
    c. activation of genes by promoter insertion upstream of a currently unexpressed gene
    d. All of the above are possible effects of transposable elements.

7. Which of the following is not true about Hfr X F⁻ matings?
    a. The recipient can become F⁺ (or Hfr) if the mating lasts long enough for the entire bacterial chromosome to be transferred.
    b. The recipient usually remains F⁻ because the connection usually breaks before the entire bacterial chromosome can be transferred.
    c. The recipient may become F′ if more than half of the plasmid is transferred.
    d. All of the above are true about Hfr × F⁻ matings.

8. Transfer of bacterial genes by an F′ plasmid is often called
    a. transduction.
    b. sexduction.
    c. transfection.
    d. F-duction.

9. Which of the following has not been used to map chromosomal locations of bacterial genes?
    a. Hfr × F⁻ conjugation
    b. F′ × F⁻ conjugation
    c. F⁺ × F⁻ conjugation
    d. All of the above have been used to map bacterial genes.

10. Which of the following cannot be used to map the distance between any two genes on the bacterial chromosome?
    a. the frequency of co-conjugation during Hfr × F⁻ matings
    b. the frequency of cotransformation
    c. the frequency of cotransduction during generalized transduction
    d. the frequency of cotransduction during specialized transduction

11. Which of the following represents the best description of host restriction?
    a. the inability to take up an exogenote during transformation
    b. the inability to integrate an exogenote into the host chromosome
    c. the degradation of an exogenote by host nucleases
    d. the inability to express the genes located on an exogenote

12. Which of the following is not a possible fate for an exogenote?
    a. integration into the host chromosome
    b. expression of the genes and replication of the exogenote without integration into the host chromosome
    c. survival of the exogenote without integration or replication
    d. All of the above are possible fates for an exogenote.

TRUE/FALSE

_____ 1. Bacteria that are partial diploids, containing nonintegrated, transduced DNA, are called abortive transductants.

_____ 2. A plasmid that can exist independent of the host chromosome but that cannot be integrated into the host chromosome is called an episome.

_____ 3. The fertility factor is a conjugative plasmid that is particularly efficient at initiating conjugation with appropriate recipient cells.

_____ 4. Each antibiotic resistance gene is carried on a separate plasmid. Therefore, for a bacterium to be resistant to five different antibiotics, it must carry five different plasmids simultaneously.

_____ 5. Nonconjugative plasmids can move between bacteria during conjugation if such conjugation is initiated by another (conjugative) plasmid.

_____ 6. Multiple-drug resistance plasmids are usually produced when a single plasmid accumulates several transposons, each carrying one or more antibiotic resistance genes.

_____ 7. Transposable elements have only been found in procaryotes and do not appear to play a major role in eucaryotic genetics.

_____ 8. In an $F^+ \times F^-$ mating, the recipient becomes F^+ after the mating has been completed.

_____ 9. In an $F^+ \times F^-$ mating, the donor becomes F^- after the mating has been completed.

_____ 10. Hfr $\times$ F^- conjugation is probably the most efficient natural mechanism of gene transfer between bacteria.

_____ 11. All resistance plasmids are nonconjugative.

_____ 12. Host restriction refers to ability of some organisms to degrade exogenotes that enter the cell.

CRITICAL THINKING

1. Transposable elements and retroviruses have certain structural features in common; therefore, this also means that certain of their effects on their respective hosts are also similar. Discuss these similarities and show how the study of transposable elements has contributed to our understanding of retrovirus-mediated carcinogenesis.

2. Hfr $\times$ F^- matings have been used to map the bacterial genome. Demonstrate with diagrams how this is done. Why is more than one Hfr strain needed in order to generate a complete map of the bacterial chromosome?

ANSWER KEY

Terms and Definitions

1. o, 2. g, 3. e, 4. k, 5. i, 6. m, 7. f, 8. c, 9. p, 10. u, 11. l, 12. s, 13. b, 14. r, 15. q, 16. j, 17. n, 18. h, 19. t, 20. d, 21. a

Fill in the Blank

1. general; reciprocal; homologous 2. site-specific; nonhomologous 3. exogenote; endogenote; merozygote 4. self-replication; recombination; replicated copy; replicative transposition 5. F$^+$; F$^-$; reciprocal 6. pilus formation; plasmid transfer; conjugation 7. transformation; competent; competent; calcium chloride 8. assembly; generalized; specialized 9. Hfr; interrupted mating

Multiple Choice

1. d, 2. b, 3. a, 4. c, 5. d, 6. d, 7. c, 8. b, 9. c, 10. d, 11. c, 12. d

True/False

1. T, 2. F, 3. T, 4. F, 5. T, 6. T, 7. F, 8. T, 9. F, 10. T, 11. F, 12. T

15 Recombinant DNA Technology

CHAPTER OVERVIEW

This chapter focuses on practical applications of microbial genetic principles discussed in previous chapters. Although we have been altering the genetic makeup of organisms for centuries and nature has been doing it even longer, only recently have we been able to manipulate the DNA directly using genetic engineering or recombinant DNA technology. The potential benefits of these techniques are great and affect such diverse areas as medicine, agriculture, and industry. However, the use of these techniques is not without risks, and these risks must be considered in any discussion of this technology.

CHAPTER OBJECTIVES

After reading this chapter you should be able to:

- discuss the use of recombinant DNA technology to genetically engineer various organisms
- discuss the key role played by restriction endonucleases in genetic engineering
- discuss how plasmids, phages, and cosmids are used as vectors for insertion and expression of foreign genes in an organism
- discuss the use of both procaryotes and eucaryotes as target organisms for foreign gene insertion
- discuss the contributions already made by the use of this technology
- discuss the risks and the ethical problems associated with the use of this technology

CHAPTER OUTLINE

I.   Historical Perspectives
    A.   Arber and Smith (late 1960s) discovered restriction endonucleases, which cleave DNA at specific sequences
    B.   Boyer (1969) first isolated the restriction endonuclease *Eco*RI
    C.   Baltimore and Tëmin (1970) independently discovered reverse transcriptase; this enzyme can be used to construct a DNA copy, called complementary DNA (cDNA), of any RNA molecule
    D.   Jackson, Symons, and Berg (1972) generated the first recombinant DNA molecules
    E.   Cohen and Boyer (1973) produced the first recombinant plasmid (vector), which was introduced into and replicated within a bacterial host
    F.   Southern (1975) developed the blotting procedure for detecting specific DNA fragments, which uses radioactive DNA hybridization probes; this is useful in isolating particular genes of interest
    G.   Maxam and Gilbert (and independently, Sanger) (1976–1978) developed procedures for rapidly sequencing DNA molecules
    H.   Nonradioactive, enzyme-linked probes can now replace the earlier radioactive probes; they are faster and safer, but may be less sensitive
II. Synthetic DNA
    A.   Can be made by adding one nucleotide at a time to a growing chain; this takes about 40 minutes per nucleotide

B. In site-directed mutagenesis, a small synthetic oligonucleotide containing the desired sequence change is used as a primer for DNA polymerase, which then replicates the remainder of the target gene, and produces a new gene copy with the desired mutation; this can then be introduced into a new host

III. The polymerase chain reaction (PCR)
  A. Synthetic DNAs with sequences identical to those flanking the target sequence (100 to 5,000 base pairs in length) are used as primers; this is followed by successive cycles with a heat-stable DNA polymerase; using this technique, billions of copies of the target sequence can be produced in one to two hours
  B. New procedures allow RNA to be used as a template to produce and amplify complementary DNA (cDNA)
  C. PCR is now being used in forensic science as part of the DNA fingerprinting process

IV. Preparation of Recombinant DNA
  A. Isolating and cloning fragments
    1. DNA fragments are generated by shearing or by restriction endonuclease cleavage
    2. DNA fragments are separated electrophoretically
    3. Desired fragment is located by the Southern blotting technique; there must be a way of specifically identifying the desired fragment
    4. Fragment is inserted into plasmid vector
    5. Plasmid is inserted into bacterium by transformation
    6. Alternatively, fragments can be inserted into phage particles and introduced into bacterium by infection (transduction)
    7. Bacteria are grown, replica plated onto nitrocellulose filters, lysed, reacted with a hybridization probe specific for the desired gene, and then the desired clone is identified
  B. Alternative strategies
    1. Production of a genomic library
      a. The DNA is fragmented by endonuclease cleavage
      b. All resulting fragments are cloned
      c. The clone containing the desired fragment is identified and amplified
      d. The plasmid is extracted, and the fragment is purified
    2. A synthetic DNA can be produced and then cloned
  C. Gene probes
    1. cDNA probes can be used if the gene is expressed in certain tissue; the mRNA is obtained and reverse transcriptase is used to produce the cDNA probe
    2. Synthesized probes that are 20 nucleotides long or longer can be made in the laboratory if the amino acid sequence (a partial sequence will suffice) is known
    3. Previously cloned genes can be used as probes if they have sufficient sequence homology
  D. Isolating and purifying cloned DNA
    1. An appropriate colony is picked from a master plate and then propagated
    2. The plasmid (or phage) DNA is extracted and purified by gradient centrifugation
    3. The desired fragment is cleaved from the DNA with restriction endonuclease and then separated by electrophoresis

V. Cloning Vectors—small, well-characterized DNA molecules that contain at least one replication origin and that can be replicated within the appropriate host
  A. Plasmids
    1. Easy to isolate and purify
    2. Can be introduced into bacteria by transformation
    3. Often bear antibiotic resistance genes that can be used to select recombinants
  B. Phage vectors are more conveniently stored for long periods, and they contain insertion sites that do not interfere with replication when foreign DNA is inserted
  C. Recombinant phage DNA (rather than complete phage particles) can also be used directly, a process known as transfection; however, this approach is less efficient

133

D. Cosmids are plasmids with lambda phage cos sites and can be packaged into lambda capsids; they can be transmitted similarly to phages but can exist in the cell like a plasmid; these can be used for larger pieces of DNA

E. Shuttle vectors are used to transfer genes between very different organisms; they usually contain one replication origin for each host

F. Artificial chromosomes (yeast or bacterial) have all of the elements necessary to propagate as a chromosome that can be used to clone DNA fragments from 100kb to 2000kb in length

VI. Inserting Genes into Eucaryotic Cells

A. Genes of interest can be inserted directly into animal cells by microinjection; if the genes are stably incorporated into fertilized eggs, the resulting organism is called a transgenic animal

B. Electroporation is a procedure in which target cells are mixed with DNA and are then exposed briefly to high voltage; this works with mammalian cells and plant cell protoplasts

C. The gene gun is used to shoot DNA-coated microprojectiles into plant and animal cells

VII. Expression of Foreign Genes in Bacteria

A. A promoter is necessary

B. An appropriate leader is necessary; leaders are different in procaryotes and eucaryotes

C. Introns in eucaryotic genes must be removed because bacteria cannot remove them

D. Regulatory sequences are often used to control expression

E. Other modifications have been used to increase the efficiency of expression, and to facilitate recovery of the protein product

F. Expression vectors are cloning vectors that have been modified for easy insertion and expression of foreign DNA

VIII. Applications of Genetic Engineering

A. Medical applications

1. Somatostatin, a growth regulatory hormone

2. Human growth hormone for treating pituitary dwarfism

3. Human insulin, for people allergic to bovine or porcine insulin

4. Interferon, for treatment of viral infections and possibly cancer

5. Other important polypeptides, including interleukin-2 and blood-clotting factor VIII

6. Transgenic corn and soybeans can be used to produce monoclonal antibodies

7. Genetically engineered mice can now produce fully human monoclonal antibodies

8. Synthetic vaccines (investigational)

9. Diagnostic probes for certain genetic disorders (investigational)

10. Somatic cell gene therapy for certain genetic disorders (investigational, but has been recently attempted in a patient with an immune deficiency disease with the use of a modified retrovirus)

11. The use of transgenic livestock to produce large amounts of human gene products (investigational)

12. Fusion toxins have been produced that target certain cells to deliver the toxin

B. Industrial applications

1. Improvement of bacterial, fungal, and mammalian cell strains used in industrial bioprocesses

2. Development of new strains for additional bioprocesses

3. Bacterial metabolism of petroleum products (to clean up oil spills)

4. Bacterial metabolism of other toxic materials

C. Agricultural applications

1. Introduction of new desirable traits (e.g., increased growth rate) into farm animals

2. Transfer of nitrogen fixation capabilities to nonlegume crop plants

3. Rendering plants resistant to environmental stresses

4. Rendering fruits and vegetables less susceptible to rotting so that they can be left to ripen longer; the first such food, the Flavr Savr tomato, has been approved by the FDA for sale

5. Protecting crops against frost damage

6. Making plants poisonous to insect pests

       7.   Increasing milk production in dairy cattle

IX.   Social Impact of Recombinant DNA Technology

    A.   Benefits are inherent in applications

    B.   It is possible that environmental release might trigger widespread infections

    C.   The transfer of genes from a weakened strain to a hardy one (with subsequent spread of undesirable genes) is a potential problem

    D.   Unethical use of information obtained about an individual; this is a concern within the human genome project which will improve the process of genetic screening for disorders

    E.   The unscrupulous use of technology to create biological warfare agents must be considered

    F.   Eugenic application in germ line cells could cause problems

    G.   Ecosystem disruption could result from environmental release

    H.   Guidelines for safe use have been developed and are overseen by various government agencies

    I.   Thus far, no obvious negative ecological effects have been observed, and currently, safety guidelines are being somewhat relaxed

TERMS AND DEFINITIONS

Place the letter of each term in the space next to the definition or description that best matches it.

_____ 1. Enzymes that recognize and cleave DNA at specific base pair sequences

_____ 2. A DNA copy of an mRNA that is produced by reverse transcriptase in order to be inserted into a cloning vector

_____ 3. A carrier of foreign DNA into the cloning host

_____ 4. A piece of detectably labeled nucleic acid that hybridizes with complementary DNA fragments and is used to locate them

_____ 5. The phenomenon of movement of charged molecules in an electrical field which is used to separate nucleic acid fragments (and/or proteins)

_____ 6. A set of cloned fragments that represents the entire genome of an organism

_____ 7. A plasmid that can transfer genetic material to several different organisms because it has at least one origin of replication that will function in each organism

_____ 8. A plasmid that has sequences necessary for packaging into bacteriophage lambda capsids

_____ 9. A plasmid that has all of the necessary transcription and translation start and stop signals, and that has nearby useful restriction endonuclease sites to enable the insertion of foreign DNA fragments in proper orientation

_____ 10. The process in which a high voltage electric current induces target cells to take up DNA

_____ 11. The process in which DNA is directly inserted into the target cell

_____ 12. The process in which recombinant phage DNA is taken up directly by the target cell without using complete phage particles

_____ 13. A DNA fragment that is made in the laboratory by sequential addition of one nucleotide at a time to a growing chain

_____ 14. A piece of DNA with all of the features necessary for chromosomal replication which can carry large (100kb to 2000 kb) pieces of foreign DNA into a host organism

a. artificial chromosomes
b. complementary DNA (cDNA)
c. cosmid
d. electrophoresis
e. electroporation
f. expression vector
g. genomic library
h. microinjection
i. probe
j. restriction endonuclease
k. shuttle vector
l. synthetic DNA
m. transfection
n. vector

FILL IN THE BLANK

1. Many enzymes called _____ cleave DNA at specific sequences. Often these cleavages are staggered on the two strands of the DNA, and result in _____ ends that can be used to insert foreign DNA if both vector and host DNA are cut with _____ enzyme.

2. Successful isolation of recombinant clones is dependent on the availability of suitable _____, which can be obtained in a variety of ways. Frequently _____ clones are used; these are produced by the action of the enzyme _____ on isolated mRNA molecules to make a DNA copy.

3. Some vectors have been specially designed for a specific function. For example, a vector used to transfer genes between two very different organisms is called a _____ vector. It usually contains one _____ for each host so that it can be propagated in both hosts.

4. One of the most widely used plasmid vectors is pBR322, which has resistance genes for both _____ and _____. In addition it has several restriction sites that occur only once and are useful for inserting foreign DNA. These sites may be located in a(n) _____ _____ gene, and resistance may be lost after _____ of a foreign gene. This loss can then be used to identify a _____ plasmid.

5. The problem of recombinant gene expression in host cells is overcome with the help of special cloning vectors, called _____ vectors, that contain the necessary _____ and _____ start signals. Some even contain portions of the _____, which enables the vector to have a _____ function over gene expression.

6. Genetic engineering has contributed and will continue to contribute to the fields of _____, _____, and _____, in addition to its contributions to basic research.

7. An animal that comes from a fertilized egg into which a foreign gene has been stably incorporated is called a(n) _____ animal.

8. A procedure in which target cells are mixed with DNA and briefly exposed to high _____ in order to introduce the DNA into the target cell is called _____. This has been shown to work with _____ cells and plant cell _____.

MULTIPLE CHOICE

For each of the questions below select the *one best* answer.

1. Complementary DNA (cDNA) is produced by which of the following enzymes?
   a. restriction endonucleases
   b. DNA polymerase
   c. reverse transcriptase
   d. DNA ligase
2. Any DNA molecule that is used to carry a foreign DNA into a host organism is called a
   a. plasmid.
   b. vector.
   c. probe.
   d. blot.
3. Which of the following is not effective for inserting foreign DNA into a plasmid?
   a. Cut plasmid and foreign DNA with the same enzyme to produce the same sticky ends, which can then be used to hold the fragment of foreign DNA in place for insertion.
   b. Cut plasmid and foreign DNA with an enzyme to produce blunt ends, and then add complementary tails by using terminal transferase to produce sticky ends.
   c. Cut plasmid and foreign DNA with an enzyme that produces blunt ends, and then use T4 DNA ligase to do a blunt-end ligation.
   d. All of the above are effective for inserting foreign DNA into a plasmid.
4. DNA probes can be constructed if the amino acid sequence of the protein is known. The protein amino acid sequence allows the determination of a DNA base sequence within a portion of the molecule that is
   a. exactly the same as the naturally occurring DNA sequence for that portion of the molecule.
   b. a close enough match to the naturally occurring sequence to allow hybridization of the probe to the proper fragment.
   c. usually a close match, and sometimes identical to the naturally occurring sequence.
   d. None of the above are correct.

5. Which of the following is not normally used as a cloning vector?
   a. transposons
   b. plasmids
   c. cosmids
   d. bacteriophages
6. Which of the following is(are) currently produced by recombinant DNA technology for use in humans?
   a. human growth hormone
   b. human interferon
   c. human insulin
   d. All of the above are being produced by recombinant DNA technology for use in humans.
7. Which of the following is not true about the use of nonradioactively labeled hybridization probes when compared with radioactively labeled probes?
   a. Nonradioactively labeled probes are faster to detect.
   b. Nonradioactively labeled probes are safer to use.
   c. Nonradioactively labeled probes are more sensitive than radioactively labeled probes.
   d. All of the above are true.
8. Which of the following sequences of steps cannot be used to clone a desired DNA fragment?
   a. Cleave the DNA; isolate the fragment; clone the isolated fragment.
   b. Cleave the DNA; clone all the resulting fragments; isolate a clone containing the desired fragment.
   c. Synthesize the desired fragment; clone the synthesized fragment.
   d. All of the above can be used to clone a desired DNA fragment.
9. Gene copies with known sequence alterations are normally produced by a process called
   a. polymerase chain reaction.
   b. side-directed mutagenesis.
   c. genomic library mutagenesis.
   d. None of the above are correct.

10. PCR can be used to amplify sequences found in
    a. DNA.
    b. RNA.
    c. Both (a) and (b) are correct.
    d. Neither (a) nor (b) is correct.
11. Which of the following cloning vectors can carry the least amount of foreign DNA?
    a. artificial chromosome
    b. bacteriophage
    c. cosmid
    d. plasmid
12. Which of the following cloning vectors can carry the most foreign DNA?
    a. artificial chromosome
    b. bacteriophage
    c. cosmid
    d. plasmid

TRUE/FALSE

____ 1. The Southern blotting technique was named after the person who developed the procedure, E. M. Southern.

____ 2. In electrophoresis, DNA fragments separate according to size, with the largest fragments migrating the farthest.

____ 3. Regardless of the specific technique used in recombinant DNA technology, one of the keys to successful cloning is choosing the right vector.

____ 4. Cosmids are so named because they can be used to express foreign genes in a variety of different cloning hosts.

____ 5. It is not necessary to remove introns from eucaryotic genes before cloning them in a procaryotic organism, because when the eucaryotic RNA transcript is produced in the procaryote, the intron is removed by the same posttranscriptional processing inherent in the eucaryotic cell of origin.

____ 6. The gene gun is so named because it shoots DNA-coated microprojectiles directly into the target cells.

____ 7. The FDA has approved for sale the Flavr Savr tomato which has been genetically engineered to allow it to vine-ripen longer without rotting.

CRITICAL THINKING

1. Expression vectors have been extensively modified to allow for efficient expression of recombinant genes. List as many of these modifications as possible and explain the advantage(s) that each provides for gene expression.

2. Bacterial strains that metabolize petroleum substances have been produced. These can be used to clean up oil spills. However, there has been much resistance to the use of these organisms in actual clean-up operations. What are the major objections? How would you argue against these objections? If you agree with the objections, propose modifications that could be made to overcome them.

ANSWER KEY

Terms and Definitions

1. j, 2. b, 3. n, 4. i, 5. d, 6. g, 7. k, 8. c, 9. f, 10. e, 11. h, 12. m, 13. l, 14. a

Fill in the Blank

1. restriction endonucleases; sticky (cohesive); the same 2. probes; cDNA; reverse transcriptase 3. shuttle; replication origin 4. ampicillin; tetracycline; antibiotic resistance; insertion; chimeric 5. expression; transcription; translation; lactose operon; regulatory 6. medicine; industry; agriculture 7. transgenic 8. voltage; electroporation; mammalian; protoplasts

Multiple Choice

1. c, 2. b, 3. d, 4. c, 5. a, 6. d, 7. c, 8. d, 9. b, 10. c, 11. d, 12. a

True/False

1. T, 2. F, 3. T, 4. F, 5. F, 6. T, 7. T

V   THE VIRUSES

16   The Viruses: Introduction and General Characteristics

CHAPTER OVERVIEW

Viruses are generally small, acellular entities that possess only a single type of nucleic acid and that must use the metabolic machinery of a living host in order to reproduce. Viruses have been and continue to be of tremendous importance for a variety of reasons: many human diseases have a viral etiology; the study of viruses has contributed greatly to our knowledge of molecular biology; and the blossoming field of genetic engineering is largely based on discoveries in the field of virology. This chapter focuses on the general properties of viruses, the development of the science of virology, and the methodology used in the study of virology.

CHAPTER OBJECTIVES

After reading this chapter you should be able to:

- define viruses and discuss the implications of the concepts embodied in the definition
- discuss the various requirements for culturing viruses
- discuss the methodology employed for virus purification and enumeration
- discuss the composition and arrangement(s) of viral capsids
- discuss the variety found in viral genomes (DNA or RNA, single or double stranded, linear or circular, etc.)
- describe the way in which viruses are classified

CHAPTER OUTLINE

   I.   Early Development of Virology
      A.   Many epidemics of viral diseases occurred before anyone understood the nature of the causative agents of those diseases
      B.   Edward Jenner (1798) published case reports of successful attempts to prevent disease (smallpox) by vaccination; these attempts were made, even though Jenner did not know that the etiological agent of the disease was a virus
      C.   The word *virus,* which is Latin for poison, was used to describe diseases of unknown origin, and only later came to be used to describe a particular type of disease-causing entity
      D.   Dimitri Ivanowski (1892) demonstrated that the causative agent of tobacco mosaic disease would pass through filters designed to remove bacteria; however, he thought the agent was a nonreproducing toxin

E.   Martinus Beijerinck (1898–1900), working independently of Ivanowski, showed that the causative agent of tobacco mosaic disease was still infectious after filtration (i.e., capable of reproduction); he referred to it as a filterable virus

F.   Loeffler and Frosch (1898–1900) showed that hoof-and-mouth disease in cattle was also caused by a filterable virus

G.   Walter Reed (1900) showed that yellow fever in humans was caused by a filterable virus and could be transmitted by a mosquito

H.   Ellerman and Bang (1908) showed that leukemia in chickens was caused by a filterable virus

I.   Peyton Rous (1911) showed that muscle tumors in chickens were caused by a filterable virus

J.   Frederick Twort (1915) first isolated viruses that would infect bacteria, but did not follow up on these observations

K.   Felix d'Herelle (1917) firmly established the existence of viruses that infect bacteria, and devised a method for enumerating them; he also demonstrated that these viruses could reproduce only in live bacteria

L.   W. M. Stanley (1935) crystallized the tobacco mosaic virus and showed that it was mostly (or completely) composed of protein

M.   F. C. Bawden and N. W. Pirie (1935) separated the tobacco mosaic virus particles into protein and nucleic acid components

II.   General Properties of Viruses

A.   They have a simple, acellular organization, consisting of one or more molecules of DNA or RNA enclosed in a coat of protein, and sometimes in more complex layers

B.   Both DNA and RNA do not exist together in the same virion

C.   They are obligate intracellular parasites

III.   The Cultivation of Viruses—requires inoculation of a living host

A.   Animal viruses

1.   Suitable host animals

2.   Embryonated eggs

3.   Tissue (cell) cultures—monolayers of animal cells

a.   Cell destruction can be localized if infected cells are covered with a layer of agar; the areas of localized cell destruction are called plaques

b.   Viral growth does not always result in cell lysis to form a plaque; microscopic (or macroscopic) degenerative effects can sometimes be seen; these are referred to as cytopathic effects

B.   Bacteriophages (viruses that infect bacteria) are usually cultivated in broth or agar cultures of suitable, young, actively growing host cells; broth cultures usually clear, while plaques form in agar cultures

C.   Plant viruses can be cultivated in

1.   Plant tissue cultures

2.   Cultures of separated plant cells

3.   Whole plants—may cause localized necrotic lesions or generalized symptoms of infection

4.   Plant protoplast cultures

IV.   Virus Purification and Assays

A.   Virus purification

1.   Differential centrifugation separates according to size

2.   Gradient centrifugation separates according to density or to sedimentation rate (size and density), and is more sensitive to small differences between various viruses

3.   Differential precipitation with ammonium sulfate or polyethylene glycol separates viruses from other components of the mixture

4.   Denaturation and precipitation of contaminants with heat, pH, or even organic solvents can sometimes be used

5.   Enzymatic degradation of cellular proteins and/or nucleic acids can sometimes be used because viruses tend to be more resistant to these types of treatment

B. Virus assays
    1. Particle count
        a. Direct counts can be made with an electron microscope
        b. Indirect counts can be made using methods such as hemagglutination (virus particles can cause red blood cells to clump together or agglutinate)
    2. Infectious unit counts are based on the observation that many virion particles may not be infectious
        a. Plaque assays involve plating dilutions of virus particles on a lawn of host cells; clear zones result from viral damage to the cells; results are expressed as plaque-forming units (PFU)
        b. Infectious dose assays are an end point method for determining the smallest amount of virus needed to cause a measurable effect, usually on 50% of the exposed target units; results are expressed as infectious dose (ID_{50}) or lethal dose (LD_{50})

V. The Structure of Viruses
  A. Virion size ranges from 10 nm to 400 nm
  B. Nucleocapsid—the nucleic acid plus the surrounding capsid; for some viruses this may be the whole virion; other viruses may possess additional structures as well; viral nucleocapsids are usually constructed without outside aid in a process called self-assembly
  C. Capsid—protein coat that surrounds the genome, protects the viral genetic material, and aids in transfer between host cells
    1. Helical—hollow tube with a protein wall shaped as a helix or spiral; may be either rigid or flexible
    2. Icosahedral—regular polyhedron with 20 equilateral triangular faces and 12 vertices; appears spherical
  D. Nucleic acids—genome
    1. May be either RNA or DNA, single- or double-stranded, linear or circular
    2. May have the common bases that occur in RNA or DNA, or genome may have one or more unusual bases (e.g., hydroxymethylcytosine instead of cytosine)
    3. Viruses with single-stranded RNA (ssRNA) come in several arrangements:
        a. Plus strand viruses have a genomic RNA with the same sequence as the viral mRNA; the genomic RNAs may have other features (5′ cap, poly-A tail, etc.) common to mRNA, and may direct the synthesis of proteins immediately after entering the cell
        b. Negative strand viruses have a genomic RNA complementary to the viral mRNA
        c. Segmented genomes are those in which the virion contains more than one RNA molecule; each segment is unique and frequently encodes a single protein; in some viruses segments may be packaged into more than one virion structure
  E. Viral envelopes and enzymes
    1. Envelopes are membrane structures surrounding some (but not all) viruses
        a. Lipids and carbohydrates are usually derived from the host membranes
        b. Proteins are virus specific
        c. Many have protruding glycoprotein spikes (peplomeres)
    2. Enzymes—some viruses have capsid-specific enzymes; these may be required for virus attachment or entry into the host cell; many, however, are involved in viral nucleic acid replication
  F. Viruses with capsids of complex symmetry
    1. Poxviruses are large (200 to 400 nm) with an ovoid exterior shape
    2. Some bacteriophages have complex, elaborate shapes composed of heads (icosahedral symmetry) coupled to tails (helical symmetry); the structure of the tail regions are particularly variable; such viruses are said to have binal symmetry

VI. Principles of Virus Taxonomy—Grouped according to
    A. Nature of the host—animal, plant, bacterial, insect, fungal
    B. Nucleic acid type
        1. DNA or RNA
        2. Single or double stranded
        3. Molecular weight
        4. Segmentation and the number of pieces of RNA
    C. Capsid symmetry
    D. Presence or absence of an envelope and ether sensitivity
    E. Diameter of capsid (or nucleocapsid)
    F. Number of capsomeres in icosahedral viruses
    G. Immunological properties
    H. Gene number and genomic map
    I. Intracellular location of virus replication
    J. Presence or absence of a DNA intermediate (ssRNA viruses)
    K. Type of virus release
    L. Disease caused by the virus, its special clinical features, or its mode of transmission

TERMS AND DEFINITIONS

Place the letter of each term in the space next to the definition or description that best matches it.

____ 1. The complete virus particle as it exists extracellularly	a.	capsid
____ 2. The localized areas of cellular destruction and lysis that appear as clear zones on confluent lawns of host cell growth	b.	cytopathic effects
	c.	differential centrifugation
____ 3. Microscopic or macroscopic degenerative changes or abnormalities in infected host cells or tissues	d.	envelope
	e.	gradient centrifugation
____ 4. Centrifugation of a suspension at various speeds in order to separate particles of different sizes	f.	hemagglutination
	g.	necrotic lesions
____ 5. Centrifugation of a suspension through a medium whose density varies from top to bottom; particles will separate on the basis of size and/or density	h.	nucleocapsid
	i.	peplomers
	j.	plaques
____ 6. The clumping of red blood cells in the presence of a virus suspension	k.	segmented genome
	l.	virion
____ 7. The viral nucleic acid plus the surrounding protein coat		
____ 8. The protein coat surrounding the viral nucleic acid		
____ 9. Viral genetic material that is divided into separate parts, which may be packaged together or separately; all parts are needed to establish a productive infection		
____ 10. Glycoproteins on the envelope of a virus that project outward from the surface of the virion		
____ 11. A membranous structure that surrounds the nucleocapsid of some virions		
____ 12. The localized areas of destruction occurring on plants infected by a virus		

KEY SCIENTISTS AND THE HISTORY OF VIROLOGY

Match the following scientists with their discoveries.

_____ 1. Began vaccinating humans to prevent smallpox
_____ 2. Demonstrated that the causative agent of tobacco mosaic disease would pass through a filter designed to retain bacteria
_____ 3. Demonstrated that the filterable causative agent of tobacco mosaic disease could reproduce in a susceptible host
_____ 4. Demonstrated that the causative agent of hoof-and-mouth disease in cattle was filterable and reproductively active
_____ 5. Demonstrated that the causative agent of yellow fever in humans was filterable and could be transmitted by mosquitoes
_____ 6. Demonstrated that leukemia in chickens was caused by a filterable agent
_____ 7. Demonstrated that muscle tumors in chickens were caused by a filterable agent
_____ 8. Demonstrated that bacteria could be infected by viruses
_____ 9. Firmly established the existence of bacterial viruses and developed a plaque assay for enumerating them
_____ 10. Crystallized tobacco mosaic virus and found that it was largely or completely composed of protein
_____ 11. Separated tobacco mosaic virus particles into protein and nucleic acid

a. Bawden and Pirie
b. Beijerinck
c. d'Herelle
d. Ellerman and Bang
e. Ivanowski
f. Jenner
g. Loeffler and Frosch
h. Reed
i. Rous
j. Stanley
k. Twort

FILL IN THE BLANK

1. A complete virus particle, or _____, consists of one or more molecules of _____ or _____ enclosed in a coat of _____, and sometimes in other more complex layers.
2. Viruses can be enumerated by inoculating a layer of cells and then covering the cells with a thin layer of _____ to limit _____ of the virus, so that only adjacent cells are infected by newly produced virions. As a result, localized areas of cellular destruction and lysis called _____ are formed.
3. In a plaque assay, each plaque is assumed to have arisen from the reproduction of _____ virion. Therefore, the number of plaques at a given dilution will give the number of infectious virions, or _____, which is usually _____ than the number of virus particles present.
4. Virions range in size from about 10 nm to 300-400 nm. The smallest viruses are little larger than _____, while the largest viruses are about the same size as small _____ and can be seen in the light microscope.
5. Viruses that are regular polyhedrons with 20 faces are said to have _____ symmetry, while those shaped like hollow cylinders have _____ symmetry. Some complex viruses exhibit the former type of symmetry in the head region and the latter type in the tail. These viruses are said to have _____ symmetry.
6. Many viruses have an outer membrane layer called a(n) _____ surrounding the nucleocapsid. These viruses often have glycoprotein spikes or _____ protruding from their outer surface, which may be involved in virus _____ to their host cells.

7. Viral capsids are constructed from many copies of one or a few types of protein subunits, or _____. These interact specifically with each other and associate spontaneously to form the capsid. Since the capsid is constructed without any outside aid, the process is called _____.

8. Icosahedral viruses are constructed from ring- or knob-shaped units called _____ that can either be _____, which are located at the vertices, or _____, which are located at the edges or triangular faces.

9. When biological effects are not readily quantified, an end point dilution is often useful. The normal end point used is the dose necessary to affect _____% of the cultures. If the end point measured is host damage, that dilution is called the _____ _____; if the end point is the death of the culture or organism, it is called the _____ _____.

MULTIPLE CHOICE

For each of the questions below select the *one best* answer.

1. Which of the following is not true about the word *virus?*
   a. It is of Latin origin and means *poison.*
   b. It is currently used to describe small, acellular entities that cause disease.
   c. It was used to describe disease-causing agents long before those agents were characterized.
   d. All of the above are true about the word *virus.*

2. In which of the following ways do viruses differ from living microbial cells?
   a. Viruses have acellular structure or organization.
   b. Viruses have RNA or DNA, but not both, within the same virion.
   c. Viruses are obligate intracellular parasites that cannot reproduce independently of host cells.
   d. All of the above are ways in which viruses differ from living microbial cells.

3. Centrifugation of a suspension at various speeds in order to separate particles of various sizes is called
   a. isopycnic centrifugation.
   b. density centrifugation.
   c. differential centrifugation.
   d. variable centrifugation.

4. In gradient density centrifugation, which of the following is true?
   a. Particles will continue to settle towards the bottom of the tube if the centrifugation is continued.
   b. Particles will come to rest when the density of the surrounding medium is equal to the density of the particle, even if the centrifugation is continued longer.
   c. The smallest particle will sediment the fastest.
   d. Large particles will sediment farther than small particles.

5. For viruses that have as the genome a single-stranded RNA in which the base sequence is the same as the viral mRNA, the genome is said to be
   a. plus strand.
   b. minus strand.
   c. mRNA-like.
   d. None of the above are correct.

6. An RNA genome that exists as several separate, nonidentical molecules packaged together or separately is called
   a. a diploid genome.
   b. a segmented genome.
   c. a polyploid genome.
   d. a fractionated genome.

TRUE/FALSE

_____ 1. In the extracellular phase, viruses possess few if any enzymes, and cannot reproduce independently of living cells.

_____ 2. Like bacteria and eucaryotic microorganisms, viruses can be cultured on artificial media.

_____ 3. Bacteriophages are viruses that have the typical procaryotic appearance of bacteria.

_____ 4. In viruses with helical symmetry, the RNA is wound in a spiral and lies in an interior groove formed by the protein subunits.

_____ 5. All capsids with helical symmetry tend to be rigid.

_____ 6. In viral envelopes, the lipids and carbohydrates are normal host cell components, while the proteins are coded for by viral genes.

_____ 7. The presence or absence of an envelope is not a useful characteristic in classifying viruses because any given virus may at one time have an envelope and at another time may not have an envelope.

_____ 8. In viral classification the greatest weight is given to the intracellular location of viral replication.

_____ 9. For some viruses with segmented RNA genomes, the segments may be packaged into separate virion structures.

CRITICAL THINKING

1. Viruses are thought by some scientists to be among the most primitive of living organisms. Others suggest that they should not be considered as living organisms. Take a stand on this question and defend your position. What would be the most likely arguments from those taking the opposing viewpoint? How would you counter these arguments?

2. Some viruses use as their genetic material single-stranded RNA that has the same sequence as the mRNA (plus strand viruses), while other viruses use RNA that has a sequence complementary to the mRNA (minus strand viruses). Compare these two types of viruses with respect to occurrences during the early stages of infection. What enzyme(s) would a minus strand virus need in the cell that a plus strand virus would not need? Why?

ANSWER KEY

Terms and Definitions

1. l, 2. j, 3. b, 4. c, 5. e, 6. f, 7. h, 8. a, 9. k, 10. i, 11. d, 12. d

Key Scientists and the History of Virology

1. f, 2. e, 3. b, 4. g, 5. h, 6. d, 7. i, 8. k, 9. c, 10. j, 11. a

Fill in the Blank

1. virion; DNA; RNA; protein 2. agar; spread; plaques 3. a single (one); plaque-forming units (PFU); less 4. ribosomes; bacteria 5. icosahedral; helical; binal 6. envelope; peplomers; attachment 7. protomers; self-assembly 8. capsomers; pentamers; hexamers 9. 50; infectious dose (ID_{50}); lethal dose (LD_{50})

Multiple Choice

1. d, 2. d, 3. c, 4. b, 5. a, 6. b

True/False

1. T, 2. F, 3. F, 4. T, 5. F, 6. T, 7. F, 8. F, 9. T

17 The Viruses: Bacteriophages

CHAPTER OVERVIEW

This chapter focuses on the characteristics of the bacterial viruses, or bacteriophages. It begins with their classification and then details the infectious cycle of those DNA viruses that cause destruction (lysis) of host cells. RNA phages are discussed briefly, and the chapter concludes with information about phages that can integrate their DNA into the host chromosome and thereby set up a stable residence within the host cell. These phages are called temperate phages, and the process is referred to as lysogeny.

CHAPTER OBJECTIVES

After reading this chapter you should be able to:

- describe the four phases of the viral life cycle
- discuss the differences between DNA phages and RNA phages in terms of their life cycles and their interactions with their hosts
- discuss the establishment and maintenance of lysogeny by temperate phages

CHAPTER OUTLINE

   I.   Classification of Bacteriophages—the most important criteria are phage morphology and nucleic acid properties
       A.   Morphology
           1.   Tailless icosahedral
           2.   Viruses with contractile tails
           3.   Viruses with noncontractile tails
           4.   Filamentous viruses
       B.   Nucleic acid properties
           1.   DNA or RNA
           2.   Single stranded (ss) or double stranded (ds)
   II.   Reproduction of DNA Phages: The Lytic Cycle—culminates with the host cell bursting and releasing virions
       A.   The one-step growth experiment
           1.   Reproduction is synchronized
           2.   Bacteria are infected and then diluted so that the released phages will not immediately find new cells to infect
           3.   The released phages are then enumerated
           4.   Several distinct phases are observed
               a.   Latent period—no release of virions detected; represents the shortest time required for virus reproduction and release; the early part of this period is called the eclipse period, and during this period no infective virions can be found even inside infected cells
               b.   Rise period (burst)—rapid lysis of host cells and release of infective phages

   c. Plateau period—no further release of infective virions
  5. Burst size—the number of infective virions released per infected cell
 B. Adsorption to the host cell and penetration
  1. Viruses attach to specific receptor sites (proteins, lipopolysaccharides, teichoic acids, etc.)
  2. Receptor variation is at least partly responsible for host range specificity
  3. Binding is probably due to electrostatic interactions and is influenced by pH and the presence of ions such as Mg^{2+} and Ca^{2+}
  4. Many viruses inject DNA into the host cell, leaving an empty capsid outside
  5. The tail tube of T-even bacteriophages may interact with the plasma membrane to form a pore through which the DNA passes
 C. Synthesis of phage nucleic acids and proteins
  1. Host synthesis of DNA, RNA, and protein is halted
  2. Some virus-specific mRNAs (early mRNA) are transcribed using host RNA polymerase
  3. Early proteins, made at the direction of these mRNAs, may:
   a. Take over host cells
   b. Degrade host DNA
   c. Replicate the viral nucleic acid
   d. Alter host RNA polymerase to preferentially transcribe viral genes by altering promoter recognition
  4. Viral DNA is replicated
   a. Alternate bases (if needed) are synthesized; these are sometimes used to protect the phage DNA from host enzymes (restriction endonucleases) that would otherwise degrade the viral DNA and thereby protect the host
   b. Replication of viral DNA—the mechanism of single-stranded DNA phage replication is different from that of double-stranded DNA phage replication; single-stranded DNA phages usually require the formation of a double-stranded replicative form (RF), which in turn directs the synthesis of mRNA and the new genome
 D. The Assembly of phage particles
  1. Capsid proteins are synthesized at the direction of late mRNAs (made after viral nucleic acid replication)
  2. Noncapsid (scaffolding) proteins needed for assembly and/or lysis are also made at the direction of late mRNAs
  3. Assembly proceeds either sequentially or by subassemblies, which are then put together
  4. Different phages package viral DNA differently
   a. Some phages build capsid around the DNA
   b. Some phages insert DNA into preformed (but as yet incomplete) capsid structures
 E. Release of phage particles
  1. Enzymes damage the cytoplasmic membrane
  2. Other enzymes damage the cell wall
  3. A few phages (e.g., filamentous fd phages) are released without lysing the host cell, which secretes the phage through the plasma membrane (into which phage coat proteins have been inserted)
III. Reproduction of RNA Phages
 A. RNA replicase—the virus must provide an enzyme for replicating the RNA genome because the host does not produce an enzyme with this capability
 B. RNA is usually plus stranded (+) and can act similarly to mRNA in directing the synthesis of the replicase during an initial step after penetration
 C. +ssRNA is then converted to ±dsRNA, the replicative form
 D. Replicative form is then used as a template for production of multiple copies of the genomic (and messenger) +ssRNA
 E. Capsid proteins are made, and +ssRNA is packaged into new virions
 F. One or more lysis proteins then function to release the phage

G. Only one dsRNA phage has so far been discovered (ϕ6)
   1. It infects *Pseudomonas phaseolicola*
   2. It possesses a membranous envelope

IV. Temperate Bacteriophages and Lysogeny
   A. Temperate phages are capable of lysogeny, a nonlytic relationship with their hosts
   B. In lysogeny, the viral genome (called a prophage) is integrated into the host DNA and is replicated with it; it does not kill (lyse) the host cell; the cells are said to be lysogenic (or are called lysogens)
   C. In the lysogenic state, the host cell cannot be superinfected by a virus of the same type (i.e., it has immunity to superinfection)
   D. It may switch to the lytic cycle at some later time; this process is called induction
   E. Conditions at the time of infection may determine whether the virus will establish a lytic infection or lysogeny
   F. Establishment of lysogeny (bacteriophage lambda)
      1. Two sets of promoters are available to host RNA polymerase
      2. A repressor protein may be made from genes adjacent to one of these promoters
      3. If this repressor binds to its target operator before the other promoter is used, then that promoter is blocked and lysogeny is established
      4. If genes associated with that second promoter are expressed before the repressor can bind to the operator, then the lytic cycle is established
      5. For lambda and most temperate phages, if lysogeny is established, the viral genome integrates into the host chromosome; however, some temperate phages can establish lysogeny without integration
      6. Induction (the termination of lysogeny and entry into the lytic cycle) will occur if the level of the repressor protein decreases; this is usually in response to environmental damage to the host DNA
   G. Lysogenic conversion is a change that is induced in the host phenotype by the presence of a prophage, and that is not directly related to the completion of the viral life cycle; examples include:
      1. Modification of lipopolysaccharide structure in infected *Salmonella*
      2. Production of diphtheria toxin only by lysogenized strains of *Corynebacterium diphtheriae*

TERMS AND DEFINITIONS

Place the letter of each term in the space next to the definition or description that best matches it.

_____ 1. The term that describes the time immediately after infection during which no infective virions are released

_____ 2. The term that describes the time immediately after infection during which no infective viruses can be found even inside the infected cell

_____ 3. The term that describes time during which there is a rapid lysis of host cells and release of infective virions

_____ 4. The number of infective virions produced per infected cell

_____ 5. The place on the surface of a host cell where a phage can attach

_____ 6. The mRNA that is made before viral nucleic acid has been replicated

_____ 7. The mRNA that is made after viral nucleic acid has been replicated

_____ 8. An enzyme that cleaves DNA at specific points, thereby destroying it

_____ 9. A base sequence that is repeated at both ends of a DNA molecule

_____ 10. Viruses that can only establish a lytic infection

_____ 11. An infection that kills the cell and causes it to burst open and release virus particles that have been assembled within

_____ 12. Viruses that can either establish a lytic infection or establish lysogeny

_____ 13. An infection that does not kill the host and that maintains the viral genome in a dormant state; this dormant genome is replicated when the host genome is replicated

_____ 14. A change in the phenotype of a lysogenized cell that is not directly related to the completion of the virus life cycle

_____ 15. A protein encoded within the lambda genome that establishes and maintains the lysogenic state

_____ 16. A dsDNA or dsRNA that is produced by a phage with a single-stranded genome, and that acts as a template for the synthesis of mRNA and genomic nucleic acid

_____ 17. The switching of an infected bacterial lysogen to the active production of viral progeny

_____ 18. An infected bacterium carrying a dormant prophage in the lysogenic state

a. burst size
b. early mRNA
c. eclipse period
d. induction
e. lambda repressor
f. late mRNA
g. latent period
h. lysogen
i. lysogenic conversion
j. lysogeny
k. lytic infection
l. receptor site
m. replicative form
n. restriction endonuclease
o. rise period
p. temperate phages
q. terminal redundancy
r. virulent phages

FILL IN THE BLANK

1. In a one-step growth experiment, a _____ period during which there is no release of virions immediately follows virus addition. Early in this period, no infective virions are found even inside the infected cells; this is called the _____ period. Subsequently, a rapid lysis of host cells and release of infective virions occur, which is called the _____ period. Finally, a plateau is reached and no more virions are released.

2. Virus-specific mRNA that is synthesized before viral nucleic acid is replicated is called _____ mRNA, while that produced after viral replication is called _____ mRNA.

3. The DNA of T-even phages contains _____ instead of cytosine. This helps to protect it from _____ enzymes, which would otherwise degrade the viral DNA and destroy it, thereby preventing infection.

4. Several T4 gene products are needed for cell lysis. One directs the synthesis of _____, which attacks the cell wall peptidoglycan. Another damages the bacterial _____, allowing the first product to reach the peptidoglycan.

5. Phages that are capable only of the lytic infectious cycle are called _____ phages, while those that are capable of both lytic infection and lysogeny, a dormant state, are called _____ phages. The bacteria carrying these dormant phages are called _____.

6. The latent form of the genome that exists when a phage establishes lysogeny is called a _____. This can be _____ into the host chromosome or may exist independently.

7. A temperate phage may cause a change in the _____ of the host that is not directly related to the completion of its life cycle. This process is called _____. For example, only lysogenized bacteria of the species *Corynebacterium diphtheriae* will produce the _____ that causes the disease diphtheria.

8. Under some conditions, *E. Coli* cells infected with lambda phage will change from the lysogenic state and enter the lytic cycle. This is called _____ and is triggered by a drop in _____ levels due to environmental damage to host DNA, which stimulates the production of the recA protein. The recA protein may act as a _____ enzyme, cleaving the _____ and thereby inactivating it; alternatively, recA may bind to the _____ and stimulate it to proteolytically cleave itself.

9. Bacteria that are infected by a temperate phage and have entered into a lysogenic relationship cannot be infected by a virus of the same type; this is referred to as _____ to _____.

10. Binding of a virus to its host cell receptor is probably due to _____ interactions and is influenced by _____ and the presence of ions such _____ and _____.

MULTIPLE CHOICE

For each of the questions below select the *one best* answer.

1. Which of the following is least important in classifying bacteriophages?
   a. phage morphology
   b. host range
   c. type of nucleic acid (DNA or RNA)
   d. strandedness of nucleic acid (single- or double-stranded)

2. Which of the following do not serve as phage receptor sites?
   a. lipopolysaccharides
   b. teichoic acids
   c. proteins
   d. All of the above serve as phage receptor sites.

3. In T-even phages, which of the following makes the initial contact with the appropriate receptor site?
   a. tail fiber
   b. base plate
   c. collar
   d. tail tube

153

4. Virus-specific enzymes produced before viral nucleic acid replication perform which of the following functions?
   a. degrade host DNA
   b. modify host RNA polymerase to recognize viral promoters
   c. produce any unusual bases required by the virus for DNA replication
   d. All of the above are functions performed by virus-specific enzymes.
5. The sequence of genes in each T4 virus within a population is the same but starts with a different gene at the 5′ end. If each of these linear pieces is coiled into a circle, the gene sequences are identical. Therefore, the T4 DNA is said to be
   a. a linear circle.
   b. linearly permuted.
   c. circularly permuted.
   d. linearly circular.
6. Which of the following is not a function of the replicative form (dsDNA) of the ssDNA phage φX174?
   a. synthesis of more RF copies
   b. synthesis of minus strand DNA
   c. synthesis of plus strand DNA
   d. synthesis of mRNA

7. Which of the following is not a translation product of late mRNA?
   a. phage structural proteins
   b. phage assembly proteins that are not incorporated in the capsid
   c. phage proteins needed to replicate the phage nucleic acid
   d. phage release proteins
8. The number of infective virions released from a single infected cell is called the
   a. infective dose.
   b. burst size.
   c. multiplicity of infection.
   d. burst plateau.
9. The molecule responsible for the establishment and maintenance of lysogeny in cells infected with bacteriophage lambda is called the
   a. lactose repressor.
   b. lambda repressor.
   c. lambda lysogeny protein.
   d. lysogeny maintenance protein.
10. Bacteriophage φ6 which infects *Pseudomonas phaseolicola,* is the only known bacteriophage with a
   a. dsRNA genome.
   b. membranous envelope.
   c. Both of the above are correct.
   d. None of the above are correct.

TRUE/FALSE

_____ 1. In the one-step growth experiment, diluting the culture after initial infection prevents the spread of infection to additional cells by virions released from the originally infected cells.

_____ 2. When the single-stranded DNA phage φX174 infects a cell, transcription must take place before replication can occur.

_____ 3. In T4 infection, the first complete infective virions appear about 15 minutes after infection.

_____ 4. The single-stranded genome of RNA phages serves both as a template for its own replication and as mRNA.

_____ 5. The multiplicity of infection (MOI) can affect whether cells will be lysogenized or lytically infected. Generally, a low MOI will favor lysogeny, while a high MOI will favor lytic infection.

_____ 6. Filamentous fd phages are released by secretion through the host plasma membrane, leaving the host relatively undamaged and able to continue to release more phage particles.

_____ 7. The tail tube of a complex bacteriophage may interact with the plasma membrane to form a pore through which the DNA passes.

_____ 8. Noncapsid proteins that aid in the assembly of virion structures are referred to as scaffolding proteins.

CRITICAL THINKING

1. Consider the results of a one-step growth experiment presented in the following graph. Label the brackets on the figure and explain what each represents. Under ideal conditions, how much time is associated with each period, and how many phage particles are released?

(a) _____

(b) _____

(c) _____

(d) _____

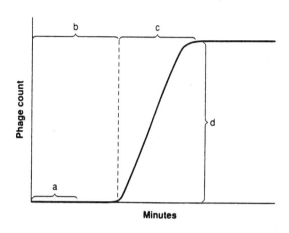

2. When the T4 genome is packaged into its head from long multigenome concatemers, the packaged DNA is about 2% longer than the complete genome. Using diagrams and an arbitrary gene set (A-M), show how the linear T4 DNA is circularly permuted.

ANSWER KEY

Terms and Definitions

1. g, 2. c, 3. o, 4. a, 5. l, 6. b, 7. f, 8. n, 9. q, 10. r, 11. k, 12. p, 13. j, 14. i, 15. e, 16. m, 17. d, 18. h

Fill in the Blank

1. latent; eclipse; rise (burst) 2. early; late 3. hydroxymethylcytosine; restriction 4. lysozyme; plasma membrane 5. virulent; temperate; lysogens 6. prophage; integrated 7. phenotype; lysogenic conversion; toxin 8. induction; lambda repressor; protease repressor; repressor 9. immunity; superinfection 10. electrostatic; pH; Ca^{2+}; Mg^{2+}

Multiple Choice

1. b, 2. d, 3. a, 4. d, 5. c, 6. b, 7. c, 8. b, 9. b, 10. c

True/False

1. T, 2. F, 3. T, 4. T, 5. F, 6. T, 7. T, 8. T

18   The Viruses: Viruses of Eucaryotes

CHAPTER OVERVIEW

This chapter focuses on the characteristics of viruses that infect eucaryotes. Animal (mammalian) viruses are emphasized because they are causative agents of many human diseases. Other viruses, such as plant viruses and insect viruses, are also discussed. The chapter concludes with a discussion of infectious agents that are even simpler than viruses, the viroids and prions.

CHAPTER OBJECTIVES

After reading this chapter you should be able to:

- compare and contrast viruses that infect eucaryotes with those that infect procaryotes
- describe the various ways that viruses of eucaryotes can harm their host organisms
- discuss the establishment of chronic or slow virus infections
- discuss the mechanisms that have been proposed for virus involvement in the establishment of certain cancers
- discuss the importance of plant viruses and the technical difficulties that have hindered rapid progress in this area
- discuss fungal and algal viruses
- discuss the potential use of insect viruses for pest control
- discuss the nature and significance of viroids and prions

CHAPTER OUTLINE

I.   Classification of Animal Viruses—the most important criteria are virus morphology, nucleic acid properties, and genetic relatedness
    A.   Morphology—size, shape, and presence or absence of envelope
    B.   Nucleic acid properties
        1.   DNA or RNA
        2.   Single stranded or double stranded (ss or ds)
        3.   Size and/or segmentation
    C.   Genetic relatedness—nucleic acid hybridization and sequencing
II.   Reproduction of Animal Viruses
    A.   Adsorption of virions
        1.   Attach to specific receptor sites; usually cell surface glycoproteins that are required by the cell for normal cell functioning
        2.   May be species-specific or tissue-specific, or may have a broad multispecies host range
        3.   Many viral receptors are part of the immunoglobulin superfamily
        4.   Viral surface glycoproteins and/or enzymes may mediate virus attachment to the cellular receptor molecules

B. Penetration and uncoating vary with different viruses
1. Injection similar to that of bacteriophages may be used by some viruses
2. Envelope fusion with the cytoplasmic membrane with resultant deposition of the nucleocapsid core within the cell is used by other viruses
3. Engulfment within coated vesicles (endocytosis) may occur with most viruses; lysosomal enzymes and low endosomal pH often trigger the uncoating process
C. Replication and transcription in DNA viruses
1. Host synthesis of DNA, RNA, and protein may be halted, unaffected, or even stimulated
2. Viral DNA replication usually occurs in the nucleus; however, in poxviruses it occurs in the cytoplasm
3. Transcription usually uses host RNA polymerase (poxviruses are an exception)
4. Gene expression may be divided into early and late phases
5. Some viruses use overlapping genes to package more information into a very small genome
D. Replication and transcription in RNA viruses
1. RNA viruses are more diverse in their reproductive strategies than are DNA viruses
2. Plus-strand viruses use genome RNA as mRNA
3. Minus-strand viruses produce replicative form (dsRNA), using a virion-associated transcriptase, and then produce mRNA
4. Double-stranded RNA viruses use a virion-associated transcriptase for replication and mRNA production
5. Retroviruses make a dsDNA copy (called proviral DNA) using the enzyme reverse transcriptase
   a. The proviral DNA is integrated into the host chromosome
   b. The integrated proviral DNA can then direct the synthesis of mRNA
   c. Sometimes these viruses can change the host cells into tumor cells
E. Synthesis and assembly of virus capsids
1. Capsid proteins are synthesized by host cell ribosomes under the direction of the late genes
2. Empty procapsids are produced
3. Nucleic acid is inserted
4. Enveloped virus nucleocapsids are assembled similarly, except for poxvirus nucleocapsids
F. Virion release
1. Naked viruses are released when host cell lyses
2. Enveloped viruses are usually released by the following mechanisms:
   a. Virus-encoded proteins are incorporated into plasma membrane
   b. Nucleocapsid buds outward, forming the envelope during release
3. Herpesvirus envelope formation usually involves the host's nuclear envelope rather than the plasma membrane; other membrane structures can also be used by certain viruses
4. Poxviruses use actin cytoskeleton microfilaments to propel them through the plasma membrane and thereby escape without destroying the host cell
III. Cytocidal Infections and Cell Damage—damage may or may not result in cell death; if death occurs the infection is cytocidal; mechanisms of host cell damage may include:
A. Inhibition of host DNA, RNA, and protein synthesis
B. Lysosome damage, leading to release of hydrolytic enzymes into the cell
C. Plasma membrane alteration leading to host immune system attack on the cell or to cell fusion
D. Toxicity from high protein concentrations
E. Formation of inclusion bodies that may cause direct physical disruption of cell structure
F. Chromosomal disruptions
G. Malignant transformation to a tumor cell

IV. Persistent, Latent, and Slow Virus Infections
   A. In persistent (chronic) infections, the virus reproduces at a slow rate without causing disease symptoms
   B. In latent infections, the virus stops reproducing and remains dormant for a period before becoming active again
   C. Slow virus infections are those that cause progressive, degenerative diseases with symptoms that increase slowly over a period of years
   D. Defective interfering (DI) particles are usually produced by deletion mutations; they cannot reproduce but they slow normal virus reproduction, thereby reducing host damage and establishing a chronic infection
V. Viruses and Cancer
   A. Viruses may cause cancer by a variety of mechanisms
      1. Virus may carry one or more cancer-causing genes (oncogenes)
      2. Viruses may insert a promoter or enhancer next to a cellular oncogene (an unexpressed cellular gene that regulates cell growth and reproduction), causing an abnormal expression of this gene and thereby deregulating cell growth
      3. Viruses may produce a regulatory protein, which in turn activates a cellular oncogene
   B. Viruses and Human Cancers
      1. Epstein-Barr virus (EBV)—a herpesvirus that may cause:
         a. Burkitt's lymphoma; found mostly in central and western Africa
         b. Nasopharyngeal carcinoma; found in southeast Asia
         c. Infectious mononucleosis; found in the rest of the world
         d. Evidence suggests that host infection with malaria is necessary for EBV to cause Burkitt's lymphoma; this is supported by the low incidence of Burkitt's lymphoma in the U.S. where there is almost no malaria
      2. Hepatitis B virus may be associated with one form of liver cancer
      3. Human papillomavirus has been linked to cervical cancer
      4. Human T-cell lymphotropic viruses (the retroviruses HTLV-1 and HTLV-2) are associated with adult T-cell leukemia and hairy-cell leukemia, respectively
   C. Viral etiology of human cancers is difficult to establish because Koch's postulates could only be satisfied by experimenting on humans
VI. Plant Viruses
   A. Virion morphology does not differ significantly from that of animal viruses or bacteriophages
   B. Plant virus taxonomy—almost all are RNA viruses and are classified on the basis of nucleic acid type, strandedness, capsid symmetry, size, and the presence or absence of an envelope
   C. Plant virus reproduction (using tobacco mosaic virus as an example)
      1. The virus uses either a cellular or a virus-specific RNA replicase; the evidence is not clear
      2. The virus produces proteins, which then spontaneously assemble
      3. The virus causes many cytological changes, such as the formation of inclusion bodies and the degeneration of chloroplasts
      4. Viral spread is through the plant vascular system
   D. Transmission of plant viruses is sometimes difficult because of the tough walls that cover plant cells
      1. Some may enter only cells that have been mechanically damaged
      2. Some are transmitted through contaminated seeds, tubers, or pollen
      3. Soil nematodes can transmit viruses while feeding on roots
      4. Some may be transmitted by parasitic fungi
      5. Most important agents of transmission are insects such as aphids or leafhoppers that feed on plants

VII. Viruses of Fungi and Algae
    A. Fungal viruses
        1. Higher fungi are infected by dsRNA viruses that are most often latent rather than cytopathic viruses
        2. Lower fungi are infected by dsRNA or dsDNA viruses that cause lysis of infected cells
    B. Algal viruses have been detected in electron micrographs, but have not been well studied

VIII. Insect Viruses
    A. Members of at least seven virus families are known to infect insects
    B. Infection is often accompanied by formation of granular or polyhedral inclusion bodies
    C. May persist as latent infections
    D. Current interest in most insect viruses focuses on their use as a possible means of biological pest control; they have several advantages over chemical toxins:
        1. They are invertebrate-specific and, therefore, should be safe
        2. They have a long shelf life and high environmental stability
        3. They are well suited for commercial production because they reach high concentrations in infected insects

IX. Viroids and Prions
    A. Viroids
        1. Circular ssRNA molecules
        2. No capsids
        3. Cause diseases in plants
        4. Do not act as mRNAs
        5. Mechanism that produces symptoms of disease is unknown
        6. May give rise to latent infections
    B. Prions
        1. Proteinaceous infectious particles
        2. No apparent nucleic acid, only a protein called PrP has been identified
        3. Genetic activity is unknown, but the existence of a small, untranslated nucleic acid that interacts with the host cell has not been ruled out; such an infectious agent containing this small nucleic acid coated by PrP would be called a virino
        4. Cause progressive, degenerative central nervous system disorders
            a. Scrapie in sheep and goats
            b. Bovine spongiform encephalopathy (mad cow disease)
            c. Kuru (found only in the Fore, an eastern New Guinea tribe that practice ritual cannibalism)
            d. Creutzfeldt-Jakob, fatal familial insomnia and Gerstmann-Strassler-Scheinker Syndrome are all human diseases caused by prions
        5. Mechanism of pathogenesis may involve a conformational change in the PrP to an abnormal form

TERMS AND DEFINITIONS

Place the letter of each term in the space next to the definition or description that best matches it.

_____ 1. The stage in which a virus attaches to the host cell surface
_____ 2. The stage in which all or part of the virion enters the host cell
_____ 3. The stage in which the capsid is removed and the viral genome released
_____ 4. A sequence of bases on a nucleic acid genome that can encode up to three proteins with the same sequence by using different reading frames
_____ 5. An enzyme that makes a DNA copy of an RNA genome
_____ 6. Term for viruses that do not have a membranous envelope surrounding the nucleocapsid
_____ 7. An infection that results in cell death
_____ 8. Clusters of subunits or virions within the host nucleus or cytoplasm
_____ 9. Viruses that cause progressive degenerative diseases in which several years pass before symptoms develop
_____ 10. Abnormal new cell growth and reproduction due to a loss of regulation
_____ 11. Genes whose expression or abnormal expression leads to the development of cancer
_____ 12. A cancer-causing gene that is introduced into a cell as a result of a viral infection
_____ 13. A cancer-causing gene that is a normal cellular gene, but whose abnormal expression may lead to the onset of cancer; this abnormal expression may be the result of infection by a virus
_____ 14. Small, infective, circular RNA molecules that have no protein capsids, and that infect plants
_____ 15. Small, infective, proteinaceous particles that have no apparent nucleic acid genome, and that cause progressive degenerative diseases in man and animals
_____ 16. Hypothetical, small, infectious particles that may contain a small, untranslated nucleic acid coated by PrP (prion protein)
_____ 17. Virus particles that cannot replicate and which slow normal virus reproduction thereby reducing host damage and establishing a chronic infection

a. adsorption
b. cellular oncogene
c. cytocidal infection
d. defective interfering (DI) particles
e. inclusion bodies
f. naked virions
g. neoplasia
h. oncogenes
i. overlapping genes
j. penetration
k. prions
l. reverse transcriptase
m. slow viruses
n. uncoating
o. viral oncogenes
p. virinos
q. viroids

CHARACTERISTICS OF ANIMAL VIRUSES

Complete the table by filling in the missing information for each virus family:

Virus	Envelope (yes or no)	Nucleic Acid Type (DNA or RNA)	Nucleic Acid Strandedness (single or double)
1. Adenovirus			
2. Arenavirus			
3. Baculovirus			
4. Bunyavirus			
5. Coronavirus			
6. Herpesvirus			
7. Iridovirus			
8. Orthomyxovirus			
9. Papovavirus			
10. Paramyxovirus			
11. Parvovirus			
12. Picornavirus			
13. Poxvirus			
14. Reovirus			
15. Retrovirus			
16. Rhabdovirus			
17. Togavirus			

FILL IN THE BLANK

1. The mechanisms of _____ and _____ must vary with the type of virus because viruses differ so much in structure and mode of reproduction. For example, _____ viruses probably enter differently from naked viruses.
2. Some virions attach to coated pits on the host cell membrane that are coated on the _____ side with the protein _____. The pits then pinch off to form coated vesicles filled with virions, and these fuse with _____, which aid in virus uncoating.
3. For animal viruses with DNA as their genetic material, replication of the DNA takes place in the _____, except poxviruses, whose DNA replication occurs in the _____.
4. Picornaviruses use their genome as a giant _____ and synthesize a single _____, which is then cleaved by host and viral _____ to form the proper proteins.
5. Because their genome is _____ to the mRNA, negative strand viruses must employ a virion-associated _____ to synthesize mRNA.
6. Single-stranded RNA viruses, except _____, use a viral _____ to convert the ssRNA to a dsRNA, which is called the _____.
7. Retroviruses use an enzyme called _____ to convert their ssRNA genome to a _____ copy, which is then integrated into the host chromosome. Viral products are only formed _____ this integration has occurred.
8. Infections that have a rapid onset and last for a relatively short time are called _____ infections, while those that develop slowly and last many years with no apparent disease symptoms are called _____ infections. In some infections, the virus quits replicating and remains dormant before becoming active again. These are called _____ infections.
9. Carcinogenesis is a complex, multistep process. Often cancer-causing genes, or _____, are directly involved and may come from the cell itself or be contributed by a virus. Many of these seem to be involved in the regulation of _____ and _____.
10. Small ssRNA molecules that have no capsid and that cause disease in plants are called _____, while small proteinaceous particles that have no apparent nucleic acid and that cause progressive degenerative diseases in man and animals are called _____.

MULTIPLE CHOICE

For each of the questions below select the *one best* answer.

1. The most common type of molecule functioning as an animal virus receptor is a
    a. lipoprotein.
    b. glycoprotein.
    c. phosphoprotein.
    d. teichoic acid.
2. Poliovirus receptors are found
    a. in cells of all tissues.
    b. in spinal cord anterior horn cells only.
    c. in nasopharynx, gut, and spinal cord anterior horn cells.
    d. in gut cells only.

3. The most common type of interaction between a virus and its receptor is
    a. ionic.
    b. hydrophobic.
    c. covalent.
    d. hydrogen bonding.
4. Which of the following is not a way that animal viruses enter host cells?
    a. injection of nucleic acid
    b. fusion of envelope with plasma membrane
    c. endocytosis
    d. All of the above are ways that animal viruses enter host cells.

5. Enveloped viruses, except poxviruses, acquire their envelopes in which of the following ways?
   a. during assembly but before release
   b. during release
   c. after release
   d. during penetration

6. In some viruses, deletion mutations result in the production of altered viruses that cannot reproduce but that can slow normal virion reproduction. These are called
   a. deletion particles.
   b. replication incompetent viruses.
   c. defective interfering particles.
   d. None of the above are correct.

7. Which of the following diseases are thought to be produced by prions rather than viruses?
   a. kuru
   b. scrapie
   c. Creutzfeldt-Jakob disease
   d. All of the above are thought to be produced by prions.

8. Some cancer cells revert to a more primitive or less differentiated state. This is called
   a. anaplasia.
   b. neoplasia.
   c. metastasis.
   d. carcinogenesis.

9. Epstein-Barr virus (EBV) is one of the best-studied human cancer viruses. It causes
   a. Burkitt's lymphoma.
   b. nasopharyngeal carcinoma.
   c. infectious mononucleosis.
   d. All of the above are caused by EBV.

10. Which of the following is not a way in which viruses may cause cancer?
    a. introduction of an oncogene carried by the virus
    b. introduction of a viral promoter or enhancer next to a previously unexpressed gene that controls cellular reproduction and division
    c. production of a regulatory protein that activates cellular genes controlling cellular reproduction and division
    d. All of the above are ways in which viruses may cause cancer.

11. Which of the following is the most important agent of plant virus transmission?
    a. soil nematodes feeding on plant roots
    b. parasitic fungi
    c. pollen
    d. insects that feed on plant leaves

12. Which of the following makes some insect viruses well suited for commercial production as biological pest control agents?
    a. They have a long shelf-life and are environmentally stable.
    b. They are specific for invertebrates.
    c. They reach extremely high concentrations in larval tissue.
    d. All of the above make insect viruses suitable biological pest control agents.

13. For viruses that enter the cell by receptor-mediated endocytosis, uncoating is often triggered by
    a. lysosomal enzymes.
    b. low endosomal pH.
    c. Both (a) and (b) are correct.
    d. Neither (a) nor (b) is correct.

14. Which of the following represent the way(s) in which enveloped viruses acquire their envelopes?
    a. budding through the plasma membrane
    b. budding through internal cellular membranes
    c. For some viruses (a) is correct; for others viruses (b) is correct.
    d. Neither (a) nor (b) is correct.

TRUE/FALSE

_____ 1. The high degree of specificity makes host range the most important criterion for classifying animal viruses.

_____ 2. All viruses, except parvoviruses, use some virus-specific enzymes for replication. Therefore, some viral proteins must enter the cell with the viral nucleic acid.

_____ 3. Unlike bacteriophages, most animal viruses do not cause degradation of host DNA.

_____ 4. All of the DNA viruses use host RNA polymerase to transcribe at least early mRNA.

_____ 5. The parvovirus genome is so small that it must use overlapping genes (i.e., a single sequence of bases read in different reading frames) to encode the three proteins that are its only gene products.

_____ 6. Poxviruses have a complex infectious cycle in which partial uncoating occurs followed by early gene expression. One of these early gene products is an enzyme that completes the uncoating process prior to DNA replication.

_____ 7. RNA viruses are more uniform in their replication strategies than DNA viruses.

_____ 8. Unlike other enveloped viruses that use altered cellular membranes as their envelope source, herpesviruses use the nuclear membrane in envelope formation.

_____ 9. Some plant viruses that require insect vectors for transmission can actually be cultivated in insect cells.

_____ 10. All of the known plant viruses are RNA viruses.

_____ 11. Plant virus mRNA arises by complex processing of genomic RNA, even though the genome is plus stranded and could conceivably function as mRNA directly.

_____ 12. EBV causes fewer cancers in the U.S. compared to Africa; this is because malaria may be a required precondition for EBV-induced cancer, and there is little malaria in the U.S.

_____ 13. Prion mediated pathogenesis may involve conformational change in the prion protein (PrP) to an abnormal form.

_____ 14. Poxviruses use actin microfilaments to escape from host cells without damaging the cell.

CRITICAL THINKING

1. Compare and contrast the replicative and transcriptive strategies used by plus strand and minus strand viruses. Be sure to include retroviruses in your discussion.

2. Discuss the possible ways that prions, which have no apparent genomic nucleic acid, could infect cells and reproduce more prions. Why might this mean that we have to reexamine our concept of genetic inheritance?

ANSWER KEY

Terms and Definitions

1. a, 2. j, 3., n 4. i, 5. l, 6. f, 7. c, 8. e, 9. m, 10. g, 11. h, 12. o, 13. b, 14. q, 15. k, 16. p, 17. z

Characteristics of Animal Viruses

1. no; DNA; ds 2. yes; RNA; -ss 3. no; DNA; ds 4. yes; RNA; -ss 5. yes; RNA; +ss 6. yes; DNA; ds 7. no; DNA; ds 8. yes; RNA; -ss 9. no; DNA; ds 10. yes; RNA; -ss 11. no; DNA; +ss 12. no; RNA; +ss 13. yes; DNA; ds 14. no; RNA; ds 15. yes; RNA; +ss 16. yes; RNA; -ss 17. yes; RNA; +ss

Fill in the Blank

1. penetration; uncoating; enveloped 2. cytoplasmic; clathrin; lysosomes 3. nucleus; cytoplasm 4. mRNA; polypeptide; proteases 5. complementary; transcriptase 6. retroviruses; replicase; replicative form 7. reverse transcriptase; dsDNA; after 8. acute; persistent (chronic); latent 9. oncogenes; cell growth; differentiation 10. viroids; prions

Multiple Choice

1. b, 2. c, 3. a, 4. d, 5. b, 6. c, 7. d, 8. a, 9. d, 10. d, 11. d, 12. d, 13. c, 14. c

True/False

1. F, 2. F, 3. T, 4. F, 5. T, 6. T, 7. F, 8. T, 9. T, 10. F, 11. T, 12. T, 13. T, 14. T

Part VI

THE DIVERSITY OF THE MICROBIAL WORLD

19 Microbial Taxonomy

CHAPTER OVERVIEW

Microorganisms are tremendously diverse in size, shape, physiology and lifestyle. This chapter introduces the general principles of microbial taxonomy and presents an overview of the current classification scheme. Subsequent chapters will examine the various groups of microorganisms in greater detail.

CHAPTER OBJECTIVES

After reading this chapter you should be able to:

- discuss the rationale behind the science of taxonomy
- discuss the three domains of living organisms (*Eubacteria, Archaea,* and *Eucarya*)
- discuss the meaning of the word *species* and the basis for grouping organisms into species
- discuss the two ways (phylogenetic and phenetic) of classifying organisms
- discuss the various characteristics used in taxonomy and explain why nucleic acid sequences are probably the best indicators of microbial phylogeny and relatedness
- discuss the classification scheme(s) used in *Bergey's Manual of Systematic Bacteriology*
- discuss the dynamic nature of bacterial taxonomy and the new types of data that are contributing to the changes being made

CHAPTER OUTLINE

I. General Introduction and Overview
    A. Taxonomy is the science of biological classification
    B. Classification is the arrangement of organisms into groups (taxa)
    C. Nomenclature refers to the assignment of names to taxonomic groups
    D. Identification refers to the determination of the particular taxon to which a particular isolate belongs
    E. Systematics is the scientific study of organisms with the ultimate object of characterizing and arranging them in an orderly manner
    F. New molecular techniques are being used in classifying microorganisms but the traditional approaches still have value

II. Microbial Evolution and Diversity
   A. Fossilized remains of bacterial cells around 3.5 to 3.8 billion years old have been found in stromatolites and sedimentary rocks
   B. Stromatolites are layered or stratified rocks that are formed by incorporation of mineral sediments into microbial mats
   C. The earliest bacteria were probably anaerobic
   D. Aerobic cyanobacteria probably developed 2.5 to 3.0 billion years ago
   E. The work of Carl Woese and his collaborators suggests that organisms fall into one of three domains (empires) into which the traditional kingdoms are distributed
      1. *Eucarya*—contains all eucaryotic organisms
      2. *Bacteria (Eubacteria)*—contains procaryotic organisms with eubacterial rRNA and membrane lipids that are primarily diacyl glycerol ethers
      3. *Archaea*—contains procaryotic organisms with archaeobacterial rRNA and membrane lipids that are primarily isoprenoid glycerol diether or diglycerol tetraether derivatives
   F. Modern eucaryotic cells appear to have arisen form procaryotes about 1.4 billion years ago
   G. One hypothesis for the deveopment of chloroplasts and mitochondria involves invagination of the plasma membrane and subsequent compartmentalization of function
   H. The alternative is the endosymbiotic hypothesis which suggests the following
      1. The first event in the development of eucaryotes was the formation of the nucleus (possibly by fusion of of ancient eubacteria and archaea
      2. Chloroplasts were formed from free living photosynthetic bacteria that entered into a symbiotic relationship with the primitive eucaryote (cyanobacteria and *Prochloron* have been suggested as possible candidates)
      3. Mitochondria may have arisen by a similar process (ancestors of *Agrobacterium*, *Rhizobium*, and the rickettsias have been suggested)
   I. The endosymbiotic hypothesis has received support from the discovery of an endosymbiotic cyanobacterium that inhabits the biflagellate protist *Cyanophora paradoxa* and acts as its chloroplast; the endosymbiont is called a cyanelle
III. Taxonomic Ranks
   A. The taxonomic ranks (in ascending order) are: species, genus, family, order, class and kingdom
   B. Microbiologists often use less formal group (section) names that are descriptive (e.g., methanogens, purple bacteria, lactic acid bacteria, etc.)
   C. The basic taxonomic group is the species
   D. Bacterial species are defined on the basis of sexual reproductive compatibility (as for higher organisms) but rather are based on phenotypic and genotypic differences
      1. A bacterial species is a collection of strains that share many stable properties and differ significantly from other groups of strains
      2. A strain is a population of organisms that descends from a single organism or pure culture isolate
         a. Biovars—strains that differ biochemically or physiologically
         b. Morphovars—strains that differ morphologically
         c. Serovars—strains that differ in antigenic properties
      3. The type strain is usually the first studied (or most fully characterized) strain of a species; it does not have to be the most representative member
   E. A genus is a well-defined group of one or more species that is clearly separate from other genera
   F. The binomial system of nomenclature devised by Carl von Linne (Carolus Linnaeus) is used in which the genus name is capitalized while the specific epithet is not; both terms are italicized (e.g., *Escherichia coli*). After first usage in a manuscript the first name will often be abbreviated to the first letter (e.g., *E. coli*)

G. *Bergey's Manual of Systematic Bacteriology* focuses on the classification and biology of bacteria but is often more detailed than is necessary for identification

H. *Bergey's Manual of Determinative Bacteriology* is a single volume that is intended for use in identifying bacteria

IV. Classification Systems

    A. Natural classification—arranges organisms into groups whose members share many characteristics and reflects as much as possible the biological nature of organisms

    B. Phenetic systems group organisms together based on overall similarity

        1. Frequently a natural system based on shared characteristics

        2. Not dependent on phylogenetic analysis

        3. Use unweighted traits

        4. Best system compares as many attributes as possible

    C. Numerical Taxonomy

        1. Information about the properties of an organism is converted to a form suitable for numerical analysis

        2. Compared by means of a computer

        3. The presence or absence of at least 50 (preferably several hundred) characters should be compared

            a. Morphological, biochemical and physiological characters should be included

            b. Determine an association coefficient between characters possessed by two organisms

                (1) Simple matching coefficient -proportion that match whether present or absent

                (2) Jaccard coefficient—ignores characters that both organisms lack

            c. Arrange to form a similarity matrix

            d. Organisms with great similarity are grouped together into phenons

        4. A treelike diagram called a dendrogram is used to display the results of numerical taxonomic analysis

        5. The significance of the phenons is not always obvious but phenons with an 80% similarity often are equivalent to bacterial species

    C. Phylogenetic (phyletic) systems group organisms together based on probable evolutionary relationships

        1. Has been difficult for bacteria because of the lack of a good fossil record

        2. Direct comparison of genetic material and gene products such as rRNA and proteins overcomes this problem

V. Major Characteristics Used in Taxonomy

    A. Classical Characteristics

        1. Morphological characteristics are easy to analyze, genetically stable and do not vary greatly with environmental changes; often are good indications of phylogenetic relatedness

        2. Physiological and metabolic characteristics are directly related to enzymes and transport proteins (gene products) and therefore provide an indirect comparison of microbial genomes

        3. Ecological characteristics include life-cycle patterns, symbiotic relationships, ability to cause disease, habitat preferences and growth requirements

        4. Genetic analysis includes the study of chromosomal gene exchange through transformation and conjugation; these processes only rarely cross genera; one must take care to avoid errors that result from plasmid-borne traits

    B. Molecular Characteristics

        1. Comparison of proteins is useful because it reflects the genetic information of the organism; analysis is by:

            a. Determination of the amino acid sequence of the protein

            b. Comparison of electrophoretic mobility

   c. Determination of immunological cross-reactivity

   d. Comparison of enzymatic properties

  2. Nucleic acid base composition (G+C content)

   a. Can be determined by determination of the melting temperature (T_m) which is related to the temperature at which the two strands of a DNA molecule separate from one another as the temperature is slowly increased

   b. Taxonomically useful because variation within a genus is usually less than 10% but variation between genera is quite variable ranging from 25 to 80%

  3. Nucleic acid hybridization

   a. Determines the degree of sequence homology

   b. The temperature of incubation controls the degree of sequence homology needed to form a stable hybrid

  4. Nucleic acid sequencing

   a. rRNA gene sequences are most ideal for comparisons because they contain both evolutionarily stable and evolutionarily variable sequences

   b. Recently, complete bacterial genomes have been sequenced; direct comparison of complete genome sequences undoubtedly will become important in bacterial taxonomy

VI. Assessing Microbial Phylogeny

 A. Molecular Chronometers—based on the assumption of a constant rate of change, which is not a correct assumption; however the rate of change may be constant within certain genes

 B. Phylogenetic Trees

  1. Made of branches that connect nodes, which represent taxonomic units such as species or genes

  2. Rooted trees provide a node that serves as the common ancestor for the organisms being analyzed

  3. Developed by comparing molecular sequences and differences are expressed as evolutionary distance

  4. Organisms are then clustered to determine relatedness; alternatively, relatedness can be estimated by parsimony analysis assuming that evolutionary change occurs along the shortest pathway with the fewest changes to get from ancestor to the organism in question

 C. rRNA, DNA, and Proteins as Indicators of Phylogeny

  1. Association coefficients from rRNA studies are a measure of relatedness

  2. Oligonucleotide signature sequences occur in most or all members of a particular phylogenetic group and are rarely or never present in other groups even closely related ones; useful at kingdom or domain levels

  3. DNA similarity studies are more effective at the species and genus level

  4. Protein sequences are less affected by organism-specific differences in G+C content

  5. The three types of molecules do not always produce the same evolutionary trees

VII. The Major Divisions of Life

 A. Empires (Domains)

  1. *Eubacteria*—comprise the vast majority of procaryotes; peptidoglycan contains muramic acid; membrane lipids contain ester-linked straight-chain fatty acids

  2. *Archaea*—procaryotes that lack muramic acid and have lipids with ether-linked branched aliphatic chains, tRNAs lack thymine, RNA polymerase is distinctive, ribosomes have a different composition and shape when compared to the *Eubacteria*

  3. *Eucarya*—have a more complex membrane-delimited organelle structure

  4. Several different phylogenetic trees have been proposed relating the major domains and some trees do not even support a three-domain pattern

B. Kingdoms
   1. Five Kingdom system
      a. *Animalia*—multicellular, nonwalled eucaryotes with ingestive nutrition
      b. *Plantae*—multicellular, walled eucaryotes with photoautotrophic nutrition
      c. *Fungi*—multicellular and unicellular, walled eucaryotes with absorptive nutrition
      d. *Protista*—unicellular eucaryotes with various nutritional mechanisms
      e. *Monera* (*Procaryotae*)—all procaryotic organisms
   2. Six Kingdom system—separate *Monera* into *Eubacteria* and *Archaeobacteria*
   3. Eight Kingdom system (two empires)
      a. Separates procaryotes into *Eubacteria* and *Archaeobacteria*
      b. Redefines protists into several better-defined kingdoms

VIII. *Bergey's Manual of Systematic Bacteriology*—A detailed work that contains descriptions of all procaryotic species currently identified
   A. The First Edition of *Bergey's Manual of Systematic Bacteriology*—primarily phenetic
      1. 33 sections in 4 volumes
      2. Each section contains bacteria that share a few easily determined characteristics and bears a title that describes these properties or provides the vernacular names of the bacteria included
      3. There is considerable disagreement between the phenetic system in *Bergey's* and phylogenetic relationships as determined by a variety of means
      4. Despite limitations it is the most widely accepted system for identification of bacteria
   B. The Second Edition of *Bergey's Manual of Systematic Bacteriology*
      1. Twice the number of species with 170 newly described genera
      2. Largely phylogenetic rather than phenetic
      3. Will not be available for some time yet
      4. Pathogenic species are not grouped together but rather are scattered throughout the five volumes according to their phylogenetic relationships

IX. A Survey of Bacterial Phylogeny and Diversity—based on the 2nd edition of *Bergey's*
   A. Volume 1: The Archaea, Cyanobacteria, Phototrophs and Deeply Branching Genera
      1. *Archaea*—divided into two kingdoms
         a. *Crenarchaeota*—diverse kingdom that contains thermophilic and hyperthermophilic organisms as well as some organisms that grow in oceans at low temperatures as picoplankton
         b. *Euryarchaeota*—contains primarily mathanogenic and halophilic bacteria and also thermophilic, sulfur-reducing bacteria
      2. *Eubacteria*—complex with several small groups of phototrophs, cyanobacteria, and deeply branching eubacteria
   B. Volume 2—Gram negative proteobacteria (purple bacteria)—complex group
   C. Volume 3—Gram positive bacteria with low G+C content (<50%)
   D. Volume 4—Gram positive bacteria with high G+C content (>50-55%)
   E. Volume 5—An assortment of deeply branching phylogenetic groups that are not necessarily related to one another although all are Gram negative

TERMS AND DEFINITIONS

Place the letter of each term in the space next to the definition or description that best matches it.

_____ 1.	The science of biological classification	a. hybridization
_____ 2.	The scientific study of organisms to ultimately characterize them and arrange them in an orderly manner	b. melting temperature (*Tm*)
		c. numerical taxonomy
_____ 3.	A population of organisms that descends from a single organism	d. phenetic classification
		e. phenon
_____ 4.	A classification system based on evolutionary relationships	f. phyletic classification
_____ 5.	A classification system based on mutually similar attributes	g strain
_____ 6.	A classification system based on the general similarity of organisms, in which computers are used to calculate association coefficients	h. stromatolites
		i systematics
		j. taxon
_____ 7.	The temperature at which the two strands of a double-stranded DNA molecule will separate from each other	k. taxonomy
_____ 8.	The phenomenon in which two strands of nucleic acid associate with each other because they share some degree of sequence homology	
_____ 9.	Layered or stratified rocks that are formed by incorporation of minerals into microbial mats	
_____ 10.	Organisms with great similarity that are grouped together by numerical taxonomy methods	
_____ 11.	A generic term for the groups into which organisms are placed	

FILL IN THE BLANK

1. Bacterial strains that are characterized by biochemical or physiological differences are called _____; those that differ morphologically are called _____; and those that differ antigenically are called _____.

2. The most desirable classification system is one in which organisms are arranged into groups whose members _____ many characteristics and in which the biological nature of organisms is reflected as much as possible. These kinds of systems are called _____ systems. The two most common of these are _____ systems, based on evolutionary relatedness, and _____ systems, based on mutual similarity.

3. In numerical taxonomy a large number of characteristics are determined. After character analysis, the _____ is calculated, which is a measure of the agreement between characteristics possessed by two organisms for each pair of organisms in the group. Using this value, organisms with great mutual similarity are grouped together in _____ and separated from dissimilar organisms. The results are often summarized in a treelike diagram called a _____.

4. The _____ coefficient, the most commonly used coefficient in bacteriology, is the proportion of characteristics that match regardless of whether the attribute is _____ or _____. The _____ coefficient is calculated by ignoring any characteristics that are _____ in both organisms.

5. The classification scheme used in _____ is the most widely accepted scheme for bacteria. It is primarily a _____ scheme.

6. The 16S rRNA of most major phylogenetic groups has one or more characteristic nucleotide sequences called _____ _____ sequences. They are _____ or _____ present in other groups, even closely related ones.

7. One alternative classification scheme that has been proposed would divide all organisms into three _____ that would be above the kingdom level and into which the current kingdoms would be distributed; the _____ would include all of the eucaryotic organisms, the _____ would consist of the eubacteria, and the _____ would contain the archaeobacteria.

MULTIPLE CHOICE

For each of the questions below select the *best* answer or answers.

1. Which of the following has not been used in systematics?
   a. physiology
   b. epidemiology
   c. ecology
   d. All of the above have been used in systematics.

2. In the current system of classification, the rank immediately below the level of the kingdom is the
   a. class.
   b. division.
   c. order.
   d. family.

3. A population of organisms that descends from a single organism is a
   a. family.
   b. genus.
   c. strain.
   d. binomial system.

4. One strain, the _____ strain, is used to define the characteristics of a species. Only those strains very similar to this strain are included in a species.
   a. type
   b. prime
   c. proto
   d. comparison

5. Which of the following gene-exchange mechanisms is not useful in classification studies?
   a. transformation
   b. transduction
   c. conjugation
   d. All of the above have been useful in classification studies.

6. Which of the following is not true about the G + C content of organisms?
   a. Organisms with similar G + C content have similar base sequences.
   b. Organisms with different G + C content have dissimilar base sequences.
   c. Only if two organisms are alike phenotypically does their similar G + C content suggest relatedness.
   d. All of the above are true about the G + C content of organisms.

7. In which kingdom(s) will photoautotrophic organisms be found?
   a. *Animalia*
   b. *Plantae*
   c. *Fungi*
   d. *Protista*
   e. *Monera*

8. In which kingdom(s) will procaryotic organisms be found?
   a. *Animalia*
   b. *Plantae*
   c. *Fungi*
   d. *Protista*
   e. *Monera*

9. In which kingdom(s) will organisms with ingestive nutrition be found?
   a. *Animalia*
   b. *Plantae*
   c. *Fungi*
   d. *Protista*
   e. *Monera*

10. In the scheme proposed by Woese, those organisms currently found in the kingdom *Protista* would be found in which of the following domains or empires?
   a. *Eucarya*
   b. *Eubacteria*
   c. *Archaea*
   d. None of the above are correct.

11. Which of the following is not considered a classical characteristic for taxonomic purposes?
   a. ecological characteristics
   b. G + C content
   c. genetic analysis
   d. All of the above are classical characteristics for taxonomic purposes.

TRUE/FALSE

____ 1. The definition of a species as "a group of interbreeding or potentially interbreeding natural populations that are reproductively isolated from other groups" is satisfactory for higher organisms but not for microorganisms.

____ 2. In all cases where numerical taxonomy has been applied, the results have agreed with existing classification schemes.

____ 3. Although procaryotes do not reproduce sexually, the study of chromosomal gene exchange is sometimes useful in their classification.

____ 4. The temperature at which nucleic acids are incubated for the formation of stable hybrids determines how related the organisms must be.

____ 5. DNA-DNA hybridization is useful for closely related organisms, while DNA-RNA hybridization using tRNA or rRNA can be used for more distantly related organisms because these genes have not evolved as rapidly as most other microbial genes.

____ 6. Phylogenetic schemes and phenetic schemes for classifying bacteria generally agree with each other.

____ 7. Phenons determined by numerical taxonomy are generally equivalent and correspond to species as determined by traditional taxonomic methods.

____ 8. The earliest bacteria were probably anaerobic.

____ 9. The endosymbiotic hypothesis proposes that mitochondria and chloroplasts developed from free-living procaryotes that invaded a precursor to the eucaryotes and established a stable relationship.

____ 10. The five kingdom classification scheme that is currently popular clearly fits best with all of the known phylogenetic relationships and is therefore unlikely to be seriously challenged in the foreseeable future.

____ 11. Variation in G + C content among members of a particular genus usually varies less than 10%.

____ 12. Like the first edition, the second edition of *Bergey's Manual of Systematic Bacteriology* will use a phenetic classification scheme.

CRITICAL THINKING

1. Phylogenetic and phenetic schemes for classifying bacteria do not always agree with each other. Why not? Under what circumstances would it be more advantageous to use a phylogenetic scheme? In what situations would a phenetic scheme be better? How can this disagreement be resolved?

2. There have been several proposals to reclassify organisms into a three-domain classification system. This would primarily involve reorganizing the kingdom *Monera* into two separate domains and consolidating the eucaryotic kingdoms into a separate domain. Why is this being proposed? Take a stand on this proposal (for or against) and present the reasons why you have chosen that position.

ANSWER KEY

Terms and Definitions

1. k, 2. i, 3. g, 4. f, 5. d, 6. c, 7. b, 8.a, 9. h, 10. e, 11. j

Fill in the Blank

1. biovars; morphovars; serovars 2. share; natural; phyletic (phylogenetic); phenetic 3. association coefficient; phenons; dendrogram 4. simple matching; present; absent; Jaccard; absent 5. *Bergey's Manual;* phenetic 6. oligonucleotide signature; rarely (never); never (rarely) 7. domains; *Eucarya; Bacteria; Archaea*

Multiple Choice

1. d, 2. b, 3. c, 4. a, 5. b, 6. a, 7. b, d, e, 8. e, 9. a, d, 10. a, 11. b

True/False

1. T, 2. F, 3. T, 4. T, 5. T, 6. F, 7. F, 8. T, 9. T, 10. F, 11. T, 12. F

20   The Archaea

CHAPTER OVERVIEW

This chapter summarizes the properties of a diverse group of organisms known as the archaea. These organisms are very different from the eubacteria and from the eucaryotes. The chapter describes some of the major characteristics associated with each of the major groups of archaea.

CHAPTER OBJECTIVES

After reading this chapter you should be able to:

- discuss the morphological and physiological diversity of the archaea
- discuss the difference between the cell walls of archaea and those of eubacteria
- describe the lipid composition of archaeal cell membranes
- discuss the general genetic, molecular, and metabolic characteristics of the archaea
- discuss the restricted habitats that are typical for the archaea
- discuss the classification scheme for the archaea that will be used in the 2nd edition of *Bergey's Manual*
- discuss the unique cofactors used by methanogenic and sulfate-reducing archaeobacteria for methanogenesis
- describe the structural, chemical, and metabolic adaptations that allow the archaea to grow in extreme environments

CHAPTER OUTLINE

I.   Introduction to the Archaea
    A.   The archaea are quite diverse, both in morphology and physiology
        1.   They may stain gram-positive or gram-negative
        2.   They may be spherical, rod-shaped, spiral, lobed, plate-shaped, irregularly shaped or pleiomorphic
        3.   They may exist as single cells, aggregates or filaments
        4.   They may multiply by binary fission, budding, fragmentation, or other mechanisms
        5.   They may be aerobic, facultatively anaerobic, or strictly anaerobic
        6.   Nutritionally, they range from chemilithoautotrophs to organotrophs
        7.   Some are mesophiles, while others are hyperthermophiles that can grow above 100°C
        8.   They prefer restricted or extreme aquatic and terrestrial habitats
        9.   Recently, archaeobacteria have been found in cold environments and may constitute up to 34% of the procaryotic biomass in Antarctic surface waters
       10. A few are symbionts in animal digestive systems
    B.   Archaeal Cell Walls
        1.   Archaeal cell wall structure differs from that of eubacteria
        2.   Gram-positive archaeobacteria, like gram-positive eubacteria, have a cell wall that has a single homogeneous layer

3. Gram-negative archaeobacteria lack the outer membrane and complex peptidoglycan network associated with gram-negative eubacteria
4. The surface layer of archaeobacteria consists of protein or glycoprotein subunits
5. Archaeal cell wall chemistry is different from that of eubacteria
   a. Lacks muramic acid and D-amino acids
   b. Resistant to lysozyme and β-lactam antibiotics
   c. Some have pseudomurein, a peptidoglycan-like polymer that has L-amino acids in its cross-links and different monosaccharide subunits and linkage
   d. others have different polysaccharides
   e. Gram-negative archaeobacteria have a layer of protein or glycoprotein outside their plasma membrane
C. Archaeal Lipids and Membranes
   1. Archaeobacterial lipids differ from those of eubacteria and eucaryotes; branched hydrocarbons are attached to glycerol by ether links rather than straight-chain fatty acids attached to glycerol by ester links
   2. Other, more complex tetraether structures are also found
   3. Membranes contain polar lipids such as phospholipids, sulfolipids, and glycolipids
   4. Membranes also contain nonpolar lipids (7-30%) that are usually derivatives of squalene
   5. Membranes of extreme thermophiles are almost completely tetraether monolayers
D. Genetics and Molecular Biology
   1. The archaeal chromosome is a single, closed DNA circle like that of eubacteria, but generally is considerably smaller
   2. Archaea have few plasmids
   3. The genome of *Methanococcus jannaschii* has been completely sequenced
      a. 56% of its 1,738 genes are unlike those in eubacteria or eucarya
      b. These organisms, therefore, are as distinctive genotypically as they are in other respects
   4. Archaeal mRNA is like that of eubacteria (i.e., it may be polygenic, there are no intron-containing precursors, and its promoters are similar to those of eubacteria)
   5. Archaeal tRNAs contain modified bases not found in eubacterial tRNAs
   6. Archaeal ribosomes have a different, more variable shape than those of eubacteria or eucarya
   7. Archaeal ribosomes are the same size as eubacterial ribosomes but show antibiotic sensitivity similar to that of eucaryotic ribosomes
   8. Archaeal RNA polymerase enzymes are more similar to eucaryotic enzymes than to eubacterial enzymes
E. Metabolism
   1. Metabolic processes vary greatly among different groups
   2. Archaea do not use the Embden-Meyerhof pathway for glucose catabolism; however they frequently use a reversal of that pathway for gluconeogenesis
   3. Some (halophiles and extreme thermophiles) have a complete TCA cycle while others (methanogens) do not
   4. Archaeal biosynthetic pathways appear to be similar to those of other organisms
   5. Autotrophy is widespread with a variety of mechanisms for incorporating CO_2
F. Archaeal Taxonomy—shows great diversity
   1. The current edition of *Bergey's Manual* divides the archaeobacteria into 5 groups
   2. The new edition of *Bergey's Manual* will divide the archaea into two kingdoms
      a. *Euryarchaeota*—described in four sections
         (1) methanogens
         (2) extreme halophiles
         (3) sulfate users
         (4) extreme thermophiles with sulfur-dependent metabolism

  b. *Crenarchaeota*—divided into 3 orders
   (1) *Thermoproteales*—hyperthermophiles
   (2) *Sulfolobales*—thermoacidophilic
   (3) *Igneococcales*—hyperthermophiles that grow at neutral pH

II. Kingdom *Crenarchaeota*
 A. Many are acidophiles and are sulfur-dependent; they are extremely thermophilic
 B. Sulfur may be used as an electron acceptor in anaerobic respiration, or as an electron source by lithotrophs
 C. Almost all are strict anaerobes
 D. They grow in geothermally heated water or soils (solfatara) that contain elemental sulfur (sulfur-rich hot springs, waters surrounding submarine volcanic activity)
 E. Some (e.g., *Pyrodictum* spp.) can grow quite well above the boiling point of water (optimum @ 105°C)
 F. Some are organotrophic; others are lithotrophic
 G. There are three orders and at least twelve genera; two of the better studied genera are *Sulfolobus* and *Thermoproteus*
 H. *Sulfolobus*
  1. Gram-negative, aerobic, irregularly lobed, spherical bacteria
  2. Thermoacidophiles
  3. Cell walls lack peptidoglycan but contain lipoproteins and carbohydrates
  4. Oxygen is the normal electron acceptor, but ferric iron can also be used
  5. Sugars and amino acids may serve as carbon and energy sources; however, they often grow lithotrophically on sulfur granules in hot springs oxidizing the sulfur to sulfuric acid
 I. *Thermoproteus*
  1. Long, thin, bent or branched rods
  2. Cell wall is composed of glycoprotein
  3. Strict anaerobes; they have temperature optima from 70-97°C and pH optima from 2.5 to 6.5
  4. They grow in hot springs and other hot aquatic habitats that contain elemental sulfur
  5. They grow organotrophically
   a. They oxidize glucose, amino acids, alcohols, and organic acids
   b. They carry out anaerobic respiration with elemental sulfur as the electron acceptor
  6. They can also grow litotrophically using H_2 and S^0
  7. CO or CO_2 can serve as the sole carbon source

III. Kingdom *Euryarchaeota*
 A. The Methanogens
  1. Strict anaerobes that obtain energy by converting CO_2, H_2, formate, methanol, acetate, and other compounds to either methane or to methane and CO_2
  2. There are at least three orders and 25 genera, which differ greatly in shape, 16S rRNA sequence, cell wall chemistry and structure, membrane lipids, and other features
  3. Metabolism is unusual; members of this group contain several unique cofactors, some of which are associated with methane production
  4. They thrive in anaerobic environments rich in organic matter, such as animal rumens and intestinal tracts, freshwater and marine sediments, swamps, marshes, hot springs, anaerobic sludge digesters, and even within anaerobic protozoa
  5. They are of great potential importance because methane is a clean-burning fuel and an excellent energy source
  6. They may be an ecological problem however, because methane is a greenhouse gas that could contribute to global warming and also because methanogens can oxidize iron, which contributes significantly to the corrosion of iron pipes

B. The Halobacteria—extreme halophiles
    1. Consists of nine genera in one family, the *Halobacteriaceae*
    2. They are aerobic chemoheterotrophs with respiratory metabolism; they require complex nutrients
    3. They require at least 1.5 M NaCl and have growth optima near 3-4 M NaCl
    4. If the NaCl concentration drops below 1.5M the cell walls disintegrate
    5. They can cause spoilage of salted foods
    6. *Halobacterium salinarum* has a unique type of photosynthesis that uses a modified cell membrane (called the purple membrane) which contains the protein, bacteriorhodopsin, but no chlorophyll
    7. Two other rhodopsins act as photoreceptors that control flagellar activity to position the bacterium in the water column at a location of high light intensity, but one in which the UV light is not sufficiently intense to be lethal
    8. Halorhodopsin uses light energy to transport chloride ions into the cell to maintain an 4-5 M intracellular Kcl concentration

C. The Thermoplasms
    1. Thermoacidic cocci that lack cell walls
    2. *Thermoplasma* has an optimum temperature of 55-59°C and an optimal PH of 1 to 2
    3. At 59°C *Thermoplasma* takes the form of an irregular filament; the cells may be flagellated and motile
    4. Frequently found in coal mine refuse, in which chemolithotrophic bacteria oxidize iron pyrite to sulfuric acid and thereby produce a hot acidic environment
    5. They have no cell wall but the cell membrane is strengthened by large quantities of diglycerol tetraethers, lipopolysaccharides, and glycoproteins
    6. Their DNA is stabilized by its association with histonelike proteins, which form particles resembling eucaryotic nucleosomes
    7. *Picrophilus* also has no cell wall but has an S-layer outside the plasma membrane
    8. *Picrophilus* has large cytoplasmic cavities that are not membrane bounded
    9. *Picrophilus* is aerobic and grows between 47°C and 65°C with an optimum of 60°C; it grows only below pH 3.5, has an optimum of pH 0.7 and will even grow at or near pH 0

D. The Thermococci
    1. There are three orders, *Archaeogobales*, *Thermococcales* and *Methanopyrales*
    2. All have cell walls
    3. *Thermococcales* are strictly aerobic, reduce sulfur to sulfide, are motile by means of flagella, and have optimum growth temperatures around 88-100°C
    4. *Archaeoglobales* are gram-negative, irregular coccoid cells with walls of glycoprotein subunits; they cannot use elemental sulfur and are extremely thermophilic (optimum around 83°C); they are usually found near marine hydrothermal vents
    5. *Methanopyrales* have an optimum tempearture of 98°C and will grow at 110°C; they have been isolated from a marine hydrothermal vent; evidence suggests that they may be among the first living organisms to have developed

TERMS AND DEFINITIONS

Place the letter of each term in the space next to the definition or description that best matches it.

____ 1. A peptidoglycanlike polymer with L-amino acids instead of D-amino acids
____ 2. A strict anaerobe that obtains energy by converting CO_2, H_2, formate, methanol, acetate, and other compounds to methane
____ 3. An organism that requires at least 1.5 M NaCl in order to grow
____ 4. A modified cell membrane that carries out photosynthesis in the absence of chlorophyll
____ 5. A protein that mediates photosynthesis without chlorophyll
____ 6. An organism with a temperature growth optimum around 80°C and a pH growth optimum from 1.0 to 6.5
____ 7. Geothermally heated water or soils that contain elemental sulfur

a. bacteriorhodopsin
b. extreme halophile
c. methanogen
d. pseudomurein
e. purple membrane
f. solfatara
g. thermoacidophile

FILL IN THE BLANK

1. Gram-negative archaeobacteria lack a(n) _____ _____ and the complex _____ network normally found in gram-negative eubacteria.
2. Archaeal membranes contain polar lipids, such as _____, _____, and _____; they also contain nonpolar lipids that are derivatives of _____.
3. The membrane lipids of archaea have branched chain _____ attached to glycerol by _____ links, while the membrane lipids of eubacteria have _____ _____ attached to glycerol by _____ links.
4. Methanogens thrive in _____ environments that are rich in _____ _____.
5. Two _____ act as photoreceptors that control _____ activity to position *Halobacterium salinarum* in the water column at a location of _____ light intensity.

MULTIPLE CHOICE

For each of the questions below select the *one best* answer.

1. Which of the following is normally used by archaea as a mechanism for reproduction?
   a. binary fission
   b. budding
   c. fragmentation
   d. All of the above are correct.

2. Which of the following is the primary lipid component of the membranes of extreme thermophiles?
   a. C_{20} diethers
   b. C_{40} tetraethers
   c. sulfolipids
   d. cholesterol

3. Which of the following is true about archaeal mRNA?
   a. Polygenic mRNA has been found.
   b. mRNA splicing has not been found.
   c. Both (a) and (b) are correct.
   d. Neither (a) nor (b) is correct.
4. Which of the following is true about archaeal ribosomes?
   a. They are 70S, like eubacterial ribosomes.
   b. Their shape differs from both eubacterial and eucaryotic ribosomes.
   c. They have antibiotic sensitivities similar to those of eucaryotic ribosomes.
   d. All of the above are true.
5. Which of the following groups does not have a functional TCA cycle?
   a. methanogens
   b. extreme halophiles
   c. extreme thermophiles
   d. All of the above have a functional TCA cycle.
6. The largest group of archaea are the
   a. methanogens.
   b. halobacteria.
   c. thermoplasms.
   d. thermococci.
7. Which of the following is not true about methanogens?
   a. They are potentially of great importance because methane is an even-burning fuel and an excellent energy source.
   b. They may contribute to the greenhouse effect and global warming.
   c. They may cause corrosion in iron pipes.
   d. All of the above are true about methanogens.

8. What happens to the cell wall of *Halobacterium* if the NaCl concentration drops to about 1.5 M?
   a. The cell wall becomes more rigid.
   b. The cell wall loses permeability.
   c. The cell wall disintegrates.
   d. All of the above are correct.
9. Which of the following strengthens the cell membranes of cell wall-less archaeobacteria?
   a. diglycerol tetraethers
   b. lipopolysaccharides
   c. glycoproteins
   d. All of the above are correct.
10. Which of the following is true about *Sulfolobus?*
   a. Oxygen is the normal electron acceptor.
   b. Ferric ion can be used as an electron acceptor.
   c. Both (a) and (b) are correct.
   d. Neither (a) nor (b) is correct.
11. Which of the following is true about the archaeobacterial genus *Thermoplasma?*
   a. They are spherical at temperatures below 59°C.
   b. They are filamentous at temperatures at or above 59°C.
   c. Both (a) and (b) are correct.
   d. Neither (a) nor (b) is correct.
12. Which of the following do not have cell walls?
   a. halobacteria
   b. thermoplasms
   c. thermococci
   d. None of the above have cell walls.

TRUE/FALSE

_____ 1. Archaea are all strict anaerobes.

_____ 2. Archaea are resistant to lysozyme and β-lactam antibiotics because they lack peptidoglycan in their cell walls.

_____ 3. C_{40} tetraethers make a more rigid membrane than do C_{20} diethers.

_____ 4. The chromosomal DNA of the archaea is normally smaller than the chromosomal DNA of the eubacteria.

_____ 5. *Archaeoglobus* uses elemental sulfur as an electron acceptor.

_____ 6. Recently, archaeobacteria have been found in cold environments such as Antarctic surface waters.

_____ 7. The genome of *Methanococcus jannaschii* has been completely sequenced and has been found to be genotypically indistinct from the eubacteria.

_____ 8. *Picrophilus* has a pH optimum below 1 and can even grow at or near pH 0.

CRITICAL THINKING

1. There has been serious consideration given within the scientific community to separating the archaeobacteria from the eubacteria, placing them in a kingdom of their own. This is based on the difference between these two types of organisms. Do you agree or disagree that this should be done? Use the differences or similarities to defend your position.

2. Members of the archaea may, in the future, provide a pollution-free source of energy. Explain.

ANSWER KEY

Terms and Definitions

1. d, 2. c, 3. b, 4. e, 5. a, 6. g, 7. f

Fill in the Blank

1. outer membrane; peptidoglycan 2. phospholipids; sulfolipids; glycolipids; squalene 3. hydrocarbons; ether; fatty acids; ester 4. anaerobic; organic matter 5. rhodopsins; flagellar; high

Multiple Choice

1. d, 2. b, 3. c, 4. d, 5. a, 6. a, 7. d, 8. c, 9. d, 10. c, 11. c, 12. b

True/False

1. F, 2. T, 3. T, 4. T, 5. F, 6. T, 7. F, 8. T

21 Bacteria: The Deinococci and Nonproteobacteria Gram-Negatives

CHAPTER OVERVIEW

This chapter is devoted to nine of the more interesting and important eubacterial groups from volumes 1 and 5 of the 2nd edition of *Bergey's Manual of Systematic Bacteriology*. Though the organization and perspective of the 2nd edition is used, the description of each group in the current edition is summarized. Where appropriate the distinguishing characteristics, morphology, reproduction, physiology, metabolism, and ecology of each group is included. The taxonomy of each major group is summarized and representative species are discussed.

CHAPTER OBJECTIVES

After reading this chapter you should be able to:

- discuss the deeply branching eubacterial groups represented by the hyperthermophiles *Aquifex* and *Thermotoga*
- discuss the gram-positive deinococci
- discuss the photosynthetic bacteria that are described in volume 1 of the 2nd edition of *Bergey's Manual*
- discuss the chlamydiae and their loss of some of their metabolic independence, thereby becoming dependent on their host for their energy supply and for certain cellular constituents
- discuss gliding motility and its usefulness for organisms that digest insoluble nutrients or that move over the surfaces of solid substrata

CHAPTER OUTLINE

I. *Aquifex* and Thermotogas
    A. *Aquifex* (e.g., *Aquifex pyrophilus*) and its relatives such as *Hydrogenobacter* are members of the deepest or oldest branch of the eubacteria;
        1. Hyperthermophile with a temperature optimum of 85°C and a maximum of 95°C
        2. Autotroph that generates energy by oxidizing donors such as hydrogen, thiosulfate, and sulfur with oxygen as the acceptor
    B. *Thermotoga*—second oldest branch of the eubacteria
        1. Hyperthermophile with an optimum of 80°C and a maximum of 90°C
        2. Gram-negative rod with an outer sheath-like envelope (like a toga) that can balloon out from the ends of the cell
        3. Grows in active geothermal areas—marine hydrothermal vents and terrestrial solfataric springs)
        4. Chemoheterotroph with a functional glycolytic pathway
        5. Can grow anaerobically on carbohydrates and protein digests
        6. Consists of at least seven genera in two orders

II.   The Deinococci
   A.   One family with three genera
   B.   Spherical or rod-shaped with distinctively different 16S rRNA
   C.   Associated in pairs or tetrads
   D.   Aerobic, mesophilic, catalase positive, and usually able to produce acid from few sugars
   E.   They stain gram-positive but have a layered cell wall and an outer membrane like the gram-negative bacteria.
   F.   They have L-ornithine in their peptidoglycan; lack teichoic acid; have a plasma membrane with large amounts of palmitoleic acid rather than phosphatidylglycerol phospholipids
   G.   Extraordinarily resistant to desiccation and radiation: can survive 3-5 million rad whereas 100 rad can be lethal to humans
   H.   Can be isolated from ground meat, feces, air, fresh water, and other sources but their natural habitat is not known
   I.   Have an unusual ability to repair chromosomal damage (even fragmentation) that probably accounts for their ability to resist radiation
III.   Photosynthetic Bacteria
   A.   Three groups: purple bacteria, green bacteria, and cyanobacteria
   B.   Green and purple bacteria use anoxygenic photosynthesis
        1.   Use bacteriochlorophylls
        2.   Use H_2S and other reduced electron donors
        3.   Use wavelengths of light that enable them to grow at deeper water levels
   C.   Cyanobacteria use oxygenic photosynthesis
        1.   Use chlorophyll a
        2.   Use H_2O as an electron donor
        3.   Grow primarily at the surface of bodies of water
   D.   Divided into two groups (oxygenic and anoxygenic) in the 1st edition of *Bergey's Manual*; however, the 2nd edition divides them into six groups
   E.   The *Chloroflexi*
        1.   Has both photosynthetic and nonphotosynthetic members
        2.   *Chloroflexus* is the major representative of the photosynthetic green nonsulfur bacteria
             a.   Filamentous, gliding, thermophilic bacteria
             b.   Often isolated from neutral to alkaline hot springs
             c.   Grows in the form of orange-reddish mats
             d.   Metabolism is similar to that of the purple nonsulfur bacteria
             e.   Can carry out anoxygenic photosynthesis with organic compounds as carbon sources or can grow aerobically as a chemoheterotroph
             f.   Not closely related to any eubacterial group
        3.   *Herpetosiphon* also is included in this kingdom—anaerobic chemoorganotroph with respiratory metabolism and oxygen as the electron acceptor
   F.   *Chlorobia*
        1.   Green sulfur bacteria
        2.   Obligately anaerobic photolithoautotrophs that use hydrogen sulfide, elemental sulfur and hydrogen as electron sources
        3.   Elemental sulfur produced by sulfide oxidation is deposited outside the cell
        4.   Photosynthetic pigments are located in a ellipsoidal vesicles called chlorosomes (chlorobium vesicles) which are attached to the plasma membrane but not continuous with it
        5.   Chlorosome membrane is not a normal lipid bilayer
        6.   Chlorosomes have accessory bacteriochlorophylls but the reaction center bacteriochlorophyll is located in the plasma membrane

7. They lack flagella and are nonmotile but some species have gas vesicles to adjust their depth for adequate light and hydrogen sulfide; species without gas vesicles are found in sulfide-rich muds at the bottom of lakes and ponds.
8. Morphologically diverse; rods cocci, or vibrios; grow singly, in chains, or in clusters; grass-green or chocolate-brown in color

G. *Prochloron* and *Cyanobacteria*
1. The cyanobacteria are the largest and most diverse group of photosynthetic bacteria
2. The 1st edition of *Bergey's Manual* describes 34 genera in some detail
3. Procaryotes but the photosynthetic system resembles that of eucaryotes
    a. They have chlorophyll a and photosystem II
    b. Carry out oxygenic photosynthesis
4. Photosynthetic pigments are in thylakoid membranes lined with particles called phycobilisomes—contain the phycobilin pigment phycocyanin which transfers energy to photosystem II; some species are red-brown and contain the pigment phycoerythrin
5. They are photolithoautotrophs but some can grow slowly in the dark as chemoheterotrophs
6. Some species can carry out anoxygenic photosynthesis if in an anaerobic environment
7. They have typical procaryotic structures with a gram-negative cell wall
8. Vary greatly in shape and appearance: unicellular; colonies of many shapes; filaments called trichomes—rows of bacterial cells that are in close contact with one another over a large area
9. Many filamentous cyanobacteria can fix atmospheric nitrogen in special cells called heterocysts
10. They often use gas vesicles to move vertically in the water
11. Filamentous cyanobacteria have a gliding motility
12. Some marine species lack flagella but are able to move by an unknown mechanism
13. Reproduce by binary fission, budding, fragmentation, and multiple fission
14. Fragmentation generates small motile filaments called hormogonia
15. Some species develop akinetes, which are thick-walled resting cells that are resistant to desiccation
16. Some cyanobacteria use multiple fission to produce small reproductive cells called baeocytes
17. Some cyanobacteria form linear filaments; others produce branches or aggregates
18. Tolerant of environmental extremes; thermophilic species can grow at temperatures up to 75°C
19. Successful at establishing symbiotic relationships—photosynthetic partner in most lichen associations
20. The order *Prochlorales* is placed with the cyanobacteria in the 2nd edition of *Bergey's Manual*
21. *Prochlorales* are oxygenic phototrophic procaryotes that have both chlorophyll a and chlorophyll b but lack phycobilins
    a. Similar to, but different from, cyanobacteria
    b. Three recognized genera
        (1) *Prochloron*—extracellular symbiont on the surface or within the cloacal cavity of marine colonial ascidan invertebrates
        (2) *Prochlorothrix*—free living
        (3) *Prochlorococcus*—has a modified chlorophyll a and α-carotene rather than β-carotene

IV. The Planctomycetes
A. Order *Planctomycetales*
1. Spherical or oval, budding eubacteria with distinctive crateriform structures (pits) in their walls

2. In two genera, *Gemmata* and *Pirullela* the nuclear body is membrane bounded, something that is not seen in other procaryotes
3. The genus *Planctomyces* attaches to surfaces through a stalk and holdfast; other genera lack stalks
4. Most of these eubacteria have life cycles in which sessile cells bud to produce motile swarmer cells

B. Order *Chlamydiales*—contains a single genus *Chlamydia*
1. Nonmotile, coccoid, gram-negative eubacteria
2. Obligate intracellular parasites with a unique developmental cycle involving elementary bodies (EBs) that reorganize into reproductively specialized reticulate bodies (RBs) which then change back to infectious elementary bodies
3. Completely dependent on the host for ATP; most metabolic activity in RBs; very little metabolic activity in EBs.
4. Gram-negative-like wall but lacks muramic acid and peptidoglycan
5. Use cross-linking of outer membrane and, possibly, periplasmic proteins to achieve osmotic stability
6. Found mostly in mammals and birds but have been recently isolated from spiders, clams and freshwater invertebrates
7. Three recognized human pathogens
   a. *C. trachomatis*—trachoma, nongonococcal urethritis, and other diseases in humans and mice
   b. *C. psittaci*—causes psittacosis in humans and infects many other mammals as well; invades the respiratory and genital tracts, the placenta, developing fetuses, the eye, and synovial fluid of the joints
   c. *C. pneumoniae*—a causative agent of human pneumonia

V. The Spirochetes
A. Gram-negative, chemoheterotrophic, flexibly helical bacteria that exhibit a creeping (crawling) motility due to a structure called an axial filament
B. The axial filament (a complex of periplasmic flagella) lies in a flexible outer sheath (outer membrane) outside the protoplasmic cylinder that houses the nucleoid and cytoplasm; the function of the sheath is essential (spirochetes will die if it is removed) but unknown
C. Flagellar rotation is responsible for motility by an unknown mechanism presumably by rotating the outer sheath or flexing the cell for a crawling motion.
D. Can be anaerobic, facultatively anaerobic, or aerobic and can use a diverse array of organic molecules as carbon and energy sources
E. Ecologically diverse
1. *Spirochaeta*—anaerobic
2. *Leptospira*—aerobic
3. *Treponema*—includes the causative agent of syphillis (*T. pallidum*)
4. *Borrelia*—includes the causative agent of Lyme disease (*B. burgdorferi*)

VI. The *Bacteroides*
A. Obligate anaerobes, nonsporing, chemoheterotrophic
B. Found in oral cavity and intestinal tract of humans and other animals and the rumen of ruminants
C. Often benefit the host by degrading cellulose, pectins, and other complex carbohydrates, providing extra nutrition for the host
D. Can constitute as much as 30% of the bacteria isolated from human feces
E. Some species can be associated with disease

VII. The *Sphingobacteria*
A. Contains some of the eubacteria with gliding motility; others will be in the γ-proteobacteria
B. Contains the genera *Flexibacter*, *Cytophaga* and *Sporocytophaga*
1. Differ in morphology, life cycle and physiology

187

2. *Cytophaga*—slender rods with pointed ends
3. *Sporocytophaga*—similar to *Cytophaga* but form spherical resting cells called microcysts
4. *Flexibacter*—form long threads; unlike the other two genera, they are unable to degrade complex carbohydrates

C. Gliding motility used when in contact with a surface; leaves a slime trail; bacteria are nonmotile when in suspension

D. Many are capable of utilizing insoluble material as nutrient sources, encountering them during their gliding activity

E. Can position themselves for optimal conditions of light intensity, oxygen, hyrogen sulfide, temperature, and other factors that influence growth

F. Play a major role in the mineralization of organic matter and can damage exposed wooden structures

G. Contribute significantly to wastewater treatment

H. Some pathogenic species (e.g., *C. columnaris* causes disease in freshwater and marine fish)

TERMS AND DEFINITIONS

Place the letter of each term in the space next to the definition or description that best matches it.

_____ 1. A photosynthetic process in which water is used as an electron donor

_____ 2. A photosynthetic process that does not involve water as an electron donor

_____ 3. Ellipsoidal vesicles containing the green sulfur bacteria's photosynthetic pigments; these vesicles are attached to but are not continuous with the plasma membrane

_____ 4. Vesicles containing the photosynthetic pigments that line the thylakoid membranes of the cyanobacteria

_____ 5. Small, motile filaments of cyanobacteria

_____ 6. Dormant, thick-walled resting cells of cyanobacteria that are resistant to desiccation

_____ 7. Special cells of cyanobacteria that are able to fix atmospheric nitrogen

_____ 8. Small reproductive cells formed by those species of cyanobacteria that reproduce by multiple fission; these escape when the outer wall ruptures

_____ 9. A row of bacterial cells that are in close contact with one another over a large area

_____ 10. Structures used to move vertically in a water column

_____ 11. A complex of periplasmic flagella that mediate spirochete movement

_____ 12. Spherical resting cells produced by *Sporocytophaga*

_____ 13. Houses the cytoplam and nucleoid of spirochetes

a. akinetes
b. axial filament
c. anoxygenic photosynthesis
d. baeocytes
e. chlorosomes
f. gas vesicles
g. heterocysts
h. hormogonia
i. microcysts
j. oxygenic photosynthesis
k. phycobilisomes
l. protoplasmic cylinder
m. trichome

FILL IN THE BLANK

1. In oxygenic photosynthesis, _____ is used as the electron donor and _____ is produced. In contrast, anoxygenic photosynthesis does not involve _____, but instead reduced molecules such as _____, _____, and _____ are used as the electron source for the generation of NADH and NADPH.

2. Some filamentous cyanobacteria fix atmospheric nitrogen using special cells called _____. In order to form these special cells, the cyanobacteria must lose their nuclei, synthesize a very thick new wall, and discard photosystem II. This eliminates the ability to produce _____ from photosynthesis. This is necessary because the nitrogen-fixing enzyme known as _____ is sensitive to _____ and will only function if the internal environment is anaerobic.

3. The infectious stage of chlamydiae is called a(n) _____ body, while the reproductive stage is called a(n) _____ body.

4. *Bacteroides* inhabit the intestinal tract of mammals and benefit the host by degrading _____, _____ and other _____ carbohydrates.

MULTIPLE CHOICE

For each of the questions below select the *one best* answer.

1. Which of the following are capable of carrying out oxygenic photosynthesis?
   a. purple bacteria
   b. green bacteria
   c. cyanobacteria
   d. All of the above are capable of oxygenic photosynthesis.

2. Which of the following accumulates sulfur granules outside the cell?
   a. purple sulfur bacteria
   b. green sulfur bacteria
   c. cyanobacteria
   d. colorless sulfur bacteria

3. Which of the following activities is associated with *cytophagas*?
   a. damage to fishing gear and wood structures
   b. digestion of sewage in sewage treatment plants
   c. Both (a) and (b) are associated with cytophagas.
   d. Neither (a) nor (b) is associated with cytophagas.

4. Which of the following is the oldest branch of eubacteria?
   a. Thermotoga
   b. Aquifex
   c. Cyanobacteria
   d. Spirochaetes

5. Thermotogas generally grow in
   a. marine hydrothermal vents
   b. terrestrial solfataric hot springs
   c. Both (a) and (b) are correct.
   d. Neither (a) nor (b) are correct.

6. Which of the following is extremely radiation resistant?
   a. Deinococci
   b. Aquifex
   c. Both (a) and (b) are correct.
   d. Neither (a) nor (b) is correct.

7. Which of the following contain chlorophyll a and chlorophyll b?
   a. *Cyanobacteria*
   b. *Prochlorales*
   c. Both (a) and (b) are correct.
   d. Neither (a) nor (b) is correct.

8. In which of the following two genera of procaryotes are membrane-bounded nuclear regions observed?
   a. *Gemmata*
   b. *Pirullela*
   c. Both (a) and (b) are correct.
   d. Neither (a) nor (b) is correct.

TRUE/FALSE

_____ 1. The cyanobacteria comprise a small group of photosynthetic bacteria and do not differ greatly from one another.

_____ 2. All of the cyanobacteria have a blue-green color that comes from the pigment phycocyanin from which this group derives its name.

_____ 3. The photosynthetic partner in most lichen associations is generally a cyanobacterium.

_____ 4. A trichrome is a bacterial cell with three different photosynthetic pigments.

_____ 5. The deinococci stain gram-positive but have a layered cell wall and an outer membrane that is more like a gram-negative organism.

_____ 6. The deinococci have an unusual ability to repair chromosome damage, even fragmentation.

_____ 7. The chlorosomes of *Chlorobia* are continuous with the plasma membrane.

_____ 8. Chlamydiae are completely dependent on their host cells for ATP production.

_____ 9. The function of the flexible outer sheath in which the axial filaments of spirochaetes lay is unknown and non-essential; i.e., the bacteria will survive if it is removed.

_____ 10. Bacteroides constitute as much as 30% of the bacteria isolated from human feces.

_____ 11. Many bacteria with gliding motility are able to use insoluble material that they encounter while gliding as a nutrient source.

_____ 12. Cytophages contribute significantly to waste water treatment.

CRITICAL THINKING

1. In what ways are cyanobacteria similar to eucaryotic phototrophs? Why, then, are they classified in the kingdom *Monera* (i.e., in what ways are they more like other monerans)?

2. Compare and contrast oxygenic and anoxygenic photosynthesis in terms of both substrates and products.

ANSWER KEY

Terms and Definitions

1. j, 2. c, 3. f, 4. k, 5. h, 6. a, 7. g, 8. d, 9. m, 10. e, 11. b, 12. i, 13. 1

Fill in the Blank

1. water; oxygen; water; hydrogen sulfide; sulfur; hydrogen 2. heterocysts; oxygen; nitrogenase; oxygen 3. elementary; reticulate 4. cellulose; pectin; complex

Multiple Choice

1. c, 2. b, 3. c, 4. b, 5. c, 6. a, 7. b, 8. c

True/False

1. F, 2. F, 3. T, 4. F, 5. T, 6. T, 7. F, 8. T, 9. F, 10. T, 11. T, 12. T

22   Bacteria: The Proteobacteria

CHAPTER OVERVIEW

This chapter presents the diverse group of eubacteria known as the proteobacteria. The distinguishing characteristics of these gram-negative bacteria: morphology, physiology, metabolism, and ecology are presented. The phylogenetic relationships are discussed and representative species are examined.

CHAPTER OBJECTIVES

After reading this chapter you should be able to:

- discuss the importance of this diverse group of organisms, either as disease agents or because of their effects on the habitat
- discuss the use of *Escherichia coli* as a major experimental organism that is studied in many laboratories
- describe the diverse life styles and metabolism of members of this group of organisms
- discuss the complex structures (prosthecae, stalks, buds, sheaths, or complex fruiting bodies produced by some members of this group
- discuss the ecological impact by chemolithotrophic bacteria
- discuss the dependence of parasitic eubacteria, such as *Bdellovibrio* and the rickettsia on their hosts for energy and/or cell constituents

CHAPTER OUTLINE

I.   The Alpha Proteobacteria
    A.   Include most of the oligotrophic proteobacteria
    B.   The Purple Nonsulfur Bacteria
        1.   Like all purple bacteria, the purple nonsulfur bacteria use anoxygenic photosynthesis, possess bacteriochlorophyls a and b and have their photosystems in membranes that are continuous with the plasma membrane
        2.   Flexible in their choice of an energy source; normally they are grown anaerobically as photoorganoheterotophs, but can grow aerobically
        3.   They do not oxidize elemental sulfur to sulfate
        4.   Most have not been cultured
        5.   They are found in the mud and water of lakes and ponds with abundant organic matter and low sulfide levels
    C.   *Rickettsia* and *Coxiella*
        1.   *Rickettsia* will be in the α-proteobacteria while *Coxiella* will be in the γ-proteobacteria; however they are discussed together because of similar life styles.
        2.   These bacteria are rod-shaped, coccoid or pleiomorphic, with typical gram-negative walls and no flagella
        3.   Size varies but tends to be small (0.3 - 2.0 um)
        4.   All are parasitic or mutualistic

5. Rickettsias enter the host by phagocytosis; escape phagosome, and then reproduce in the cytoplasm by binary fission; in contrast *Coxiella* remains in the phagosome after fusion with a lysosome and reproduces within the resulting phagolysosome. Eventually the host cell bursts releasing new organisms
6. Metabolism is different
    a. It lacks the glycolytic pathway, and does not use glucose as an energy source;
    b. Oxidize glutamate and tricarboxylic acid cycle intermediates (e.g., succinate)
7. Order contains many important pathogens
    a. *R. prowazekii* and *R. typhi*—typhus fever
    b. *R. ricketsii*—Rocky Mountain Spotted Fever
    c. *Coxiella burnetii*—Q fever
8. They are also important pathogens in dogs, horses, sheep and cattle

D. The *Caulobacteraceae* and *Hyphomicrobiaceae*
1. Have one of three distinguishing features
    a. Prostheca—an extension of the cell, including the plasma membrane, that is narrower than the mature cell
    b. Stalk—a nonliving appendage produced by cells and extending from it
    c. Reproduction by budding—parental cell retains its identity and progeny are much smaller than the parental cell
2. *Hyphomicrobium*—chemoheterotrophic, aerobic, budding bacteria that frequently attach to solid objects in freshwater, marine and terrestrial environments
    a. Mature cell produces a hypha or prostheca that elongates
    b. The nucleoid divides and a copy moves into the hypha while a bud forms at its end
    c. The bud matures, produces one to three flagella, and a septum divides the bud from the hypha
    d. The bud is released as an oval- to pear-shaped swarmer cell
    e. Distinctive in its nutrition and physiology—grows on ethanol, acetate and one-carbon molecules such as methanol, formate, and formaldehyde
    f. May be as much as 25% of the total bacterial population in oligotrophic or nutrient-poor freshwater habitats
3. *Caulobacter*
    a. May be polarly flagellated rods or may possess prostheca and holdfast, by which they attach to solid substrata
    b. Usually found in low-nutrient freshwater and marine habitats, but are also present in soil
    c. Often adhere to bacteria, algae, and other microorganisms and may absorb nutrients released by their hosts
    d. Prostheca differs from that of *Hyphomicrobium* in that it lacks cytoplasmic components and is composed almost totally of plasma membrane and cell wall
    e. Reproduction involves formation of a single flagellum at the end opposite the prostheca; asymmetric transverse fission forms a swarmer cell that swims off; when the swarmer comes to rest, it forms a new prostheca at the flagellar end and loses the flagellum; whole cycle takes only 2 hours

E. Family *Rhizobiaceae*
1. Contains the genera *Rhizobium* and *Agrobacterium*
2. *Rhizobium*
    a. gram-negative, motile rods, often containing poly-β-hydroxybutyrate granules
    b. become pleiomorphic under adverse conditions
    c. grow symbiotically within root nodules cells of legumes as nitrogen-fixing bacteroids

3. *Agrobacterium*
   a. not capable of nitrogen fixation
   b. transforms infected plant cells into autonomously proliferating tumors
   c. *A. tumefaciens* (best studied) causes crown gall disease by means of a tumor inducing (Ti) plasmid
F. Nitrifying Bacteria
   1. *Bradyrhizobiaceae—Nitrobacter* and *Nitrococcus*
   2. *Nitrosomonadaceae—Nitrosomonas, Nitrosococcus,* and *Nitrosospira*
   3. All are aerobic, gram-negative organisms without endospores ansd able to oxidize either ammonia or nitrite
   4. May be rod-shaped, ellipsoidal, spherical, spirillar or lobate with either polar or peritrichous flagella
   5. *Nitrobacter* and *Nitrococcus* oxidize nitrite to nitrate
   6. *Nitrosomonas, Nitrosospira* and *Nitrosococcus* oxidize ammonia to nitrite
   7. If two genera such as *Nitrobacter* and *Nitrosomonas* grow together in a niche, ammonia is converted to nitrate (nitrification)
   8. Nitrate is readily used by plants but is also easily leached from the soil or denitrified to nitrogen gas
II. The Beta Proteobacteia
A. Overlap alpha-proteobacteria metabolically but tend to use substances that diffuse from organic decomposition in the anaerobic zone of habitats
B. Considerable metabolic diversity
C. Order *Neisseriales*
   1. Members of the genus *Neisseria* are nonmotile, aerobic, gram-negative cocci that most often occur in pairs with adjacent sides flattened
   2. May have capsules and fimbriae
   3. Chemoorganotrophic, oxidase-positive, and almost always catalase-positive
   4. *Neisseria gonorrhoeae*—causative agent of gonorrhea
   5. *Neisseria meningitidis*—one of the causative agents of bacterial meningitis
D. Order *Burkholderiales*
   1. The order contains four families, three of them with well-known genera
   2. *Burkholderiaceae*—contains seven genera, formed from the genus *Pseudomonas* based on rRNA data, including the genus *Burkholderia*
      a. Gram-negative, aerobic, nonfermentative, non-spore-forming, mesophilic, straight rods
      b. All but one species are motile with a single flagellum or a tuft of polar flagella
      c. Catalase-positive, often oxidase-positive
      d. Use poly-β-hydroxybutyrate as their carbon reserve
      e. *B. cepacia* is very active in recycling organic materials; is a plant pathogen; can cause disease in hospital patients due to contaminated equipment and medications; a particular problem with cystic fibrosis patients
   3. *Alcaliginaceae*—contains the genus *Bordetella*
      a. Gram-negative, aerobic coccobacilli
      b. chemoorganotroph, with respiratory metabolism, that requires organic sulfur and nitrogen (in the form of amino acids) for growth
      c. Mammalian parasite that multiplies in respiratory epithelial cells
      d. *B. pertussis*—nonmotile, encapsulated species that is the causative agent for whooping cough
   4. *Comamonadaceae*
      a. Contains twelve genera with diverse characteristics

    b. Some have a sheath, a hollow tubelike structure surrounding a chain of cells
        (1) Help bacteria attach to surfaces
        (2) Help to obtain nutrients from slowly running water as it flows past
        (3) Help protect against predators
    c. *Sphaerotilus* forms long chains and prefers slowly running freshwater polluted
        with sewage or industrial waste; can form tangled masses that interfere with
        activated sludge tanks used in sewage treatment
    d. *Leptothrix* deposits large amounts of iron and manganese oxides in its sheath; this
        seems to protect it and allows it to grow in the presence of high concentrations of
        soluble iron compounds
  E. Order *Nitrosomonadaceae*
    1. Includes the nitrifying bacteria, *Nitrosomonas, Nitrosococcus,* and *Nitrosospira*,
        discussed earlier
    2. *Gallionella*—a stalked chemolithotroph
    3. *Spirillaceae*—*Spirillum* and *Thiobacillus*; the latter is one of the best studied
        chemolithotrophs and the most prominent of the colorless sulfur bacteria
III. The Gamma Proteobacteria
  A. Largest subgroup of proteobacteria, including several deeply branching groups
  B. The 2nd edition of *Bergey's Manual* divides this group into 13 orders, 19 families, and
      around 130 genera
  C. The Purple Sulfur Bacteria
    1. Divided into two families—*Chromatiaceae* and *Ectothiorhodospiraceae* in the order
        Chromatiales
        a. *Ectothiorhodospira* has red, polarly flagellated, spiral-shaped cells that deposit
            sulfur globules externally; internal photosynthetic membranes are organized as
            lamellar stacks
        b. The family *Chromatiaceae* contains twelve genera
    2. Strict anaerobes and usually photolithoautotrophs; deposit sulfur granules internally
    3. Hydrogen may serve as an electron donor
    4. *Thiospirillum, Thiocapsa,* and *Chromatium* are typical and are usually found in
        anaerobic, sulfide-rich zones of lakes
  D. Order *Thiotrichales*
    1. Contains some of the colorless sulfur bacteria that are nonphotosynthetic, nonfruiting,
        gliding bacteria
    2. Two of the best studied gliding genera are *Beggiatoa* and *Leucothrix*
    3. *Beggiatoa*
        a. Microaerophilic and grows in sulfide-rich habitats
        b. They lack a sheath
        c. Metabolically verstile; can oxidize hydrogen sulfide to sulfur (deposited internally)
            and can oxidize sulfur to sulfate; can also grow heterotrophically with acetate as a
            carbon source and some may incorporate CO_2 autotrophically
    4. *Leucothrix*
        a. Aerobic chemoorganotroph, marine, attached to solid substrates by a holdfast
        b. Complex lifestyle in which dispersal is by formation of gonidia
        c. *Thiotrix* is a related genus that forms sheathed filaments and is chemolithotrophic;
            oxidizes hydrogen sulfide and depositis sulfur granules internally
  E. Order *Methylococcales*
    1. Contains rods, vibrios, and cocci that use methane and methanol as their sole carbon
        and energy source (methylotrophs)
    2. Contains five genera including *Methylococcus* (spherical, nonmotile) and *Methylomonas*
        (straight, curved, or branched rods with a single polar flagellum)
    3. Complex arrays of intracellular membranes

4. Methane is oxidized to methanol and then to formaldehyde which can then be assimilated into cell material

F. Order *Pseudomonadales*
1. The genus *Pseudomonas* is the most important in this order and contains straight or slightly curved rods; motile by polar flagella, lack a sheath or prosthecae
2. Aerobic chemoheterotrophs
3. Over 70 species in the genus *Pseudomonas*
4. Around 25 species have recently been removed and at least seven new genera have been formed
5. Great impact
   a. Mineralization of a wide variety of organic compounds; useful in sewage treatment
   b. Important experimental subjects
   c. Some are major animal & plant pathogens
   d. Some involved in the spoilage of refrigerated milk because they can grow at 4C and degrade lipids and proteins
6. The genus *Azotobacter* are large ovoid bacteria that are motile by peritrichous flagella; aerobic, catalase positive; fix nitrogen nonsymbiotically; widespread in soil and water

G. Order *Vibrionales*
1. Contains one family that are gram-negative, straight or curved rods with polar flagella
2. Oxidase-positive and use D-glucose as their sole or primary carbon and energy source
3. Aquatic with widespread distribution in freshwater and marine habitats
4. Five genera: *Vibrio, Photobacterium, Salinivibrio, Listonella,* and *Allomonas*
5. Pathogens include *V. cholerae*—cholera, *V. parahaemolyticus*—gastroenteritis after eating contaminated seafood, and *V. anguillarum*—a fish pathogen
6. Some, *V. fischeri* and at least two species of *Photobacterium* are among the few marine bacteria capable of bioluminescence; some bioluminescent species live symbiotically in the luminous organs of fish while others are free-living

H. Order *Enterobacteriales*
1. One family containing over 35 genera
2. Degrade sugars by Embden-Meyerhof pathway
3. The majority (e.g., *Escherichia, Proteus, Salmonella* and *Shigella*) carry out mixed acid fermentation while others (e.g., *Enterobacter, Serratia, Erwinia* and *Klebsiella*) carry out butanediol fermentation
4. Gram-negative, peritrichously flagellated or nonmotile, facultatively anaerobic, straight rods with simple nutritional requirements
5. Similarity in appearance among enteric bacteria requires the use of biochemical tests for identification
6. Rapid commercial identification systems (e.g., Enterotube, API 20-E) are based on these tests
7. *Escherichia coli* is probably the best studied bacterium and experimental organism of choice for many microbiologists; intestinal tract inhabitant; indicator organism for water quality (fecal contamination)
8. Several important human pathogens
   a. *Salmonella*—typhoid fever and gastroenteritis
   b. *Shigella*—bacillary dysentery
   c. *Klebsiella*—pneumonia
   d. *Yersinia*—plague

I. Order *Pasteurellales*
1. Small, nonmotile, normally oxidase-positive with complex nutritional requirements
2. Parasitic in vertebrates
3. Four genera: *Pasteurella, Haemophilus, Actinobacillus* and *Lonepinella*

196

4. Best known for the diseases they cause
    a. *P. multilocida*—fowl cholera
    b. *P. haemolytica*—pneumonia in cattle, sheep and goats (e.g. "shipping fever" in cattle)
    c. *H. influenzae*—major human pathogen that causes a variety of diseases, including meningitis in children

IV. The Delta Proteobacteria
  A. Small collection of genera but with diverse morphological and physiological characteristics
  B. Chemoorganotrophs placed in two orders and ten families
  C. Order *Desulfovibrionales*
    1. Gram-negative, dissimilatory sulfate- or sulfur-reducing bacteria; strict anaerobes; use elemental sulfur or sulfur compounds as electron acceptors during anaerobic respiration
    2. *Desulfovibrio*—sulfate reducing
    3. *Desulfuromonas*—uses only elemental sulfur
    4. Important in sulfur cycling
    5. Thrive in muds, polluted lake sediments, methane digesters
    6. Negative impact on industry because of their primary role in the anaerobic corrosion of iron in pipelines, heating systems, and other structures
    5. *Bdellovibrio*—gram-negative curved rods with polar flagella
      a. Preys on other gram-negative bacteria
      b. Alternates between a nongrowing predatory phase and an intracellular reproductive phase
      c. Complex lifestyle
        (1) high-speed collision with its prey
        (2) entry involves boring a hole through the host cell wall by a combination of mechanical and enzymatic action; flagellum lost during penetration
        (3) inhabits space between cell wall and plasma membrane
        (4) inhibits host DNA, RNA, and protein synthesis
        (5) disrupts plasma membrane so cell constituents leak out
        (6) grows into long filament and then divides into many smaller flagellated progeny
  D. Order *Myxococcales*
    1. The myxobacteria are gram-negative, aerobic soil bacteria with gliding motility
    2. Form fruiting bodies and dormant myxospores
    3. Divided into four families
    4. Micropredators or scavengers that lyse bacteria and yeasts by secretion of an array of digestive enzymes
    5. Use the released peptides and amino acids as primary carbon, nitrogen, and energy source
    6. Lifestyle resembles that of cellular slime molds
    7. Aggregate and differentiate into fruiting bodies when nutrient supply is exhausted
    8. Myxospores are frequently enclosed in walled structures called sporangioles or sporangia
      a. dormant
      b. desiccation-resistant
      c. fruiting bodies aid dispersal of myxospores
    9. Colonial lifestyle aids digestion by providing higher enzyme concentration than any individual bacterium could do
    10. Found in neutral soils, decaying plant material, and animal dung
    11. Most abundant in warm areas but will grow in the arctic tundra

V. The Epsilon Proteobacteria
   A. Smallest of the proteobacteria groups
   B. Slender, gram-negative rods that can be straight, curved, or helical
   C. One order with two families and two important genera
      1. *Campylobacter*—contains both pathogenic and nonpathogenic species
         a. *C. fetus*—reproductive disease and abortions in cattle and sheep; can cause septicemia and enteritis in humans
         b. *C. jejuni*—causes abortion in sheep and enteritis diarrhea in humans
      2. *Helicobacter*
         a. isolated from stomachs and upper intestines of humans, dogs, cats and other mammals
         b. *H. pylori*—cause of gastritis and peptic ulcer disease; produces large quantities of urease and urea hydrolysis appears to be associated with their virulence.

TERMS AND DEFINITIONS

Place the letter of each term in the space next to the definition or description that best matches it.

___ 1.	A nonliving appendage produced by a cell and extending from it.	a. methylotroph
___ 2.	An extension of the cell, including the plasma membrane and cell wall that is narrower than the cell.	b. prostheca
___ 3.	A hollow tubelike structure that surrounds a chain of cells.	c. sheath
___ 4.	An organism that uses methane or methanol as its sole carbon and energy source.	d. stalk

THE PROTEOBACTERIA AND THE DISEASES THEY CAUSE

Match the proteobacterial species listed and the primary disease caused by each of them.

___ 1.	bacterial dysentery	a.	*Bordetella pertussis*
___ 2.	bubonic plague	b.	*Campylobacter jejuni*
___ 3.	cholera	c.	*Coxiella burnettii*
___ 4.	enteritis diarrhea	d.	*Helicobacter pylori*
___ 5.	gastroenteritis	e.	*Klebsiella pneumoniae*
___ 6.	gonorrhea	f.	*Neisseria gonorrhoeae*
___ 7.	meningitis	g.	*Neisseria meningitidis*
___ 8.	peptic ulcer disease	h.	*Rickettsia rickettsii*
___ 9.	pneumonia	i.	*Salmonella typhi*
___ 10.	Q-fever	j.	*Shigella dysenteriae*
___ 11.	Rocky Mountain spotted fever	k.	*Vibrio cholerae*
___ 12.	typhoid fever	l.	*Vibrio parahaemolyticus*
___ 13.	whooping cough	m.	*Yersinia pestis*

FILL IN THE BLANK

1. *Rickettsia* and *Coxiella* both enter the cell by _____. However, _____ escape from the phagosome before reproducing by binary fission, while _____ remains in the phagosome after fusion with a lysosome and reproduces within the resulting structure.
2. *Hyphomicrobium* are chemoheterotrophic, aerobic, budding bacteria that frequently attach to solid objects in _____, _____, and _____ environments.
3. Reproduction of the genus *Caulobacter* involves _____ transverse fission to form a _____ cell that swims off. When this cell comes to rest, it loses its _____ and forms a new _____. The whole process takes only about _____ hours.
4. Members of the order *Desulfovibrionales* are gram-negative, disimilatory sulfate- or sulfur-reducing bacteria that use _____ sulfur or sulfur _____ *as acceptors during* _____ respiration.
5. *Bdellovibrio* preys on other _____ bacteria. They have a complex lifestyle that involves their inhabiting the space between the _____ and _____ of their host.
6. The _____ lifestyle of the myxobacteria aids digestion of other bacteria and yeasts by providing a _____ enzyme concentration than could be attained by any individual bacterium.

MULTIPLE CHOICE

For each of the questions below select the best answer or answers.

1. Members of the genus *Pseudomonas* do which of the following?
   a. mineralization of organic compounds
   b. spoilage of refrigerated milk
   c. Both (a) and (b) are correct.
   d. Neither (a) nor (b) is correct.

2. Mineralization means which of the following?
   a. breakdown of organic materials to inorganic materials
   b. release of various minerals from various ores
   c. use of minerals as energy sources
   d. None of the above is correct.

3. *Hyphomicrobium* may constitute as much as _____ of the total bacterial population in oligotrophic (nutrient poor) freshwater habitats:
   a. 10%
   b. 25%
   c. 40%
   d. 60%

4. The process of converting ammonia to nitrate is referred to as:
   a. nitrogen fixation
   b. nitrification
   c. ammonification
   d. denitrification

5. The sheath of the *Comamonadaceae* performs which of the following functions?
   a. helps bacteria attach to surfaces
   b. helps obtain nutrients from slowly running water as it flows past
   c. helps protect against predators
   d. All of the above are correct.

6. The largest subgroup of the proteobacteria is the:
   a. alpha proteobacteria
   b. beta proteobacteria
   c gamma proteobacteria
   d. delta proteobacteria

7. Which of the following is true of *Beggiatoa*.
   a. They can oxidize hydrogen sulfide to sulfur
   b. They can oxidize sulfur to sulfate
   c. They can grow heterotrophically with acetate as their carbon source
   d. All of the above are true of *Beggiatoa*.

8. Which of the following genera contain bacteria that are capable of bioluminescence?
   a. *Vibrio*
   c. *Photobacteria*
   c. Both (a) and (b) are correct.
   d. Neither (a) nor (b) is correct.

TRUE/FALSE

___ 1. The enterobacteria, in the order *Enteriobacteriale*s are widely diverse in appearance; however, they can be easily distinguished from one another by morphological criteria.

___ 2. The purple nonsulfur bacteria are so named because they do not oxidize elemental sulfur to sulfide.

___ 3. Rickettsias do not use glucose as an energy source.

___ 4. *Escherichia coli* is not a good indicator of fecal contamination of water supplies because it cannot be readily detected.

___ 5. Members of the *Desulfovibrionales* have a positive impact on industry because they protect iron-containing pipelines fron corrosion.

___ 6. Myxobacteria lyse bacteria and yeasts by secretion of digestive enzymes. They then use the resulting peptides and amino acids as their primary carbon, nitrogen, and energy source.

CRITICAL THINKING

1. At one time rickettsias, like the chlamydiae, were thought to be viruses rather than bacteria. What was the basis for this misconception? Explain why it is a misconception.

2. Explain the basis of rapid commercial identification systems such as the Enterotube or the API 20-E system. Why was it necessary and desirable to develop these types of rapid identification systems?

ANSWER KEY

Terms and Definitions

1. d, 2. b, 3. c, 4.a

The Proteobacteria and the Diseases They Cause

1. j, 2. m, 3. k, 4. b, 5. l, 6. f, 7. g, 8. d, 9. e, 10. c, 11. h, 12. i, 13. a

Fill in the Blank

1. phagocytosis; *Rickettsia; Coxiella*; lysosome 2. freshwater; marine; terrestrial 3. asymmetric; swarmer; flagellum; prostheca; two 4. elemental; compounds; electron; anaerobic 5. cell wall; plasma membrane 6. colonial; higher

Multiple Choice

1. c, 2. a, 3. b, 4. c, 5. d, 6. c, 7. d, 8. c

True/False

1. F, 2. T, 3. T, 4. F, 5. F, 6. T

23   Bacteria: The Low G + C Gram-Positives

CHAPTER OVERVIEW

This chapter describes the different approaches to Gram-positive organisms taken by the 1st and 2nd editions of *Bergey's Manual* and then focuses on the mycoplasmas, *Clostridium* and its relatives, and the bacilli and lactobacilli.

CHAPTER OBJECTIVES

After reading this chapter you should be able to:

- discuss the difference in classification of gram-positives that is used in the 1st and 2nd editions of *Bergey's Manual*
- discuss the variation in peptidoglycan structure that is useful in identifying specific groups
- discuss the various roles of these organisms: harmless, free-living saprophytes, pathogens, importance in the food and dairy industry

CHAPTER OUTLINE

I. Class *Mollicutes* (The Mycoplasmas)
    A. The new edition will have one order and three families
    B. Mycoplasmas lack cell walls and cannot synthesize peptidoglycan precursors
        1. Penicillin resistant
        2. Susceptible to lysis by osmotic shock and detergent treatment
    C. Smallest bacteria capable of self-reproduction
    D. Most are nonmotile but some can glide along liquid-covered surfaces
    E. Most species require sterols (unusual for bacteria)
    F. Usually facultative anaerobes but a few are strict anaerobes
    G. G + C content ranges from 23 to 41%
    H. Can be saprophytes, commensals or parasites
    I. Recently, the complete genome of *M. genitalium* has been sequenced
    J. Metabolism is not particularly unusual although they are deficient in several biosynthetic pathways
    K. Widespread—can be isolated from plants, animals, the soil and even compost piles
    L. Serious contaminant of mammalian cell cultures; difficult to detect; difficult to eliminate
    M. In animals they colonize mucous membranes and joints and are often associated with diseases of the respiratory and urogenital tracts
    N. Pathogenic species include:
        1. *M. mycoides*—bovine pleuropneumonia in cattle
        2. *M. gallisepticum*—chronic respiratory disease in chickens
        3. *M. pneumoniae*—primary atypical pneumonia in humans

4. *M hominis* and *Ureaplasma urealyticum*—pathogenic in humans
5. Spiroplasmas—pathogenic in insects, ticks, and a variety of plants

II. Low G + C Gram Positive Bacteria in *Bergey's Manual*
   A. First edition treats low G + C gram positives phenotypically
      1. Classified on the basis of cell shape, clustering and arrangement of cells, presence or absence of endospores, oxygen relationships, fermentation patterns, peptidoglycan chemistry, etc.
      2. Peptidoglycan structure varies considerably
         a. Some contain meso-diaminopimelic acid cross-linked through its free amino group to the carboxyl group of the terminal D-alanine of the adjacent chain
         b. Others contain lysine cross-linked by interpeptide bridges
         c. Others contain L,L-diaminopimelic acid and have one glyvcine as the interpeptide bridge
         d. Pathogenic corynebacteria use ornithine to cross-link between positions 2 and 4 of the peptide chains rather than positions 3 and 4 as used by the other forms
         e. Other cross-links and differences in cross-link frequency also contribute to variation in structure
         f. These variations are characteristic of particular groups and are therefore taxonomically useful
      3. Bacterial endospores are complex structures that allow survival under adverse conditions
         a. Recently viable endospores have been found that may be over 25 million years old
         b. Sporeformers are distributed widely but found mainly in soil
   B. Second edition takes a phylogenetic approach dividing the low G + C gram positives into two groups: the clostridia and relatives, and the bacilli and lactobacilli; sporeformers are found in both groups

III. The Clostridia and Relatives
   A. Contains one class with two orders and at least eight families
   B. The largest genus is *Clostridium*
      1. Obligate anaerobes, sporeformers, do not carry out dissimilatory sulfate reduction
      2. Over 100 species in distinct phylogenetic clusters
      3. Responsible for many cases of food spoilage, even in canned foods (e.g., *C. botulinum*)
      4. Other major disease causing clostridia include:
         a. *C. perfringens*—gas gangrene
         b. *C. tetani*—tetanus
      5. Some are of industrial value (e.g., *C. acetobutylicum*—used to manufacture butanol)
   C. *Epulopiscium*—giant bacteria
   D. *Desulfotomaculum*
      1. Anaerobic, endospore-forming genus that reduces sulfate and sulfite to hydrogen sulfide during anaerobic respiration
      2. Stains gram-negative (placed with gram-negatives in 1st edition) but actually has a gram-positive type cell wall with a lower than normal peptidoglycan content
   E. *Heliobacterium* and *Heliophilum*—are photosynthetic using bacteriochlorophyll g, but no intracytoplasmic photosynthetic membranes (pigments are in the plasma membrane)
   F. *Veillonella*
      1. Anaerobic, chemoheterotrophic cocci with complex nutritional requirements;
      2. Parasites of homeothermic animals
      3. Part of the normal microflora of the mouth, the gastrointestinal tract, and urogenital tract of humans and other animals
      4. Classified with gram-negatives in the 1st edition but are actually more closely related to low G + C gram-positives

IV. The Bacilli and Lactobacilli
   A. Order *Bacilliales*
      1. *Bacillus*
         a. Largest genus in the order
         b. Gram-positive, endospore-forming, chemoheterotrophic rods
         c. Usually motile with peritrichous flagella
         d. Usually aerobic, sometimes facultative, and catalase positive
         e. Some new genera have been created: *Alicyclobacillus* and *Paenibacillus*
         f. Many species are of considerable importance
            (1) Some produce antibiotics such as bacitracin, gramicidin, and polymyxin
            (2) *B. cereus* causes food poisoning
            (3) *B. anthracis* causes anthrax
            (4) *B. thuringiensis* and *B. sphaericus* are used as insecticides
      2. *Thermoactinomyces*
         a. Thermophilic; forms single spores on both its aerial and substrate mycelia
         b. Commonly found in damp haystacks, compost piles, and other high-temperature habitats
         c. Very heat-resistant endospores—can survive 90°C for 30 minutes
         d. *T. vulgaris*—causative agent for farmer's lung disease, an allergic respiratory disease in agricultural workers
      3. *Caryophanon*—strict aerobe, catalase positive, motile by peritrichous flagella
      4. *Staphylococcus*
         a. Facultatively anaerobic, catalase positive, oxidase negative
         b. Ferment glucose anaerobically
         c. Normally associated with skin, skin glands, and mucous membranes
         d. *S. epidermidis*—skin resident; sometimes responsible for endocarditis and for infections of patients with lowered resistance (wound infections, surgical infections, and urinary tract infections)
         e. *S. aureus*—found on nasal membranes, skin, and in the gastrointestinal and urinary tracts; associated with boils, abscesses, wound infections, pneumonia, toxic shock syndrome and food poisoning; produces coagulase which causes blood to clot
   B. Order *Lactobacilliales*
      1. Largest genus is *Lactobacillus* with nearly 80 species
      2. Can be rods, and sometimes coccobacilli, that lack catalase and cytochromes, are usually facultative or microaerophilic, produce lactic acid, and have complex nutritional requirements
      3. *Lactobacillus*
         a. Can carry out heterolactic or homolactic acid fermentation
         b. Grow optimally between pH 4.5 and pH 6.4
         c. Found on plant surfaces and in dairy products, meat, water, sewage, beer, fruits, and many other materials
         d. Normal microflora of mouth, intestinal tract, and vagina; usually not pathogenic
         e. Used in the production of fermented vegetable foods, beverages, sour dough, hard cheeses, yogurt, and sausages
         f. Responsible for spoilage of beer, milk, and meat
      4. *Leuconostoc*
         a. Facultatively anaerobic; elongated or elliptical shape; clustered in pairs or chains
         b. Lack catalase or cytochromes; heterolactic fermentation
         c. Important in wine production, fermentation of vegetables such as cabbage and cucumbers, manufacture of buttermilk, butter and cheese; tolerate high sugar concentrations and, therefore, grow in heavy syrup

5. *Streptococcus*
   a. Pairs or chains in liquid media; does not form endospores, nonmotile
   b. Homolactic fermentation; produces lactic acid but no gas
   c. Most are facultative anaerobes; a few are obligate anaerobes
   d. Distinguished by hemolysis reactions
      (1) α-hemolysis—incomplete with greenish zone
      (2) β-hemolysis—complete with clear zone—no greening
   e. Also distinguished serologically and by a variety of biochemical and physiological tests
   f. Genus is large with 38 species in four groups: pyogenic streptococci, oral streptococci, anaerobic streptococci, and other streptococci
   g. Many bacteria originally in this genus have been moved to the genus *Enterococcus* (18 species) or to the genus *Lactococcus* (8 species)
   h. Species of practical importance include:
      (1) *S. pyogenes*—causes streptococcal sore throat, acute glomerulonephritis and rheumatic fever
      (2) *S. pneumonia*—causes lobar pneumonia
      (3) *S. mutans*—associated with dental caries
      (4) *E. faecalis*—opportunistic pathogen that can cause urinary tract infections and endocarditis
      (5) *L. Lactis*—used in the production of buttermilk and cheese
6. *Listeria*—aerobic or facultative, catalase positive, peritrichously flagellated; *L. monocytogenes* is a human pathogen that causes listeriosis, an important food infection

TERMS AND DEFINITIONS

Place the letter of each term in the space next to the definition or description that best matches it.

_____ 1. An enzyme that catalyzes the breakdown of hydrogen peroxide to water and oxygen

_____ 2. An enzyme that causes blood plasma to clot

_____ 3. Incomplete lysis of red blood cells with formation of a greenish ring around the bacterial colony

_____ 4. Complete lysis of red blood cells with no greenish zone formed

a. α-hemolysis
b. β-hemolysis
c. catalase
d. coagulase

FILL IN THE BLANK

1. *Staphylococcus* _____ is a common skin resident that is sometimes responsible for endocarditis and for infections of patients with lowered resistance. The most important human staphylococcal human pathogen, however, is _____, which is associated with boils, abscesses, wound infections, pneumonia, toxic shock syndrome, and food poisoning. The latter organism differs from other staphylococcal species in that it is positive for the enzyme _____, which causes blood plasma to clot.

2. One way of distinguishing the streptococci is by determining their effect on red blood cells. Some are _____ and only partially lyse red blood cells, leading to the formation of a _____ ring around the growing colony. Others are _____, which is characterized by a zone of complete killing or lysis without a marked color change.

3. Lactobacilli are not normally pathogenic but are found as normal inhabitants of the _____, the _____, and the _____.

4. Members of the genera _____ and _____ are photosynthetic bacteriochlorophyll *g*. However, they have no _____ photosynthetic membranes. Instead the photosynthetic pigments are located in the _____ membrane.

5. Members of the genus *Veillonella* are part of the normal microflora of the _____, the _____ tract and the _____ tract of humans and other animals.

6. *Thermoactinomyces vulgaris* is commonly found in damp _____ and _____ piles. It is the causative agent for _____ disease, an allergic respiratory disease in agricultural workers.

MULTIPLE CHOICE

For each of the questions below select the *one best* answer.

1. Which of the following does not involve a contribution by members of the genus *Leuconostoc?*
    a. fermentation of vegetables like cabbage (sauerkraut) and cucumbers (pickles)
    b. manufacturing of buttermilk, butter, and cheese
    c. manufacturing of beer
    d. All of the above involve a contribution by *Leuconostoc.*

2. Which of the following can be attributed to members of the genus *Bacillus?*
    a. They produce certain useful antibiotics.
    b. They are the causative agents of some food poisonings.
    c. They have been used as a biological insecticide.
    d. All of the above are properties attributed to members of the genus *Bacillus.*

206

3. Which of the following is caused by *Clostridium perfringens*?
   a. gas gangrene
   b. tetanus
   c. botulism
   d. None of the above are caused by this organism.
4. Which of the following involve(s) lactobacilli in their production?
   a. sauerkraut and pickles
   b. hard cheeses
   c. yogurt
   d. All of the above involve lactobacilli in their production.
5. Which of the following statements is correct?
   a. *Streptococcus lactis* has been renamed *Lactococcus lactis*.
   b. *Streptococcus faecalis* has been renamed *Enterococcus faecalis*.
   c. Both (a) and (b) are correct.
   d. Neither (a) nor (b) is correct.
6. The complete genome sequence of which of the following organisms has been determined?
   a. *Mycobacterium leprae*
   b. *Mycoplasma genitalium*
   c. *Micrococcus luteus*
   d. None of the above
7. Which of the following is a reason why *Mycoplasmas* are a major problem when working with mammalism cell cultures?
   a. They are difficult to detect.
   b. They are difficult to eliminate.
   c. Both (a) and (b) are correct.
   d. Neither (a) nor (b) is correct.
8. Primary atypical pneumonia in humans is caused by which of the following?
   a. *Streptococcus pneumoniae*
   b. *Mycoplasma pneumoniae*
   c. *Clostridium pneumoniae*
   d. None of the above

TRUE/FALSE

____ 1. When the members of the gram-positive cocci are examined both phenetically and phyletically, it is found that the two ways of classifying these organisms show a close match.
____ 2. The streptococci carry out homolactic fermentation, producing lactic acid but no gas.
____ 3. *Leuconostoc* will tolerate a higher sugar content (unlike most organisms) and is, therefore, a major problem in sugar refineries. It can also cause food spoilage of foods that are baked in heavy syrup.
____ 4. Lactobacilli are alkalophilic and prefer conditions that are slightly alkaline for optimal growth.
____ 5. Mycoplasmas are serious contaminants of mammalian cell cultures.
____ 6. *Clostridium botulinum* is responsible for many cases of food spoilage even in canned foods.

CRITICAL THINKING

1. In the production of yogurt, two organisms are used. The first to grow is *Streptococcus thermophilus* and the second is *Lactobacillus bulgaricus*. This is true even though the two are inoculated into the milk simultaneously. Explain. In your explanation be sure to consider the properties of lactobacilli and how these may contribute to the sequence of events.

2. Tetanus, which is caused by *Clostridium tetani,* is of serious concern in deep puncture wounds. However, it is seldom a problem with surface lacerations. Using your knowledge of the properties of the genus *Clostridium,* explain these observations.

ANSWER KEY

Terms and Definitions

1. c, 2. d, 3. a, 4. b

Fill in the Blank

1. *epidermidis; S. aureus;* coagulase 2. α-hemolytic; greenish; β-hemolytic 3. mouth; intestinal tract; vagina 4. *Heliobacterium; Heliophilum,* intracytoplasmic; plasma 5. mouth; gastrointestinal; urogenital 6. haystacks; compost; farmer's lung

Multiple Choice

1. c, 2. d, 3. a, 4. d, 5. c, 6. b, 7. c, 8. b

True/False

1. F, 2. T, 3. T, 4. F, 5. T, 6. T

24 Bacteria: The High G + C Gram-Positives

CHAPTER OVERVIEW

This chapter surveys the general characteristics of the actinomycetes and other organisms that are classified as high G + C gram-positives in the 2nd edition of *Bergey's Manual*. The actnomycetes are filamentous bacteria that form branching hyphae and asexual spores.

CHAPTER OBJECTIVES

After reading this chapter you should be able to:

- describe the filamentous actinomycetes
- describe the morphology and arrangement of spores, cell wall chemistry, and the types of sugars present in cell extracts within the actinomycetes
- discuss the roles of actinomycetes in the mineralization of organic compounds and in the production of antibiotics
- describe the important human pathogens contained in the genera *Corynebacterium* and *Mycobacterium*

CHAPTER OUTLINE

I.   General Properties of the Actinomycetes
    A.   Form substrate mycelia
    B.   Septa divide the mycelia into long cells (20 um and longer), each containing several nucleoids
    C.   Some form a tissue-like mass called a thallus
    D.   They may have aerial mycelia that for conidospores on the end of the filament, or that form sporangiospores within a sporangium
    E.   Spores are not heat resistant but withstand desiccation well
    F.   They are generally nonmotile, but the spores may be flagellated
    G.   Cell wall types vary
    H.   Cell wall type, sugars in extracts, morphology and color of mycelia and sporangia, G + C content, membrane phospholipid composition, and spore heat resistance are all important in classifying these organisms
    I.   Comparison of 16S rRNA sequences and pulse-field electrophoresis of large DNA fragments produced by restriction endonuclease digestion are also used for classification purposes
    J.   Widely distributed in soils; degrade a number of organic compounds; important in mineralization processes
    K.   They produce most of the medically important, naturally synthesized antibiotics
    L.   A few species are pathogenic in humans, other animals and plants

II. High G + C Gram-Positive Bacteria in *Bergey's Manual*
  A. The 2nd edition will reclassify the high G + C gram positives using 16S rRNA sequences
  B. Many genera will be grouped with the actinomycetes in the class *Actinobacteria* which will be divided into orders, suborders and families
  C. This chapter will focus on the order *Actinomycetales* which is divided into 10 suborders
  D. The order *Bfidobacteriales* also will be briefly described
III. Suborder *Actinomycineae*
  A. Contains the genera *Actinomyces*, *Arcanobacterium,* and *Mobiluncus*
  B. Straight or slightly curved rods and slender filaments with true branching
  C. Facultative or obligate anaerobes; require CO_2 for best growth
  D. Cell walls contain lysine but not diaminopimelic acid
  E. *A. bovis* causes lumpy jaw in cattle
  F. Involved in ocular disease and periodontal disease in humans
IV. Suborder *Micrococcineae*
  A. Nine genera with many species
  B. *Micrococcus*
    1. Catalase-positive cocci that occur in pairs, tetrads or irregular clusters
    2. Usually nonmotile
    3. Often, yellow, red, or orange pigmented
    4. Widespread in soil, water, and on mammalian skin
    5. Not usually pathogenic
  C. *Arthrobacter*
    1. Aerobic, catalase-positive rods with respiratory metabolism and lysine in its peptidoglycan
    2. When growing in exponential phase they are rods that reproduce by a snapping division
    3. In stationary phase they change to a coccoid form
    4. Most important habitat is soil
  D. *Dermatophilus*
    1. Forms packets of motile spores with tufts of flagella
    2. Facultative anaerobe
    3. Mammalian parasite responsible for a skin infection called streptothrichosis
V. Suborder *Corynebacterineae*
  A. Contains five families with several important genera
  B. *Corynebacterium*
    1. Aerobic and facultative species; catalase-positive; straight to slightly curved rods, often with tapered ends
    2. Remain partially attached after snapping division resulting in angular arrangements
    3. Form metachromatic granules
    4. Cell walls contain meso-diaminopimelic acid
    5. Some species are harmless soil and water saprophytes
    6. Many are animal and human pathogens (e.g., *C. diphtheriae*—causative agent of diphtheria in humans
  C. *Mycobacterium*
    1. Straight or slightly curved rods
    2. Aerobic and catalase-positive; grow slowly
    3. Cell walls contain waxes with 60-90 carbon mycolic acids—makes them acid-fast (i.e. basic fuchsin dye cannot be removed with acid-alcohol treatment
    4. Some are free-living saprophytes
    5. They are best known as human and animal pathogens
      a. *M. bovis*—tuberculosis in cattle and other ruminants
      b. *M. tuberculosis*—tuberculosis in humans
      c. *M. leprae*—causes leprosy in humans

210

D. *Nocardia*
    1.   Develops a substrate mycelium that readily breaks into rods and coccoid elements
    2.   Some develop aerial mycelia
    3.   They are found in soil and aquatic habitats
    4.   Involved in biodegradation of rubber joints in water and sewage pipes
    5.   Some species (e.g., *N. asteroides*) are opportunistic pathogens causing nocardiosis—lungs are most often affected but the central nervous system and other organisms can be involved

VI. Suborder *Micromonosporineae*
    A.   Referred to as Actinoplanetes
    B.   Extensive substrate mycelia; aerial mycelia are absent or rudimentary
    C.   Form conidiospores within a sporangium that extends above the surface of the substratum
    D.   Genera vary in arrangement and development of spores
    E.   Found in soil and freshwater habitats and occasionally in the ocean
    F.   Soil dwellers play an important role in plant and animal decomposition
    G.   Some produce antibiotics such as gentamicin

VII. Suborder *Propionibacterineae*
    A.   The genus *Propionibacterium* contains pleiomorphic, nonmotile rods that are often club shaped; cells may also be coccoid or even branched; single cells, short chains, or in clumps
    B.   Facultatively anaerobic or aerotolerant; ferment sugars to produce propionic acid
    C.   Found on skin and in the digestive tract of animals; also in dairy products such as cheese
    D.   Contributes to the production of Swiss cheese
    E.   *P. acne* is involved in the development of body odor and acne vulgaris

VIII. Suborder *Streptomycineae*
    A.   Only one genus, *Streptomyces*
    B.   An enormous genus with around 500 species
    C.   Aerial mycelia divide in a single plane to form chains of nonmotile conidiospores
    D.   Ecologically and medically important
    E.   Natural habitat is soil where they represent from 1-20% of the organisms present
    F.   Impart the characteristic odor of moist earth by producing volatile substances such as geosmin
    G.   Metabolically flexible; major contributors to mineralization
    H.   Best known for the synthesis of a vast array of antibiotics useful in medicine and research
    I.   Only *S. somaliensis* is known to be pathogenic in humans; causes actinomycetoma, an infection of subcutaneous tissues that produces swelling, abscesses and even bone destruction

IX. Suborder *Streptosporangineae*
    A.   Contains 3 families and 12 genera
    B.   *Maduromycetes*
        1.   They have the sugar madurose (3-O-methyl-D-galactose) in cell extracts
        2.   Aerial mycelia produce pairs or short chains of spores and the substrate mycelia are branched
        3.   Some genera form sporangia
    C.   *Thermomonospora*
        1.   Show considerable variation in morphology and lifestyle
        2.   Produce single spores on the aerial mycelium or on both the aerial and the substrate mycelium
        3.   *Nocardiopsis* has a substrate mycelium that fragments like the nocardioforms but it has a different cell wall chemistry

X. Suborder *Frankineae*
    A. The genus *Geodermatophilus* has motile spores and is an aerobic soil organism
    B. The genus *Frankia*:
        1. Forms nonmotile sporangiospores in a sporogenous body
        2. Grows in symbiotic relationship with at least 8 families of higher nonleguminous plants
        3. Microaerophilic and able to fix atmospheric nitrogen
    C. The genus *Sporichthya* lacks a substrate mycelium but use holdfasts to anchor to the substratum; grow upward to form aerial mycelia that release motile, flagellated conidia in the presence of water
XI. Order *Bfidobacteriales*
    A. Contains one family and 3 genera
    B. *Falcivibrio* and *Gardnerella* are found in the human genitourinary tract; Gardnerella may be a major cause of vaginitis
    C. *Bfidobacterium* is best studied
        1. Nonmotile, nonsporing, gram-positive rods of varied shapes taht are slightly curved and clubbed; often they are branched
        2. Rods can be single cells, in clusters or in V-shaped pairs
        3. They are anaerobic and ferment lactose to produce acetic and lactic acids but no carbon dioxide
        4. Found in the mouth and intestinal tract of warm-blooded animals, in sewage, and in insects
        5. *B.bfidus* is a pioneer colonizer of the human intestinal tract, particularly when babies are breast fed
        6. Some infections of humans have been reported but does not appear to be a major cause of disease

TERMS AND DEFINITIONS

Place the letter of each term in the space next to the definition or description that best matches it.

_____ 1. A tissuelike mass of cells formed by actinomycetes
_____ 2. Asexual spores held on the ends of filaments
_____ 3. Asexual spores located within a structure at the end of the filament
_____ 4. A volatile substance produced by members of the genus *Streptomyces* that imparts the characteristic odor of moist earth
_____ 5. Branching network of filaments that rises above the agar growth medium
_____ 6. Branching network of filaments below the surface of the agar medium
_____ 7. A means of separating large DNA fragments in an electric field.

a. aerial mycelia
b. conidiospores
c. geosmin
d. pulse-field electrophoresis
e. sporangiospores
f. substrate mycelia
g. thallus

FILL IN THE BLANK

1. The actinomycetes usually grow on and into the substratum to form _____ mycelia. Asexual conidia, or _____, are held on the ends of filaments of _____ mycelia that sometimes form. These spores are not resistant to _____ but are resistant to _____, and therefore, these organisms are considerably adaptive.
2. Soil-dwelling members of the *Micromonosporineae* (actinoplanetes) may have an important role in the decomposition of _____ and _____ material.
3. Streptomycetes are important both _____ and _____. They play a major role in the _____ of organic material and are probably best known for the synthesis of _____, which are useful in medicine and research.
4. Members of the genus *Nocardiopsis* have a mycelium that _____ like the nocardioforms. However they have a different cell wall chemistry.
5. Members of the suborder *Actinomycineae* are involved in _____ disease and _____ disease in humans.
6. Members of the genus *Mycobacterium* are said to be _____ (i.e., basic fuchsin dye cannot be removed with acid-alcohol treatment). This is because the cell wall contains _____ with 60 to 90 carbon _____ acids.
7. Members of the genus *Sporichthya* lack _____ mycelia. Instead, they use _____ to anchor to the substratum.

MULTIPLE CHOICE

For each of the questions below select the *one best* answer.

1. Which of the following is *not* a reason for studying the actinomycetes?
   a. They contribute to the mineralization of organic material.
   b. They produce the majority of useful, naturally synthesized antibiotics.
   c. Some members are pathogenic.
   d. All of the above are reasons for studying actinomycetes.

2. *Micromonosporineae* produce which of the following antibiotics?
   a. micromycin
   b. gentamicin
   c. penicillin
   d. streptomycin

3. Members of the genus *Micrococcus* are frequently pigmented which of the following colors?
   a. red
   b. yellow
   c. orange
   d. All of the above.

4. Which of the following is correct about the genus *Arthrobacter*?
   a. They are rod-shaped during exponential growth.
   b. They are coccoid in stationary phase.
   c. Both (a) and (b) are correct.
   d. Neither (a) nor (b) is correct.

5. Diptheria is caused by a member of which of the following genera?
   a. *Corynebacterium*
   b. *Mycobacterim*
   c. *Propionibacterium*
   d. None of the above

6. Which of the following diseases is(are) caused by members of the genus *Mycobacterium*?
   a. tuberculosis
   b. leprosy
   c. Both (a) and (b) are correct.
   d. Neither (a) nor (b) is correct.

7. Which of the following is correct about members of the genus *Propionibacterium*?
   a. They contribute to the production of Swiss cheese.
   b. They cause acne vulgaris and contribute to the development of body odor.
   c. Both (a) and (b) are correct.
   d. Neither (a) nor (b) is correct.

TRUE/FALSE

_____ 1. Most actinomycetes are not motile, but when present, motility is confined to flagellated spores.

_____ 2. The actinomycetes differ from the nocardioforms in that the filaments of the actinomycetes do not readily fragment into rods and coccoid elements as the nocardioform filaments do.

_____ 3. No streptomycetes are known to be pathogenic for humans.

_____ 4. In addition to being opportunistic pathogens, some species of Nocardia cause biodeterioration of rubber joints in water and sewer pipes.

_____ 5. *Streptomyces* can constitute up to 20% of the organisms found in the soil.

_____ 6. Of the nearly 500 species in the genus *Streptomyces*, only *S. somaliensis* is known to be pathogenic in humans.

CRITICAL THINKING

1. Actinomycetes and nocardioform bacteria both form filaments. However, they differ in a number of ways, both morphologically and in terms of their practical characteristics. Discuss these differences and their significance.

2. Discuss the major ecological and medical contributions of the genus *Streptomyces*.

ANSWER KEY

Terms and Definitions

1. g, 2. b, 3. e, 4. c, 5. a, 6. f, 7. d

Fill in the Blank

1. substrate; conidiospores; aerial; heat; desiccation 2. plant; animal 3. ecologically; medically; mineralization; antibiotics 4. substrate; fragments 5. ocular; periodontal 6. acid-fast; waxes; mycolic 7. substrate; holdfasts

Multiple Choice

1. d, 2. b, 3. d, 4. c, 5. a, 6. c, 7. c

True/False

1. T, 2. T, 3. F, 4. T, 5. T, 6. T

25 The Fungi (Eumycota), Slime Molds, and Water Molds

CHAPTER OVERVIEW

This chapter discusses the characteristics of the members of the kingdom *Fungi*. The diversity of these organisms is described, and their ecological and economic impact is discussed. In addition, certain protists—the slime molds and water molds, which resemble fungi—are also presented in this chapter.

CHAPTER OBJECTIVES

After reading this chapter, you should be able to:

- discuss the distribution of fungi and their roles in the environment
- discuss the morphological characteristics of the fungi
- describe the external digestion of organic matter by fungi
- explain the formation of both asexual and sexual spores for reproduction
- discuss the five major types of organisms in this kingdom—zygomycetes, ascomycetes, basidiomycetes, deuteromycetes, and chytrids—and the basis upon which fungi are assigned to these categories
- discuss the slime molds and water molds, and their resemblance to fungi, even though they are phylogenetically distinct

CHAPTER OUTLINE

   I.   Introduction
       A.   Fungi—eucaryotic, spore-bearing organisms with absorptive metabolism and no chlorophyll that reproduce sexually and asexually
       B.   Mycologists—scientists who study fungi
       C.   Mycology—the study of fungi
       D.   Mycotoxicology—the study of fungal toxins and their effects on various organisms
       E.   Mycoses—diseases in animals caused by fungi
       F.   Belong to the kingdom *Fungi* within the domain Eucarya
       G.   The kingdom is a monophyletic group known as the *eumycota* (true fungi)
  II.   Distribution
       A.   Primarily terrestrial with a few freshwater and marine organisms
       B.   Many are pathogenic in plants or animals
       C.   Form associations with plant roots (mycorrhizae) or with algae or cyanobacteria (lichens)
 III. Importance
       A.   Beneficial
           1.   Decomposers—break down organic material and return it to environment
           2.   Industrial fermentation—bread, wine, beer, cheese, tofu, soy sauce, steroid manufacture, antibiotic production, and the production of the immunosuppressive drug cyclosporine
           3.   Research—fundamental biological processes can be studied in simple eucaryotic organism

B. Detrimental
  1. Major cause of plant diseases
  2. Cause of many animal and human diseases
IV. Structure
  A. Thallus—body or vegetative structure of a fungus
  B. Chitin—nitrogen-containing polysaccharide consisting of N-acetyl glucosamine residues; found in the cell wall
  C. Yeast—unicellular fungus with single nucleus; reproduces asexually by budding, or sexually by spore formation; daughter cells may separate after budding or may aggregate to form colonies
  D. Mold—a fungus with long, branched, threadlike filaments
  E. Hyphae—the filaments of a mold; may be coenocytic (i.e., have no cross walls within the hyphae) or septate (i.e., have cross walls)
  F. Mycelia—bundles or tangled masses of hyphae
  G. Dimorphism—a property of some fungi, which change from the yeast (Y) form (within an animal host) to the mold (M) form (in the environment); this is referred to as the YM shift; the reverse relationship exists in plant-associated fungi
V. Nutrition and Metabolism
  A. Most fungi are saprophytes, securing nutrients from dead organic material
  B. Fungi secrete hydrolytic enzymes that promote external digestion
  C. They are chemoheterotrophic—use organic materials as sources of carbon, electrons, and energy
  D. Glycogen is the primary storage polysaccharide
  E. Most are aerobic (some yeasts are facultatively anaerobic); obligate anaerobic fungi are found in the rumen of cattle
VI. Reproduction
  A. Asexual reproduction—there are several possible mechanisms, including:
    1. Transverse fission
    2. Budding of vegetative cells or spores
    3. Direct spore production
      a. Hyphal fragmentation—component cells behave as arthrospores or chlamydiospores (if enveloped in thick cell wall before separation)
      b. Sporangiospores are produced in sporangium (sac) at the end of an aerial hypha (sporangiophore)
      c. Conidiospores are unenclosed spores produced at the tip or on the sides of aerial hypha
      d. Blastospores are produced when a vegetative cell buds off
  B. Sexual reproduction
    1. Involves the union of compatible nuclei
    2. Some are self-fertilizing (male and female gametes produced on the same mycelium (homothallic), while others require outcrossing between different but sexually compatible mycelia (heterothallic)
    3. Zygote formation proceeds by one of several mechanisms
      a. Fusion of gametes (haploid)
      b. Fusion of gamete-producing bodies (gametangia)
      c. Fusion of hyphae
      d. Immediate fusion of nuclei and cytoplasm
      e. Delayed fusion of nuclei—one cytoplasm with two haploid nuclei (dikaryotic stage)
    4. Zygotes can develop into spores (zygospores, ascospores, or basidiospores)
VII. Characteristics of the Fungal Divisions
  A. Division *Zygomycota*—zygomycetes
    1. Most are saprophytes; a few are plant and animal parasites
    2. Coenocytic hyphae—no crosswalls
    3. Haploid nuclei
    4. Sporangiospores (asexual)

5. Zygospores (sexual)—tough, thick-walled zygotes that can remain dormant when the environment is too harsh for growth
6. Representative member: *Rhizopus stolonifer*—bread mold (also grows on fruits and vegetables)
    a. Normally reproduces asexually
    b. Reproduces sexually by fusion of gametangia if food is scarce or environment is unfavorable
    c. Zygospores (diploid) are produced and remain dormant until conditions are favorable
    d. Meiosis often occurs at time of germination
7. These are used in the production of foods, antibiotics, coloring agents, and other useful products

B. Division *Ascomycota*—ascomycetes
1. Members of this division cause food spoilage, mildew, chestnut blight, and Dutch elm disease
2. They include many yeasts, edible morels, and truffles, as well as the pink bread mold *Neurospora crassa*
3. The mycelia are septate
4. They produce multinucleate conidiospores (asexual)
5. Ascospores (sexual) haploid spores are located in a sac (ascus)
6. Thousands of asci may be packed together in a cup-shaped ascocarp

C. Division *Basidiomycota*—basidiomycetes
1. Includes smuts, jelly fungi, rusts, shelf fungi, stinkhorns, puffballs, toadstools, mushrooms, and bird's nest fungi
2. Basidia are produced at the tips of the hyphae, in which the basidiospores will develop
3. Basidiospores are held in fruiting bodies called basidiocarps
4. Usefulness—many are decomposers; some mushrooms serve as food (some are poisonous); one is the causative agent of cryptococcosis; and some are plant pathogens

D. Division *Deuteromycota*—deuteromycetes (commonly called Fungi Imperfecti)
1. This is a classical division based on fungi that lack a sexual reproductive phase, or fungi for which a sexual reproductive phase has not been observed; more recently molecular systematics places the *Deuteromycota* among their closest relatives in the *Eumycota* and eliminates the *Deuteromycota* as a separate division
2. Most are terrestrial; a few are freshwater or marine organisms
3. Most are saprophytes or plant parasites; some are parasitic on other fungi; some trap and consume nematodes; and some are human parasites, causing ringworm, athlete's foot, histoplasmosis, and other diseases
4. Useful activities include producing antibiotics, giving aromas to cheeses, and fermenting soy sauce

E. Division *Chytridiomycota*—simplest of true fungi
1. Do not form true mycelia
2. Consist of a multinucleate mass that resembles mycelia
3. Form a thallus containing flagellated zoospores
4. Parasites and pathogens of algae, other fungi, terrestrial and aquatic plants

VIII. Slime Molds and Water Molds—resemble fungi in appearance and life-style, but their cellular organization, reproduction, and life cycles are more closely related to protists
A. Division *Myxomycota*—plasmodial slime molds
1. The multinucleated protoplasm (plasmodium) lack a cell wall and can, therefore, exhibit amoeboid movement
2. Feed by phagocytosis
3. Form ornate fruiting bodies when food and/or moisture are in short supply
4. Form spores with cellulose cell walls that are environmentally resistant

5. Germination produces either
   a. Myxamoeba—nonflagellated
   b. Swarm cells—flagellated
6. These will fuse to form a diploid zygote to begin cycle again
B. Division *Acrasiomycota*—cellular slime molds
   1. The vegetative stage is amoeboid cells called myxamoeba
   2. Form pseudoplasmodia when food is scarce by aggregating and secreting a slimy sheath around themselves
   3. Become sedentary and differentiate into prestalk and prespore cells
   4. Form sorocarps that mature to sporangia, which produce spores
   5. Released spores will later germinate to form haploid amoebae to begin the cycle again
C. Division *Oomycota*—oomycetes (water molds)
   1. Resemble fungi, but cell walls are composed of cellulose, not chitin
   2. Produce a relatively large egg cell that is fertilized by a small sperm cell or an even smaller antheridium
   3. Usually saprophytic in freshwater environments

TERMS AND DEFINITIONS

Place the letter of each term in the space next to the definition or description that best matches it.

_____ 1. The study of fungi
_____ 2. The study of the poisonous substances released by fungi and their effects on various organisms
_____ 3. The vegetative structure of a fungus
_____ 4. Filaments in which the protoplasm streams freely, uninterrupted by cross walls
_____ 5. Filaments with cross walls, each wall having one or more pores through which protoplasmic streaming can take place
_____ 6. Asexual spores produced within a sac
_____ 7. Asexual spores produced without a sac
_____ 8. Asexual spores produced by budding from a vegetative cell
_____ 9. Sexual spores that are diploid and surrounded by a tough coating
_____ 10. Sexual spores that are haploid and produced within a sac
_____ 11. Sexual spores that are diploid and produced within a club-shaped sac
_____ 12. Large, multinucleated mass of protoplasm that exhibits amoeboid movement, leaving a slime trail
_____ 13. Large aggregate of amoeboid cells that moves as a unit, leaving a slime trail
_____ 14. Gamete-producing bodies that can fuse to form a zygote during sexual reproduction of some fungi
_____ 15. Sexual spores that are flagellated and produced by the chytrids

a. ascospores
b. basidiospores
c. blastospores
d. coenocytic hyphae
e. conidiospores
f. gametangia
g. mycology
h. mycotoxicology
i. plasmodium
j. pseudoplasmodium
k. septate hyphae
l. sporangiospores
m. thallus
n. zoospores
o. zygospores

FILL IN THE BLANK

1. A unicellular fungus that has a single nucleus and reproduces either asexually by budding or sexually by producing spores is called a(n) _____, while a(n) _____ consists of long, threadlike filaments called _____ that aggregate in bundles to form _____. Some fungi can alternate between the two states and are said to be _____.

2. Like some bacteria, fungal cells secrete _____ that promote _____ outside themselves. The nutrients released are then _____ across the plasma membrane.

3. Fungi use organic material as a source of _____, _____, and _____ for both catabolic and anabolic processes and therefore are said to be _____. Their storage polysaccharide is primarily _____.

4. Asexual spores that are produced in a saclike structure known as a(n) _____ are called _____. The saclike structure is located at the end of an aerial hypha known as a(n) _____. Asexual spores that are not enclosed in a sac, but that are produced at the tips or sides of hyphae are called _____, while spores that are produced by budding from a vegetative cell are called _____.

5. Edible truffles are in the division _____, while the edible mushrooms are in the division _____. Poisonous mushrooms are in the division _____.

6. The plasmodial slime molds are in the division _____, while the cellular slime molds are in the division _____. The water molds are in the division _____.

7. The plasmodial slime molds form a fruiting body when _____ and/or _____ is in short supply. They produce and release spores, and when the conditions are again favorable, the spores germinate to release either nonflagellated amoeboid _____ or flagellated _____ cells that feed and are haploid.

8. The oomycetes, or water molds, even though phylogenetically distinct, resemble fungi in appearance, and consist of finely branched filaments or _____.

9. The _____ occurs when dimorphic fungi switch from the _____ form (single cells) to the _____ form (hyphae).

10. Self-fertilizing fungi in which male and female gametes are produced on the same mycelia are said to be _____, while those that require outcrossing between different but sexually compatible mycelia are said to be _____.

MULTIPLE CHOICE

For each of the questions below select the *one best* answer.

1. Which of the following is not a deleterious effect attributed to fungi?
   a. spoilage of food
   b. deterioration of leather
   c. production of the immunosuppressive compound cyclosporine
   d. Dutch elm disease
2. Hyphae that are multinucleated with no cross walls to prevent protoplasmic streaming are called
   a. septate.
   b. coenocytic.
   c. nonseptate.
   d. protoplasmic.

3. Dimorphism is an advantage to fungi for which of the following reasons?
   a. It allows them to fend off certain defensive host responses.
   b. It allows them to reproduce more rapidly during adverse environmental conditions.
   c. It gives them a competitive edge over other organisms in obtaining nutrients from the environment.
   d. All of the above are correct.

4. In addition to the direct production of asexual spores, fungi can produce asexually by which of the following mechanisms?
   a. transverse fission of a parental cell
   b. fragmentation of hyphae whereby the component cells behave as spores
   c. budding of either somatic (vegetative) cells or spores
   d. All of the above are correct.
5. In most cases when sexual gametes of fungi fuse, the cytoplasm fuses first, and the fusion of the nuclei is delayed. This leads to a stage in which there is one cell containing two haploid nuclei. This is referred to as the
   a. dikaryotic stage.
   b. dinucleated stage.
   c. monocytoplasmic stage.
   d. None of the above are correct.
6. In which of the following divisions will the coenocytic hyphae be found?
   a. Zygomycetes
   b. Ascomycetes
   c. Basidiomycetes
   d. None of the above are correct.
7. In which of the following are the sexual spores haploid?
   a. Zygomycetes
   b. Ascomycetes
   c. Basidiomycetes
   d. Deuteromycetes

8. In which of the following will you find the common bread mold *Rhizopus stolonifer*?
   a. Zygomycetes
   b. Ascomycetes
   c. Basidiomycetes
   d. Deuteromycetes
9. In which of the following will most of the yeasts be found?
   a. Zygomycetes
   b. Ascomycetes
   c. Basidiomycetes
   d. Deuteromycetes
10. When the individual amoeboid cells of the cellular slime molds aggregate, they secrete a sheath around the entire mass of cells. This mass, called a _____, then moves, leaving a slime trail behind it.
    a. plasmodium
    b. pseudopodium
    c. pseudoplasmodium
    d. swarming cell
11. Because of their lack of an observable sexual reproductive cycle, the Deuteromycetes are also referred to as
    a. Fungi incompleti.
    b. Fungi imperfecti.
    c. Fungi asexuali.
    d. Fungi havnofuni.

TRUE/FALSE

____ 1. Most fungi are aerobic or facultatively anaerobic. However, a few are obligate anaerobes.
____ 2. Many fungal spores are responsible for the bright colors and fluffy texture of the molds that produce them.
____ 3. Fungal spores are light and can therefore remain suspended in the air for long periods of time.
____ 4. In all fungi that have been observed to have a sexual reproduction cycle, male and female gametes are produced on separate hyphae and must then find each other for fertilization to occur.
____ 5. The ascomycetes and the basidiomycetes both have septate hyphae. However, septal ultrastructure is a reliable way of distinguishing between the two.
____ 6. Chytrids do not have true mycelia; instead they have a multinucleate mass that resembles mycelia.

CRITICAL THINKING

1. Some fungi are placed in the deuteromycetes when they are first studied, and then are changed into another category. Explain. Is it possible that the deuteromycetes do not represent a real category of fungi? Explain.

2. In various classification schemes, the slime molds and water molds have been classified as either fungi or protists. This text classifies them as protists, but describes them in the chapter concentrating on fungi. Explain why this apparent confusion exists, and also explain the basis on which the decision was made to classify them as protists.

ANSWER KEY

Terms and Definitions

1. g, 2. h, 3. m, 4. d, 5. k, 6. l, 7. e, 8. c, 9. o, 10. a, 11. b, 12. i, 13. j, 14. f, 15. n

Fill in the Blank

1. yeast; mold; hyphae; mycelia; dimorphic 2. enzymes; digestion; absorbed 3. energy; electrons; carbon; heterotrophic; glycogen 4. sporangium; sporangiospores; sporangiophore; conidiospores; blastospores 5. *ascomycota; basidiomycota; basidiomycota* 6. *myxomycota; acrasiomycota; oomycota* 7. food; moisture; myxamoeba; swarm 8. hyphae 9. YM shift; yeast, mold (mycelial) 10. homothallic; heterothallic

Multiple Choice

1. c, 2. b, 3. d, 4. d, 5. a, 6. a, 7. b, 8. a, 9. b, 10. c, 11. b

True/False

1. T, 2. T, 3. T, 4. F, 5. T, 6. T

26   The Algae

CHAPTER OVERVIEW

This chapter discusses the characteristics of a diverse polyphyletic group of organisms known as the algae. They range from single cells to multicellular organisms over 75 meters in length. They are found in oceans and freshwater environments and are the major producers of oxygen and organic material. A few algae live in moist soil and other terrestrial environments. They do not constitute a unique kingdom; instead they are found within two of the five kingdoms previously discussed. An overview of their characteristics is presented, followed by discussion of each of the major groups of algae.

CHAPTER OBJECTIVES

After reading this chapter you should be able to:

- discuss the various habitats in which algae are found
- discuss the various morphological characteristics of the algae and the taxonomic relationships of this diverse polyphyletic group of organisms
- discuss asexual and sexual reproduction of the algae
- discuss the various classical divisions of algae and the differences in photosynthetic pigments and cell structures exhibited within these divisions

CHAPTER OUTLINE

I.   Introduction
    A.   Algae—plants or protists that lack roots, stems, and leaves, but that have chlorophyll and other pigments for carrying out oxygenic photosynthesis
    B.   Phycologists (algologists)—scientists who study algae
    C.   Phycology (algology)—the study of algae
II. Distribution of Algae—primarily aquatic, with a few terrestrial organisms growing on moist surfaces
    A.   Planktonic—suspended in the aqueous environment
       1.   Phytoplankton—algae and other small aquatic plants
       2.   Zooplankton—animals and other nonphotosynthetic protists
    B.   Benthic—attached and living on the bottom of a body of water
    C.   Neustonic—living at the air-water interface
    D.   Some algae are endosymbionts in protozoa, mollusks, worms, corals, and plants
    E.   Some associate with fungi to form lichens
III. Classification of Algae
    A.   Belong to seven divisions within two different kingdoms
    B.   Primary classification is based on cellular and not organismal properties
       1.   Cell wall (if present) chemistry and morphology
       2.   Storage of food and photosynthetic products
       3.   Types of chlorophyll molecules and accessory pigments

4. Number of flagella and their insertion location
5. Morphology of cells and/or thallus (body)
6. Habitat
7. Reproductive structures
8. Life history patterns
   C. Molecular systems have reclassified the algae as polyphyletic with diverse origins and associations
IV. Ultrastructure of the Algal Cell
   A. Surrounded by a thin, rigid cell wall (some have an outer matrix also)
   B. When present, flagella are the locomotory organelles
   C. Chloroplasts have thylakoids (sacs) that are the site of photosynthetic light reactions
   D. Chloroplasts have a dense proteinaceous pyrenoid that is associated with the synthesis and storage of starch
   E. The nucleus has a typical nuclear envelope with pores
V. Algal Nutrition—can be either autotrophic or heterotrophic
   A. Autotrophic—require only light and inorganic compounds for energy; use CO_2 as carbon source
   B. Heterotrophic—use external organic materials as source of energy and carbon
VI. Structure of the Algal Thallus (Vegetative Form)
   A. Unicellular
   B. Colonial
   C. Filamentous
   D. Membranous
   E. Tubular
VII. Algal Reproduction
   A. Asexual—occurs only with unicellular algae
      1. Fragmentation—thallus breaks up and each fragment forms a new thallus
      2. Spores formed in ordinary vegetative cell or in sporangium
         a. Zoospores are flagellated motile spores
         b. Aplanospores are nonmotile spores
      3. Binary fission—nuclear division followed by cytoplasmic division
   B. Sexual—occurs in multicellular and unicellular algae
      1. Oogonia—relatively unmodified vegetative cells in which eggs are formed
      2. Antheridia—specialized structures in which sperm are formed
      3. Zygote—fusion of sperm and egg
VIII. Characteristics of the Algal Divisions
   A. *Chlorophyta* (green algae)
      1. Contain chlorophylls *a* and *b,* and carotenoids; store carbohydrate as starch; cell walls are made of cellulose
      2. Live in fresh and salt water, in soil, on and within other organisms
      3. Have a variety of body types—unicellular, colonial, filamentous, membranous, and tubular
      4. Exhibit both asexual and sexual reproduction
      5. Molecular classification places these with the land plants
      6. *Chlamydomonas*—a representative unicellular green alga
         a. Microscopic, rounded, with two flagella at anterior end
         b. Single haploid nucleus
         c. A large chloroplast with conspicuous pyrenoid for starch production and storage
         d. Stigma (phototactic eyespot)
         e. Contractile vacuoles act as osmoregulators
         f. Asexual reproduction (zoospores) and sexual reproduction
      7. Protothecosis—a human and animal disease caused by the green alga, *Prototheca moriformis*

B. *Charophyta* (stoneworts/brittleworts)
   1. Abundant in fresh and brackish waters; worldwide distribution
   2. Some species precipitate calcium and magnesium carbonate from water to form a limestone covering (helps preserve members as fossils)
C. *Euglenophyta* (euglenoids)
   1. Same chlorophylls (*a* and *b*) as *Chlorophyta* and *Charophyta*
   2. Found in fresh and brackish waters, and in moist soils
   3. Molecular classification indicates the euglenoids to be closely associated with the amoeboflagellates (flagellated protozoa)
   4. *Euglena*
      a. Elongated cell bounded by a plasma membrane
      b. Stigma located near an anterior reservoir
      c. A pellicle (articulated proteinaceous strips lying side-by-side) maintain a cellular shape that is somewhat flexible yet rigid enough to prevent excessive alterations
      d. A large contractile vacuole collects water and empties it into the reservoir for osmotic regulation
      e. Paired flagella at anterior end arise from reservoir base; only one beats to move the cell
      f. Reproduction is by longitudinal mitotic cell division
D. *Chrysophyta* (golden-brown and yellow-green algae and diatoms)
   1. Molecular classification associates these with the *stramenophiles*
   2. Contain chlorophylls *a* and c_1/c_2, and the carotenoid fucoxanthin
   3. Some lack cell walls, some have intricately patterned scales on the plasma membrane; diatoms have a distinctive two-piece wall of silica called a frustule
   4. Zero, one, or two flagella (of equal or unequal length)
   5. Unicellular or colonial
   6. Reproduction is usually asexual, but occasionally sexual
   7. Diatoms are photosynthetic, circular, or oblong cells with overlapping silica shells
      a. Epitheca—larger half
      b. Hypotheca—smaller half
   8. Diatomaceous earth—deposits of empty diatom shells
      a. Used in detergents, polishes, paint removers, decolorizers, deodorizers, and fertilizers
      b. Filtering agent
      c. Component in soundproofing and insulating materials
      d. Paint additive to increase night visibility of license plates and signs
E. *Phaeophyta* (brown algae)
   1. Multicellular seaweeds; some species have the largest linear dimensions known in the eucaryotic world
   2. Have branched filaments and more complex arrangements; some (kelps) are differentiated into flattened blades, stalks, and holdfast organs that anchor them to rocks
   3. *Sargassum* forms huge floating masses
   4. Contain chlorophylls *a* and *c;* carotenoids include fucoxanthin, violaxanthin, and β-carotene
   5. Molecular classification associates these with the *stramenophiles*
F. *Rhodophyta* (red algae)
   1. Some are unicellular, but most are multicellular, filamentous seaweeds
   2. Comprises most of the seaweeds
   3. Contain phycoerythrin (red pigment) and phycocyanin (blue pigment), and can therefore live in deeper waters
   4. Their cell walls include a rigid inner part composed of microfibrils and a mucilaginous matrix consisting of sulfated polymers of galactose
   5. One of the polymers, agar, is widely used as a gelling agent in bacterial culture media
   6. Many also deposit calcium carbonate in their cell walls and contribute to coral reef formation

225

G. *Pyrrhophyta* (dinoflagellates)
   1. They are unicellular, photosynthetic protists
   2. Most are marine but a few are freshwater dwellers
   3. Some are responsible for phosphorescence in ocean waters
   4. Two perpendicular flagella cause organisms to spin
   5. Contain chlorophylls *a* and *c,* carotenoids, and xanthophylls
   6. Zooxanthellae—symbiotic dinoflagellates that have lost their cellulose plates and flagella and that live within the cells of other organisms
   7. Responsible for toxic red tides; toxin is ingested by shellfish (which are unaffected by it) and passed to those who consume the shellfish

TERMS AND DEFINITIONS

Place the letter of each term in the space next to the definition or description that best matches it.

_____ 1.	Term that describes algae that are suspended in the aqueous environment	a. antheridia
_____ 2.	Term that describes algae that are attached and living on the bottom of the body of water	b. aplanospores
		c. benthic
_____ 3.	Term that describes algae that live at the water-atmosphere interface	d. epitheca
		e. frustule
_____ 4.	A flexible, gelatinous layer outside the cell wall that is similar to a bacterial capsule	f. hypotheca
		g. neustonic
_____ 5.	A dense, proteinaceous area of the chloroplast that is associated with the synthesis and storage of starch	h. oogonia
		i. outer matrix
_____ 6.	Asexual spores that are flagellated and, therefore, motile	j. pellicle
_____ 7.	Asexual spores that are not flagellated and, therefore, nonmotile	k. planktonic
		l. pyrenoid
_____ 8.	Unmodified vegetative cells that are the site of egg formation	m. scales
		n. stigma
_____ 9.	Special structures that are the site of sperm formation	o. zoospores
_____ 10.	A rhodopsin-containing eyespot that aids in phototaxic responses	p. zooxanthellae
_____ 11.	An articulated, proteinaceous structure inside the plasma membrane that is rigid but somewhat flexible	
_____ 12.	Intricately patterned coverings on the plasma membrane of some algae	
_____ 13.	The distinctive two-piece wall (valve) of silica found on diatoms	
_____ 14.	The larger of the two pieces of the diatom valve	
_____ 15.	The smaller of the two pieces of the diatom valve	
_____ 16.	Symbiotic dinoflagellates found in other marine organisms	

FILL IN THE BLANK

1. Aquatic plants are collectively referred to as _____, and the study of these organisms is called _____ or _____; while the scientists who study these organisms are referred to as _____ or _____.

2. Plankton consists of free-floating, mostly microscopic, aquatic organisms. If these organisms are algae and other plants, they are called _____; if they are mostly animals and other nonphotosynthetic organisms, they are called _____.

3. Algae that reproduce sexually form eggs in unmodified vegetative cells called _____, while they form sperm in special structures known as _____.

4. Many of the algae have a structure called an eyespot, or _____ (which contains molecules of _____) that aids the organism in _____, or movement toward a light source.

5. The red algae contain two phycobilin pigments: _____, which is red, and _____, which is blue. When the first is present in high concentrations, the organisms have the characteristic red appearance from which the division gets its name. These pigments allow these organisms to live in _____ waters.

MULTIPLE CHOICE

For each of the questions below select the *one best* answer.

1. In which of the following kingdoms would you not find organisms classified as algae?
   a. *Plantae*
   b. *Fungi*
   c. *Protista*
   d. Algae are found in all of the above.

2. Which of the following is not a mechanism used by algae for asexual reproduction?
   a. fragmentation
   b. spore formation
   c. binary fission
   d. All of the above are mechanisms used by algae for asexual reproduction.

3. Asexual spores that are not flagellated and are, therefore, not motile are called
   a. zoospores.
   b. zygospores.
   c. aplanospores.
   d. gametospores.

4. Which of the following is true about toxic red tides?
   a. The toxin produced (saxitoxin) paralyzes the striated respiratory muscles of vertebrates.
   b. The toxin causes an economically devastating kill of commercially important shellfish beds.
   c. The toxin is produced by members of the *Rhodophyta* (red algae), from which the condition gets its name.
   d. All of the above are true about toxic red tides.

5. Diatom shells have been used for which of the following?
   a. abrasives in detergents and polishes
   b. filtering agents
   c. soundproofing and insulating material
   d. All of the above are correct.

6. Which of the following groups of algae contain agar, a substance widely used as a gelling agent in bacterial culture media?
   a. *Chlorophyta* (green algae)
   b. *Phaeophyta* (brown algae)
   c. *Rhodophyta* (red algae)
   d. *Pyrrophyta* (dinoflagellates)

TRUE/FALSE

___ 1. There are no algae that are known to cause disease in humans.

___ 2. Extinct stoneworts and brittleworts have been preserved as fossils because they can cover themselves with calcium and magnesium carbonate, which they precipitate from the water.

___ 3. The red algae are classified in the plant kingdom. However, unlike other plants, these organisms have centrioles that function during mitosis.

___ 4. *Euglena* uses a contractile vacuole and an anterior reservoir for osmotic regulation.

___ 5. While most algae are photoautotrophic, some chemoheterotrophs do exist among this diverse group of organisms.

___ 6. The algae are polyphyletic meaning that they are associated with multiple kingdoms with independent origins and do not, therefore, represent a single evolutionary branch of development.

CRITICAL THINKING

1. The authors state that the term *algae* "no longer has any formal significance in classification schemes." Explain the meaning of this statement. If it is true, why is an entire chapter devoted to a discussion of algae?

2. The cyanobacteria were once included in the algae, but no longer are included in the most widely accepted classification scheme. Do you agree or disagree? Defend your answer.

ANSWER KEY

Terms and Definitions

1. k, 2. c, 3. g, 4. i, 5. l, 6. o, 7. b, 8. h, 9. a, 10. n, 11. j, 12. m, 13. e, 14. d, 15. f, 16. p

Fill in the Blank

1. algae; algology; phycology; algologists; phycologists 2. phytoplankton; zooplankton 3. oogonia; antheridia 4. stigma; rhodopsin; phototaxis 5. phycoerythrin; phycocyanin; deep(er)

Multiple Choice

1. b, 2. d, 3. c, 4. a, 5. d, 6. c

True/False

1. F, 2. T, 3. F, 4. T, 5. T, 6. T

27 The Protozoa

CHAPTER OVERVIEW

This chapter discusses the characteristics of those protists that are commonly referred to as the protozoa. These protists exhibit a variety of locomotion mechanisms and demonstrate the great adaptive potential of unicellular eucaryotic organisms. Using molecular methods the protozoa have been shown to be polyphyletic. In addition to a discussion of their general features, and the vast array of their niches and habitats, individual coverage of some representative examples is given.

CHAPTER OBJECTIVES

After reading this chapter you should be able to:

- describe the various habitats, types of locomotory ability, and the specialized organelles of protozoa
- discuss the characteristics of the seven phyla in which protozoa are classified
- explain the reproductive strategies employed by protozoa
- discuss the various arrangements of nuclei that are found in protozoa
- describe the various feeding mechanisms used by protozoa

CHAPTER OUTLINE

I.   Introduction
    A. Protozoa are unicellular, eucaryotic protists that are usually motile
    B. Protozoologists are scientists that study protozoa
    C. Protozoology is the study of protozoa
    D. Protozoa have diverse, non related, polyphyletic origins
II.   Distribution—primarily in moist habitats
    A. Found in freshwater, marine, and moist terrestrial environments
    B. Most are free living, but some are parasitic in plants and animals
III. Importance
    A. Serve as an important link in food chains and food webs (zooplankton)
    B. Important in the study of molecular biology because they use the same metabolic pathways as multicellular eucaryotes
    C. Causative agents of some important diseases in humans and other animals
IV.   Morphology
    A. Ectoplasm is the gelatinous cytoplasm just inside the plasma membrane; it provides some rigidity and shape through the formation of a pellicle
    B. Endoplasm is the more fluid cytoplasm in the interior of the cell
    C. The pellicle consists of the plasma membrane and the structures immediately beneath it
    D. The macronucleus is associated with trophic activities and regenerative processes
    E. The micronucleus controls reproductive activities by sequestering genetic material for exchange during reproduction

F. Vacuoles
   1. Contractile vacuoles are osmoregulatory
   2. Phagocytic vacuoles are sites of food digestion
   3. Secretory vacuoles usually contain enzymes for specific functions, such as excystation
G. Some protozoa are anaerobic (e.g., *Trichonympha* lives in the gut of termites)
   1. They have no mitochondria or cytochromes, and they have an incomplete TCA cycle
   2. They contain hydrogenosomes—small membrane-delimited organelles containing a unique electron transfer system that uses protons as terminal electron acceptors to form molecular hydrogen
V. Nutrition—chemoheterotrophic
A. In holozoic nutrition, nutrients are acquired by phagocytosis
B. Some ciliates have a specialized structure, called a cytosome, for phagocytosis
C. In saprozoic nutrition, nutrients are acquired by pinocytosis, diffusion, or carrier-mediated transport (facilitated diffusion or active transport)
VI. Encystment and Excystment
A. Encystation is the development of a resting stage structure called a cyst
   1. The cyst is a dormant form marked by the presence of a wall and by the reduction of metabolic activity to a very low level
   2. Functions of cysts
      a. Protect against adverse changes in the environment
      b. Function as sites for nuclear reorganization and cell division
      c. Serve as a means of transfer from one host to another for parasitic species
B. Excystation is the escape of vegetative forms, called trophozoites, from the cyst; it is usually triggered by a return to favorable environment, such as entry into a new host for parasitic species
VII. Locomotory Organelles—a few protozoa are nonmotile; most use one of three major types of locomotory organelles
A. Pseudopodia—cytoplasmic extensions
B. Cilia—filamentous extensions (short)
C. Flagella—filamentous extensions (long)
VIII. Reproduction
A. The most common method of asexual reproduction is binary fission, which involves mitosis followed by cytokinesis
B. The most common type of sexual reproduction is conjugation, an exchange of gametic nuclei between paired protozoa of complementary mating types
IX. Classification
A. Protozoa are currently classified as a subkingdom of protists although Cavalier-Smith has proposed elevating the protozoa to the status of a kingdom
B. There are seven phyla within this subkingdom
C. Classification is based primarily on types of nuclei, mode of reproduction, and mechanism of locomotion
D. More recently, molecular classification schemes suggest that the protozoa do not exist as an evolutionary taxon, but rather that the protozoa are polyphyletic and are found at all evolutionary levels
X. Representative Types
A. *Sarcomastigophora*
   1. Single type of nucleus
   2. Flagella (subphylum *Mastigophora*)—zooflagellates
      a. Some are free living
      b. Some cause parasitic disease
         (1) *Trichomonas vaginalis* causes a sexually transmitted disease
         (2) *Giardia lamblia* causes a gastrointestinal disorder

    (3) *Trypanosoma brucei* and other trypanosomes (also called hemoflagellates) are
      important blood pathogens
   c. Kinetoplastids are a small group of mastigophorans whose mitochondrial DNA is in a
     special region called the kinetoplast
  3. Pseudopodia (subphylum *Sarcodina*)
   a. Amoeboid organisms
   b. Some form cysts
   c. Reproduction is usually by simple asexual binary fission
   d. Some have a loose-fitting shell (test)
   e. Foraminiferans and radiolarians are primarily marine amoebae with a few occurring in
     fresh or brackish water
   f. There are some symbionts (parasites)
    (1) *Endamoeba blattae* is common in the intestines of insects
    (2) *Entamoeba histolytica* causes a parasitic disease in humans
B. *Labyrinthomorpha*
  1. They are spindle-shaped or spherical, nonamoeboid, vegetative cells
  2. Most members are marine organisms, and are either saprozoic or parasitic on algae
  3. Move by gliding motion on mucous tracks
C. *Apicomplexa*—sporozoans
  1. Lack locomotor organelles, except the male gametes and the zygotes (ookinetes)
  2. Apical complex—a unique arrangement of fibrils, tubules, vacuoles, and other organelles at
    one end of the cell
   a. Consist of one or two polar rings at the apical end
   b. The conoid—spirally arranged fibers adjacent to the polar rings
   c. Subpellicular microtubules radiate from the polar rings and probably serve as support
     elements
   d. Rhoptries extend to the plasma membrane and secrete their contents at the cell surface
     (probably aids in host cell penetration)
   e. Micropores take in nutrients
  3. Complex life cycle involves two different hosts (usually mammal and mosquito)
  4. Undergo an alternation of haploid and diploid generations
  5. Undergo schizogony, a rapid series of mitotic events producing a large number of small
    infective organisms through the formation of uninuclear buds
  6. Sexual reproduction involves the formation of a thick-walled oocyst after fertilization;
    meiosis within this structure then produces haploid infective spores
  7. This group incudes some very important parasitic disease causers
   a. *Plasmodium*—malaria
   b. *Cryptosporidium*—cryptosporidiosis
   b. *Toxoplasma*—toxoplasmosis
   c. *Eimeria*—coccidiosis
D. *Microspora*
  1. Obligate intracellular parasites
  2. Abundant and pathogenic in insects
  3. Increased interest in their use as biological pest control
  4. Recently, five genera have been implicated in human diseases in immunosuppressed patients
    and in AIDS patients
E. *Acetospora*—spores lack polar caps or polar filaments, parasitic in mollusks
F. *Myxozoa*—resistant spore with one to six coiled polar filaments; parasitic on freshwater and
  marine fish; can cause a major economic problem in cultured salmon
G. *Ciliophora*
  1. Use cilia as locomotory organelles (longitudinal rows or spirals)
  2. Can move forward or backward by control of ciliary movement

3. Food processing is somewhat complex
   a. Food enters the cytostome; it is brought in by the action of the surrounding cilia
   b. The food passes to phagocytic vacuoles that fuse with lysosomes, where digestion occurs
   c. After digestion the vacuoles fuse with a special region of the pellicle, called the cytoproct, which empties the cell's waste material to the outside
4. Most have two types of nuclei
   a. Micronucleus—diploid; functions in mitosis and meiosis
   b. Macronucleus—polyploid for some genes; maintains routine cellular functions
5. Asexual reproduction is by transverse binary fission
6. Sexual reproduction usually is by conjugation
7. Some ciliates are harmless commensals, while others are disease-causing parasites

TERMS AND DEFINITIONS

Place the letter of each term in the space next to the definition that best matches it.

_____ 1. A series of organisms, each feeding on the preceding one
_____ 2. A complex interlocking series of organisms with a hierarchy of feeding relationships
_____ 3. The vegetative form of a protozoan
_____ 4. The dormant form of a protozoan
_____ 5. Semisolid or gelatinous cytoplasm just beneath the plasma membrane
_____ 6. Fluid cytoplasm in the interior of protozoa
_____ 7. The single site of phagocytosis in some ciliated protozoa
_____ 8. Formation of the cyst stage from the vegetative stage
_____ 9. Formation of the vegetative stage from the cyst stage
_____ 10. Cytoplasmic extensions used for locomotion and food capture
_____ 11. Long filamentous extensions used for locomotion
_____ 12. Short filamentous extensions used for locomotion
_____ 13. The process whereby there is an exchange of gametes between paired protozoa of complementary mating types
_____ 14. The loose-fitting shell around some amoeboid organisms
_____ 15. An arrangement of fibrils, microtubules, vacuoles, and other organelles at one end of the cell
_____ 16. A region of the pellicle of ciliates where phagocytic vacuoles empty their contents after food digestion has taken place
_____ 17. Specialized region found in some zooflagellates where their mitochondrial DNA is stored

a. apical complex
b. cilia
c. conjugation
d. cyst
e. cytoproct
f. cytosome
g. ectoplasm
h. encystation
i. endoplasm
j. excystation
k. flagella
l. food chain
m. food web
n. kinetoplast
o. pseudopodia
p. test
q. trophozoite

FILL IN THE BLANK

1. In some species of protozoa, the cytoplasm immediately under the plasma membrane is gelatinous and is termed _____. The plasma membrane and structures immediately beneath it are referred to as the _____. The more fluid portion of the cytoplasm in the interior of the cell is referred to as the _____.

2. One or more vacuoles are usually found in the cytoplasm of protozoa. Vacuoles that function as osmoregulatory organelles are called _____ vacuoles and are usually found in organisms that live in a _____ environment, such as a freshwater lake. Holozoic and parasitic organisms have _____ vacuoles, which serve as the site of food digestion. Vacuoles with enzymes that function during the excystation process are referred to as _____ vacuoles.

3. Most protozoa are heterotrophic. When nutrients are acquired by phagocytosis, the organisms are said to be _____. In ciliated organisms, phagocytosis occurs at a single location called the _____. When nutrients are acquired by pinocytosis or other forms of direct transport, the organisms are said to be _____.

4. Organisms of the phylum *Apicomplexa* have an arrangement of organelles at one end of the cell; this _____ consists of one or two polar rings at one end where a cone of spirally arranged fibers called the _____ is located. Radiating from the polar rings are _____ microtubules, which probably serve as support elements; two or more structures, called _____, extend to the plasma membrane and secrete their contents at the cell surface. A structure called the _____ functions in the intake of nutrients.

5. The protozoan *Trichonymphia* lives in the gut of termites; it is _____ and, therefore, has no mitochondria, no cytochromes, and an incomplete TCA cycle. It does have a specialized organelle called a _____ that houses a unique electron transfer system which uses protons as terminal electron acceptors to form molecular hydrogen.

MULTIPLE CHOICE

For each of the questions below select the *one best* answer.

1. Which of the following is not a reason for studying protozoa?
   a. They are an important link in many food chains and food webs.
   b. They are a good model for studying eucaryotic metabolism.
   c. They cause important diseases in humans and other animals.
   d. All of the above are reasons for studying protozoa.

2. Which of the following is not a major function of cysts?
   a. They have locomotory organelles for movement through an aqueous environment.
   b. They protect against adverse changes in the environment.
   c. They play a role in reproduction.
   d. They serve as a means of transfer from one host to another in parasitic species.

3. Which of the following is a method of locomotion used by protozoa?
   a. flagella
   b. cilia
   c. pseudopodia
   d. All of the above are correct.

4. Which of the following is not a zooflagellate that is parasitic to humans?
   a. *Giardia lamblia*
   b. *Entamoeba histolytica*
   c. *Trichomonas vaginalis*
   d. *Trypanosoma cruzi*

5. Protozoans that have spindle-shaped or spherical nonamoeboid vegetative cells, and that move within a network of mucous tracks by a typical gliding motion, belong to which of the following phyla?
   a. *Apicomplexa*
   b. *Ascetospora*
   c. *Labyrinthomorpha*
   d. *Myxozoa*

6. Which of the following is not true about the phylum *Apicomplexa?*
   a. Only the male gametes have special locomotory organelles.
   b. Their complex life cycles involve an alternation of diploid and haploid generations.
   c. After fertilization, the zygote forms a thick-walled cyst called an oocyst.
   d. All of the above are true about the phylum *Apicomplexa.*

7. Which of the following is not used as a basis for classifying protozoa?
   a. types of nuclei
   b. mode of reproduction
   c. mechanism of locomotion
   d. All of the above are used to classify protozoa.

8. Which of the following diseases is not caused by members of *Apicomplexa?*
   a. malaria
   b. toxoplasmosis
   c. giardiasis
   d. coccidiosis

9. When nutrients are acquired by pinocytosis, it is referred to as _____ nutrition.
   a. holozoic
   b. saprozoic
   c. Both (a) and (b) are correct.
   d. Neither (a) nor (b) is correct.

TRUE/FALSE

_____ 1. Although most protozoa are aerobic, they often live where the partial pressure of oxygen is low; and therefore, oxidization of glucose and other nutrients is incomplete.

_____ 2. In conjugation, the micronuclei divide by meiosis to form the haploid gamete nuclei.

_____ 3. In conjugation, the parental macronuclei disappear and the daughter cells develop new macronuclei after the diploid zygote has formed.

_____ 4. The *Sarcodina,* which move by means of pseudopodia, and the *Mastigophora,* which move by means of flagella, are classified together into a single phylum.

_____ 5. Zooflagellates that are important blood pathogens of humans and other animals are also called hemoflagellates.

_____ 6. Microsporeans are of interest as biological pest control agents, although none have been approved for use yet.

_____ 7. The ciliates have two types of nuclei: the micronucleus, which is diploid, and the macronucleus, which is haploid.

_____ 8. All protozoa exhibit motility as adult organisms.

_____ 9. Like the algae, molecular methods have shown the protozoa to be polyphyletic and therefore not an evolutionary distinct taxon.

CRITICAL THINKING

1. Many protozoa use contractile vacuoles as osmoregulatory organelles. How do these organelles function to maintain osmotic balance? Why are they found in freshwater protozoa but not in marine protozoa?

2. The amoeboid organisms (subphylum *Sarcodina*) and the flagellated organisms (subphylum *Mastigophora*) are classified into the same phylum (*Sarcomastigophora*). On what basis is this done?

ANSWER KEY

Terms and Definitions

1. l, 2. m, 3. q, 4. d, 5. g, 6. i, 7. f, 8. h, 9. j, 10. o, 11. k, 12. b, 13. c, 14. p, 15. a, 16. e, 17. n

Fill in the Blank

1. ectoplasm; pellicle; endoplasm 2. contractile; hypotonic; phagocytic; secretory 3. holozoic; cytosome; saprozoic 4. apical complex; conoid; subpellicular; rhoptries; micropore 5. anaerobic; hydrogenosome

Multiple Choice

1. d, 2. a, 3. d, 4. b, 5. c, 6. d, 7. d, 8. c, 9. b

True/False

1. T, 2. T, 3. T, 4. T, 5. T, 6. F, 7. F, 8. F, 9. T

28 Symbiotic Associations: Commensalism, Mutualism, and Normal Microbiota of the Human Body

CHAPTER OVERVIEW

This chapter discusses the relationships between populations of microorganisms living in close association with each other. The discussion focuses on commensalism and mutualism. Examples of each type of relationship are presented. Additionally, a monitored or gnotobiotic condition in which the identities of all microorganisms present are known is discussed. Finally, the chapter presents an overview of the normal microorganisms associated with the human body.

CHAPTER OBJECTIVES

After reading this chapter you should be able to:

- discuss symbiosis, commensalism, and mutualism
- give some examples of endosymbiosis and ectosymbiosis
- discuss the establishment and maintenance of microbially monitored (gnotobiotic) animal colonies
- discuss the type and distribution of the microbiota (normal microflora) of the human body

CHAPTER OUTLINE

I. Introduction
    A. A symbiont is any microorganism that spends all or a portion of its life associated with another organism of a different species
    B. Symbiosis is the living together in close physical association of two or more different organisms
        1. Ectosymbiosis—organisms remain outside each other
        2. Endosymbiosis—one organism is found within the other
II. Types of Symbiosis, Functions, and Examples
    A. Commensalism—the microorganism (commensal) benefits, while the host is neither harmed nor helped
        1. The microorganism shares the same food source with the host
        2. The microorganism is not directly dependent on the metabolism of the host
        3. The microorganism causes no particular harm to host
        4. Example—the common nonpathogenic strain of *Escherichia coli* lives in the human colon; this facultative anaerobe uses oxygen creating an anaerobic environment in which obligate anaerobes (e.g. *bacteroids*) can grow. The bacteroids benefit but the *E. coli* derives no obvious benefit or harm.

B. Mutualism
    1. There is some reciprocal benefit to both partners
    2. The microorganism (mutualist) and host are metabolically dependent upon each other
    3. Syntropism is a mutually beneficial relationship in which each organism provides one or more growth factors, nutrients or substrates for the other organism; also referred to as cross-feeding or the satellite phenomenon
    4. Examples
        a. Protozoan-termite relationship—protozoa live in the guts of insects which ingest but cannot metabolize cellulose; the protozoa secrete cellulases, which metabolize cellulose, releasing nutrients that the insects can use
        b. Lichens—an association between a fungus (ascomycetes) and an alga (green algae) or cyanobacteria
            (1) Fungal partner (mycobiont) obtains nutrients from alga by hyphal projections (haustoria) that penetrate the algal cell wall
            (2) Algal partner (phycobiont) is protected from excess light intensity and is provided with water, minerals, and a firm substratum in which it can grow protected from environmental stress
            (3) Lichens have one of three characteristic morphologies
                (a) Crustose—compact and appressed to a firm substratum
                (b) Foliose—leaflike appearance
                (c) Fruticose—shrubby shape
        c. Zooxanthellae—algae harbored by marine invertebrates; reef-building (hermatypic) corals are unable to utilize zooplankton as food and rely heavily on endosymbiotic zooxanthellae; the coral pigments protect the algae from ultraviolet radiation
        d. The tube worm-bacterial relationship
            (1) Exist in hydrothermal vent communities where the vent fluids are anoxic, with high concentrations of hydrogen sulfide and temperatures up to $350°C$
            (2) The surrounding seawater reaches temperatures $10°$ to $20°C$ above the normal temperature of $2.1°C$
            (3) Free living and endosymbiotic chemolithotrophic bacteria provide the main energy source in the community through the oxidation of hydrogen sulfide
            (4) Chemolithotrophic endosymbiotic bacteria are maintained in a specialized tissue (trophosome) of giant (less than one meter in length), red, gutless tube worms
            (5) The tube worm binds hydrogen sulfide to hemoglobin and transports it to the bacteria; the bacteria use the energy from hydrogen sulfide oxidation to synthesize reduced organic material that is supplied to the tube worms
        e. Rumen ectosymbiosis—bacteria in the rumen anaerobically metabolize cellulose, and then normal digestion occurs; microorganisms produce the majority of vitamins that are needed by the ruminant; methane is also produced in the process

III. Gnotobiotic Animals—all microbial species present are known; may even be germ-free
    A. Mammalian fetuses (in utero) are free from microorganisms
    B. Gnotobiotic colonies are established by cesarean-section delivery in a germfree isolator
    C. They are maintained in a sterile environment
    D. Normal mating and delivery of gnotobiotic animals maintains the condition
    E. Gnotobiotic animals allow investigation of the interactions of animals with specific microorganisms, which are deliberately introduced
    F. Gnotobiotic animals are only superficially normal and require many nutritional supplements normally provided by the normal microbiota
    G. Gnotobiotic animals are unusually susceptible to pathogens but may be resistant to some diseases caused by protozoa that use bacteria as a food source (e.g. *Entamoeba histolytica*)

IV. Distribution of the Normal Microbiota of the Human Body
   A. Reasons to acquire knowledge of normal human microbiota and its distribution
      1. It provides greater insight into possible infections resulting from injury to these areas
      2. It gives perspective on the possible sources and significance of microorganisms isolated from an infection site
      3. It increases understanding of the causes and consequences of overgrowth of microorganisms normally absent from a specific body site
      4. It aids awareness of the role these indigenous microorganisms play in stimulating the immune response that provides protection against potential pathogens
   B. Skin
      1. Resident microbiota multiply on or in the skin
         a. They vary from one part of the body to another
         b. They experience periodic drying in a slightly acidic, hypertonic environment
      2. Transient microbiota are found on the skin for a short time and do not multiply there; they usually die in a few hours
      3. One species, *Propionibacterium acnes,* is associated with acne vulgaris
   C. Nose and nasopharynx
      1. Nose—just inside the nares; contains the same organisms as skin, including *Staphylococcus epidermidis* and *S. aureus*
      2. Nasopharynx—above the level of the soft palate; contains nonencapsulated strains of some of the same species that may cause clinical infection; other species also are found
   D. Oropharynx—between the soft palate and upper edge of the epiglottis; houses many different species
   E. Respiratory tract—no normal microbiota due to mucociliary blanket, the enzyme *lysozyme* in mucus, and the phagocytic action of alveolar macrophages
   F. Oral cavity (mouth)—contains those organisms that survive mechanical removal by adhering to gums (anaerobes) and teeth (aerobes); organisms contribute to the formation of dental plaque, dental caries, gingivitis, and periodontal disease
   G. Eye—aerobic commensals are found on the conjunctiva
   H. External ear—resembles microbiota of the skin with some fungi
   I. Stomach—most microorganisms are killed by acidic conditions unless they pass through very quickly; the number of microorganisms present increases immediately after a meal, but decreases quickly
   J. Small intestine
      1. Duodenum—few microorganisms present because of stomach acidity and inhibitory action of bile and pancreatic secretions
      2. Jejunum—*Enterococcus fecalis*, diphtheroids, lactobacilli, and *Candida albicans*
      3. Ileum—microbiota resemble that of the colon as the pH becomes more alkaline
   K. Large intestine (colon)—largest microbial population of the body
      1. Over 300 different species have been isolated from human feces
      2. Most are anaerobes or facultative organisms growing anaerobically
      3. Ratio of anaerobes to facultatives is approximately 300:1
      4. They are excreted by peristalsis, segmentation, desquamation, and movement of mucus
      5. They are replaced rapidly because of their high reproductive rate
      6. This is a self-balancing (self-regulating) microbial ecology
      7. The balance may change with: stress, altitude, starvation, diet, parasite infection, diarrhea, use of antibiotics or probiotics (microorganisms orally administered that promote health)
   L. Genitourinary tract
      1. Kidneys, ureter, and bladder are normally free of microorganisms
      2. Males—a few microorganisms are found in distal portions of the urethra
      3. Females—complex microbiota in a state of flux due to menstrual cycle; Döderlein's bacilli are primarily *Lactobacillus acidophilus* that forms lactic acid and thereby maintains the pH of the vagina and cervical os between 4.4 and 4.6

TERMS AND DEFINITIONS

Place the letter of each term in the space next to the definition or description that best matches it.

____	1.	The relationship in which two organisms share the same food source	a. commensalism
____	2.	The relationship in which two organisms are metabolically dependent on each other	b. ectosymbiotic
____	3.	The relationship in which one organism harms the other	c. endosymbiotic
____	4.	The relationship in which the two organisms remain outside of each other	d. gnotobiotic
____	5.	The relationship in which one organism lives inside the cells of the other	e. microbiota
____	6.	The population(s) of microorganisms normally associated with a particular tissue or structure	f. mutualism
____	7.	Animals that are germfree or that associate with one or more known species of microorganisms	g. parasitism
____	8.	A specialized structure in which the endosymbiotic chemolitrophic bacteria associated with tube worms live	h. probiotics
____	9.	Microorganisms that are orally administered to promote health; they take up residence in the colon and affect the microbial balance there	i. syntropism
____	10.	A mutually beneficial association in which the growth of an organism is dependent upon one or more growth factors, nutrients, or substrates provided by another organism in the same vicinity	j. trophosome

FILL IN THE BLANK

1. A relationship between any two organisms that spend all or part of their lives in association with each other is called _____. If the organisms remain outside of each other, it is called an _____; and if one organism is found within the cells of the other, it is called an _____.

2. There are several different types of symbiotic relationships. If the microorganism and the host share the same food source, but neither is metabolically dependent on the other, the relationship is called _____; if the two partners gain mutual benefit and are metabolically dependent on each other, the relationship is termed _____; and if one microorganism benefits while the host is harmed, it is called _____.

3. Lichens take on certain characteristic forms. Lichens that are compact and appressed to a substratum are termed _____; those that have a leaflike appearance are termed _____; and those that look like small shrubs are termed _____.

4. Giant tube worms live in _____ vent communities and have within a special tissue called the _____, endosymbiotic chemolithotrophic bacterial that process _____, supplied by the tube worm for energy and in turn they supply reduced _____ material to the tube worm.

5. Organisms whose microbial inhabitants are completely known are said to be _____. They are established by delivering newborns by _____ in a sterile environment and then maintaining the colony in a sterile environment.

6. Microorganisms that are normally found on the skin and that are able to multiply on or in the skin are called _____ microbiota, while those that are found on the skin temporarily and are unable to reproduce there are termed _____ microbiota.

7. In a lichen, the fungal partner is called the _____, and the algal partner is called the _____.

MULTIPLE CHOICE

For each of the questions below select the *one best* answer.

1. In lichens, the fungi send projections of their hyphae across the algal cell wall in order to obtain nutrients from the photosynthetic partner. These projections are called
   a. mycelia.
   b. hyphal extensions.
   c. haustoria.
   d. lichenthropes.

2. Which of the following is not true about the coral reef-zooxanthellae mutualism?
   a. The zooxanthellae provide most of the energy for the corals.
   b. The pigments produced by the corals protect the algae from harmful ultraviolet radiation.
   c. The rate of calcification (reef-building) is at least ten times higher in the light than in the dark.
   d. All of the above are true about the coral-zooxanthellae mutualism.

3. In a symbiotic relationship in which the microorganism benefits while the colonized animal is neither helped nor harmed, the microorganism is called the
   a. bacteriont.
   b. commensal.
   c. mutualist.
   d. host.

4. The ectosymbionts of ruminant animals are useful to their hosts in which of the following ways?
   a. They digest the cellulose of plant cell walls to provide access to the interior contents of the cells, which the ruminant can metabolize.
   b. They produce most of the vitamins needed by the ruminant.
   c. They themselves are digested as nutrients when they enter the remaining stomachs of the ruminant.
   d. All of the above are ways in which the ectosymbionts of ruminant animals are useful to their hosts.

5. The microorganisms normally associated with a particular tissue or structure are referred to as
   a. indigenous microbial populations.
   b. microbiota.
   c. microflora.
   d. All of the above are correct.

6. Why does the skin have an environment unfavorable for colonization by microorganisms?
   a. It has a slightly alkaline pH, which prevents growth of many microorganisms.
   b. It is subject to periodic drying, which is harmful to many microorganisms.
   c. It is hyperosmotic because of salt produced by sweat glands, which stresses most microorganisms.
   d. All of the above are reasons why the skin has an environment unfavorable for colonization by microorganisms.

7. Acne, which is caused by the bacterium *Propionibacterium acnes,* is more prevalent in adolescents because
   a. it only colonizes the skin during adolescence.
   b. hormones, which are produced in greater abundance during adolescence, stimulate the production of sebum, which provides an ideal environment for the bacterium.
   c. hormones, which are produced in greater abundance during adolescence, are taken up by the bacterium and increase its reproductive rate.
   d. All of the above are reasons why acne is more prevalent in adolescents.

8. Which of the following area(s) of the human body is(are) not normally free of microorganisms?
   a. respiratory tract
   b. kidneys
   c. eyes
   d. All of the above are not normally free of microorganisms.

9. Which of the following is the reason why the respiratory tract is normally free of microorganisms?
   a. the mucociliary blanket
   b. the presence of lysozyme in mucus
   c. phagocytosis by alveolar macrophages
   d. All of the above are correct.

TRUE/FALSE

____ 1. The protozoan-termite mutualism can be tested. When placed in an environment with a high concentration of oxygen (which kills the protozoan), the termite dies of starvation because it can no longer digest cellulose. When the oxygen content is low, both survive—even though there is only wood as a food source—because the protozoa process the cellulose into compounds the insect can metabolize.

____ 2. The human fetus in utero is normally free from bacteria and other microorganisms.

____ 3. The animals in gnotobiotic colonies are otherwise normal in all ways.

____ 4. Although gnotobiotic animals are generally more susceptible to infection by microbial and other parasitic organisms, they are almost completely resistant to dysentery caused by *Entamoeba histolytica* because they lack the bacteria that this protozoan normally uses as a food source.

____ 5. The male and female genitourinary tracts have nearly identical microbiota.

____ 6. Gnotobiotic animals are usually germ-free.

CRITICAL THINKING

1. Explain the complex endosymbiont-tube worm relationship and why hydrogen sulfide has been referred to as the "sunlight" of hydrothermal vent communities.

2. The endosymbiotic theory of organelle formation suggests that chloroplasts and mitochondria may once have been free-living bacteria that infected a primitive eucaryote and established a permanent endosymbiosis. Describe the likely sequence of events in progressing from infection to what we have today. Are there any current symbioses that support this other than the chloroplast symbiosis mentioned above? If so what are they and how do they support the theory?

ANSWER KEY

Terms and Definitions

1. a, 2. f, 3. g, 4. b, 5. c, 6. e, 7. d, 8. j, 9. h, 10. i

Fill in the Blank

1. symbiosis; ectosymbiosis; endosymbiosis 2. commensalism; mutualism; parasitism 3. crustose; foliose; fruticose 4. hydrothermal; trophosome; hydrogen sulfide; organic 5. gnotobiotic; cesarean section 6. resident; transient 7. mycobiont; phycobiont

Multiple Choice

1. c, 2. d, 3. b, 4. d, 5. d, 6. a, 7. b, 8. c, 9. d

True/False

1. T, 2. T, 3. F, 4. T, 5. F, 6. F

29 Symbiotic Associations: Parasitism, Pathogenicity, and Resistance

CHAPTER OVERVIEW

This chapter continues the discussion of symbiosis and focuses on parasitism. One of the possible consequences of parasitism—pathogenicity—is also discussed. Most higher animals, including man, possess various defense mechanisms that are employed against parasitic, pathogenic organisms. Some of these are general, nonspecific mechanisms for resistance and are discussed in this chapter. Other more specific mechanisms associated with the immune system are discussed in subsequent chapters.

CHAPTER OBJECTIVES

After reading this chapter you should be able to:

- discuss the general characteristics of parasitic symbiosis
- discuss the concepts of pathogens, disease, and infection
- describe the stages which pathogens must go through—transmission, colonization, multiplication, and interference—in order to cause disease
- discuss the general, physical, chemical, and biological barriers that organisms have for defending themselves against continual parasitic invasion

CHAPTER OUTLINE

I.   Introduction
    A.   Parasitism has evolved in nearly all groups of microorganisms
    B.   There are more parasitic organisms than nonparasitic organisms
II. Host-Parasite Relationships (Parasitism)
    A.   Parasitism—a relationship in which a symbiont harms or lives at the expense of a host
    B.   Host—the body of the animal in which the parasite lives; it provides the microenvironment that shelters and supports the growth and multiplication of the parasitic organism
    C.   Parasite—usually the smaller of the two organisms; it is metabolically dependent on the host
    D.   Ectoparasite—lives on the surface of the host
    E.   Endoparasite—lives within the host
    F.   Final host—the host on (or in) which the parasite either gains sexual maturity or reproduces
    G.   Intermediate host—a host that serves as a temporary but essential environment for parasite development
    H.   Transfer host—a host that is not necessary for development but that serves as a vehicle for reaching the final host
    I.   Reservoir host—an organism (other than a human) that is infected with a parasite that can also infect humans
    J.   Infection—the state occurring when a parasite is growing and multiplying on or within a host

K. Infectious disease—a change from a state of health as a result of an infection by a parasitic organism

L. Pathogen—any parasitic organism that produces an infectious disease

M. Pathogenicity—the ability of a parasitic organism to cause a disease

N. Virulence—the degree or intensity of pathogenicity of an organism; it is determined by three characteristics of the pathogen:
   1. Invasiveness—the ability of the organism to spread to adjacent tissues
   2. Infectivity—the ability of the organism to establish a focal point of infection
   3. Pathogenic potential—the degree to which the pathogen can cause morbid symptoms (e.g., toxigenicity)

O. The final outcome of most host-parasite relationships is dependent on three main factors:
   1. The number of pathogenic organisms present
   2. The virulence of the organism
   3. The host's defenses or degree of resistance

P. Endogenous disease—a disease caused by the host's own microbiota because the host's resistance has dropped

III. Determinants of Infectious Diseases

A. Transmissibility of the pathogen—involves initial transport to the host
   1. Direct contact—coughing, sneezing, body contact
   2. Indirect contact—soil, water, food
   3. Vectors—living organisms that transmit a pathogen
   4. Fomites—inanimate objects contaminated with a pathogen that spread the pathogen

B. Attachment and colonization (multiplication) by the pathogen
   1. Pathogen must be able to adhere to and colonize host cells and tissue; this is mediated by special molecules or structures called adhesins
   2. Pathogen must compete with normal microbiota for essential nutrients

C. Invasion of the pathogen
   1. Penetration of the host's epithelial cells or tissues
      a. Pathogen-associated mechanisms involve the production of lytic substances that may:
         (1) Attack the ground substance and basement membranes of integuments and intestinal linings
         (2) Depolymerize carbohydrate-protein complexes between cells or on cell surfaces
         (3) Disrupt cell surfaces
      b. Passive mechanisms of entry involve:
         (1) Breaks, lesions, or ulcers in the mucous membranes
         (2) Wounds, abrasions, or burns on the skin surface
         (3) Arthropod vectors that penetrate when feeding
         (4) Tissue damage caused by other organisms
         (5) Endocytosis by host cells
   2. Gaining of access to deeper tissues
   3. Penetration into the circulatory system

D. Growth and multiplication of the pathogen—pathogen must find an appropriate environment

E. Toxigenicity—the capacity of an organism to produce a toxin
   1. Intoxications—diseases that result from the entry of a specific toxin into the host
   2. Toxin—a specific substance, often a metabolic product of the organism, that damages the host in some specified manner
   3. Toxemia—symptoms caused by toxins in the blood of the host
   4. Exotoxins—soluble, heat-labile proteins produced by and released from an organism; may damage the host at some remote site
      a. Neurotoxins damage nervous tissue
      b. Enterotoxins damage the small intestine
      c. Cytotoxins do general tissue damage

5. Endotoxins—molecules that are part of the cell wall of the pathogen
   a. Released only when the microorganism, whose cell wall contains it, lyses or during bacterial multiplication
   b. One of the best characterized is lipopolysaccharide (LPS), which is part of the cell wall of gram-negative bacteria
   c. Usually capable of producing fever, shock, blood coagulation, weakness, diarrhea, inflammation, intestinal hemorrhage, and/or fibrinolysis
   d. Stimulates fever directly by acting as an exogenous pyrogen; also may stimulate fever indirectly by causing macrophages to release interleukin-1, an endogenous pyrogen
   e. May affect macrophages and monocytes by binding to special LPS-binding proteins in the plasma that, in turn, bind to receptors on the macrophages and monocytes triggering cytokine release that produces endotoxin effects
6. Leukocidins—extracellular enzymes that kill phagocytic leukocytes
7. Hemolysins—extracellular enzymes that kill erythrocytes

IV. General or Nonspecific Host Immune Defense Mechanisms—mechanisms designed to prevent infection; can be general, physical, chemical, or biological
   A. General barriers
      1. Nutrition—the more malnourished the host, the greater the susceptibility to infection
      2. Acute-phase reactants—during an acute infection by one organism, qualitative and quantitative changes in the host's blood can increase or decrease the virulence of another organism or the defense capabilities of the host
      3. Fever—elevated body temperature; usually induced by exogenous or endogenous pyrogens
         a. Stimulates leukocytes into action
         b. Enhances microbiostasis (growth inhibition) by decreasing availability of iron
         c. Enhances specific activity of the immune system
      4. Age—the very young and the very old are more susceptible
      5. Genetic factors—include host temperature; metabolic, physiologic, and anatomic differences; and food procurement abilities
   B. Physical barriers
      1. Skin and mucous membranes
         a. Skin—reasonably effective mechanical barrier
         b. Mucous membranes—mucus secretions form a protective covering that contains antibacterial substances, such as lysozyme
      2. Respiratory system—aerodynamic filtration deposits organisms onto mucosal surfaces, and mucociliary blanket transports them away from the lungs
      3. Gastrointestinal tract and enzymes
         a. Stomach—gastric acid
         b. Small intestine—pancreatic enzymes, bile, intestinal enzymes, and secretory IgA
         c. Peristalsis and loss of columnar epithelial cells help eliminate pathogens
         d. Large intestine—normal microbiota may create unfavorable environment for colonization
      4. Genitourinary tract
         a. Unfavorable environment (e.g., acid pH in the vagina)
         b. Flushing action
         c. Length of urethra in males constitutes a distance barrier
      5. The eye—flushing action, lysozyme, and other antibacterial substances
   C. Chemical barriers
      1. Gastric juices, salivary glyco-proteins, lysozyme, oleic acid on the skin, urea (already discussed)
      2. Fibronectin—glycoprotein that reacts with bacteria to promote clearance, or with host cell receptors to prevent attachment

245

3. Hormones—affect inflammatory response and immune system activity; have cyclical effects on vaginal microbiota
4. Beta-lysin and other polypeptides—may lyse or otherwise damage specific types of microorganisms
5. Interferons—respond to viruses and other inducing agents to reduce the spread of viruses to neighboring cells
6. Tumor necrosis factors (α and β)—released from monocytes or macrophages, natural killer cells or various lymphocytes mediate the inflammatory response, enhance phagocytosis, and TNFX is cytotoxic for tumor cells
7. Bacteriocins—plasmid-encoded antibacterial substances produced by the normal microbiota of the body

D. Biological barriers
1. Normal indigenous microbiota may be involved in the following ways:
   a. Bacteriocin production
   b. Competition for space and nutrients
   c. Prevention of pathogen attachment
   d. Influence on specific clearing mechanisms
2. Inflammation
   a. Response to tissue injury through the release of chemical signals (inflammatory mediators)
   b. Interaction of selectins on vascular endothelial surface and integrins on neutrophil surface; promotes neutrophil extravasation
   c. Involves neutralization and elimination of the offending pathogen, usually by phagocytosis and fibrin formation, which walls off the inflamed area
   d. Characterized by redness, heat, pain, swelling, and altered function
3. Phagocytosis
   a. Phagocytic cells arriving at the site of inflammation attach and phagocytize infecting organisms
   b. Recognition is mediated through surface receptors that allow them to attach nonspecifically to a variety of organisms
   c. May damage the pathogen by respiratory burst (formation of highly reactive, toxic oxygen products, such as superoxides and peroxides)
   d. Phagosomes may fuse with lysosomes and thereby hydrolyze the invading organism
   e. Neutrophils also release defensins, a family of broad spectrum antimicrobial peptides

TERMS AND DEFINITIONS

Place the letter of each term in the space next to the definition or description that best matches it.

_____ 1. An organism that harms or lives at the expense of another organism

_____ 2. The organism that provides the microenvironment that shelters and supports the growth of the parasitic organism

_____ 3. The state occurring when a parasite is growing and multiplying on or within a host

_____ 4. A change from a state of health as a result of infection by a pathogenic, parasitic organism

_____ 5. The ability of an organism to cause disease

_____ 6. The degree or intensity of the ability to cause disease, as indicated by case fatality rates and/or the ability of the pathogen to invade the tissues of the host

_____ 7. Transfer of pathogens from host to host with no intermediate vector or host

_____ 8. Transfer of pathogens from host to environment and then to another host

_____ 9. Organisms that transfer pathogens from one host to another

_____ 10. Inanimate objects that can be contaminated with pathogens and that can therefore be involved in the spread of those pathogens

_____ 11. A condition resulting from the multiplication of pathogens in the blood and the subsequent accumulation of toxic metabolic waste products

_____ 12. Diseases that result from the entry of specific toxins into the body of the host, even though the toxin-producing organism itself may not be present

_____ 13. The capacity of an organism to produce a toxin

_____ 14. Specific symptoms caused by toxins in the blood of the host

_____ 15. Soluble proteins that are produced by an organism and released into the host; they may damage the host at some site remote from the site of infection

_____ 16. Substances that are part of the cell wall of the pathogen and that cause toxic reactions primarily at the site of the infection

_____ 17. Extracellular enzymes produced and released by a pathogen that kill leukocytes

_____ 18. Extracellular enzymes produced and released by a pathogen that kill erythrocytes

_____ 19. A specific constituent of the infecting organism that directly triggers fever production

_____ 20. Mucous secretions and hairlike appendages of the cells lining the respiratory tract that trap invading organisms and move them away from the lungs

_____ 21. A specific substance, often a metabolic product of the organism, that damages the host in some specific manner

a. adhesins
b. defensins
c. direct contact
d. endogenous disease
e. endogenous pyrogen
f. endotoxins
g. exogenous pyrogen
h. exotoxins
i. fomites
j. hemolysins
k. host
l. indirect contact
m. infection
n. infectious disease
o. integrins
p. intoxication
q. leukocidins
r. mucociliary blanket
s. parasite
t. pathogenicity
u. selectins
v. septicemia
w. toxemia
x. toxigenicity
y. toxin
z. vector
aa. virulence

_____ 22. A disease caused by the host's own microbiota when the host's resistance is impaired

_____ 23. A substance that is released from macrophages as a result of an infection, and that triggers fever

_____ 24. Broad spectrum antimicrobial peptides released from neutrophils

_____ 25. Molecules on the vascular endothelium to which neutrophils can attach during the inflammatory response

_____ 26. Molecules on the surface of neutrophils which act as selectin receptors to mediate the attachment of neutrophils to the vascular endothelium during the inflammatory response

_____ 27. Molecules on the surface of a pathogen that mediate attachment to host cells or tissues

TYPES OF INFECTIONS ASSOCIATED WITH PARASITIC ORGANISMS

Using your own words and the textbook (table 30.1), complete the following table by describing the type of infection named in the first column.

Type	Description
Abscess	
Acute	
Bacteremia	
Chronic	
Covert	
Cross	
Focal	
Fulminating	
Generalized	
Iatrogenic	
Latent	
Localized	
Mass	
Mixed	
Nosocomial	
Opportunistic	
Overt	
Phytogenic	
Primary	
Pyrogenic	
Secondary	
Sepsis	
Septicemia	
Septic shock	
Severe sepsis	
Subclinical (inapparent or covert)	
Systemic	
Terminal	
Toxemia	
Zoonosis	

TYPES OF NONSPECIFIC DEFENSE MECHANISMS

Indicate whether each of the following is a general (G), physical (P), chemical (C), or biological (B) defense mechanism. If it serves more than one purpose, list the alternatives.

Mechanism	Type
1. Skin	___
2. Interferon	___
3. Acute-phase reactants	___
4. Gastric acid	___
5. Mucociliary blanket	___
6. Inflammation	___
7. Bacteriocins	___
8. Fever	___
9. Lysozyme	___
10. Indigenous microbiota	___
11. Bile	___
12. Phagocytosis	___
13. Nutrition	___
14. Urine	___
15. Fibronectin	___
16. Defensins	___

FILL IN THE BLANK

1. An organism that causes harm or that lives at the expense of another organism is called a(n) _____ . If the organism lives on the surface of its host, it is termed a(n) _____ ; while if it lives within its host, it is called a(n) _____ .

2. The host in which the organism attains sexual maturity or reproduces is called the _____ host; a host that serves as a temporary but essential environment for development is called a(n) _____ host; a host that is not necessary for development but is used as a vehicle for reaching a final host is called a(n) _____ host; and a nonhuman organism infected with a parasitic organism that can also infect humans is called a(n) _____ host.

3. Some pathogens release a substance that kills leukocytes called _____ .

4. Many pathogens possess mechanical, chemical, or molecular abilities to damage the host and cause disease. Based on the type of participation by the pathogens, two distinct categories of disease can be recognized. If the disease is caused by a host-parasite interaction due to the pathogen's growth, metabolism, reproduction, or tissue alterations, it is called a(n) _____ ; however, if the disease results from the entry of a specific toxin into the body that causes disease even in the absence of the toxin-producing organism, it is called a(n) _____ . The term _____ is used if specific symptoms are caused by toxins in the blood of the host.

5. Soluble proteins produced and released by an organism that act at sites remote from the site of infection are called _____ . They are usually categorized according to the symptoms they cause and the tissues they affect. Therefore, toxins that cause nervous system damage are called _____ ; toxins that cause cell death are called _____ ; and those that damage the intestinal tract are referred to as _____ .

MULTIPLE CHOICE

For each of the questions below select the *one best* answer.

1. Which of the following has no effect on the outcome of the host-parasite relationship?
   a. the number of parasites on or in the host
   b. the virulence of the parasite
   c. the defenses of the host
   d. All of the above have an effect on the outcome of the host-parasite relationship.

2. An organism that transmits a parasite from one host to another is called a
   a. fomite.
   b. vector.
   c. transmitter.
   d. carrier.

3. Specific extracellular enzymes that are released by pathogens and that kill red blood cells are called
   a. leukocidins.
   b. erythrocidins.
   c. hemolysins.
   d. beta-lysins.

4. Which of the following is not a way in which fever augments the host's defenses?
   a. It inhibits the parasite's growth by raising the temperature above the optimum growth temperature.
   b. It inhibits the parasite's growth by decreasing the availability of iron to the organism.
   c. It stimulates leukocytes into action so they can kill the organism.
   d. All of the above are ways in which fever augments the host's defenses.

5. Which of the following is not used to help protect the lungs from infection?
   a. Turbulent airflow deposits airborne pathogens on sticky mucosal surfaces.
   b. The mucociliary blanket moves trapped organisms away from the lungs by ciliary action.
   c. Coughing and sneezing forcefully expel organisms away from the lungs.
   d. All of the above are used to help protect the lungs from infection.
6. Which of the following is a factor that helps determine the virulence of a pathogen?
   a. invasiveness
   b. infectivity
   c. pathogenic potential
   d. All of the above are correct.
7. Which of the following is not a mode of action associated with endotoxins?
   a. shock
   b. paralysis
   c. diarrhea
   d. All of the above are modes of actin associated with endotoxins.

8. Phagocytosis leads to destruction of engulfed pathogens by which of the following mechanisms?
   a. lysosomal mediated hydrolysis
   b. phagosomal mediated respiratory burst
   c. Both (a) and (b) are correct.
   d. Neither (a) nor (b) is correct.
9. Endotoxin is released from a pathogen when
   a. the microorganism whose cell wall contains the endotoxin lyses.
   b. during bacterial multiplication.
   c. Both (a) and (b) are correct.
   d. Neither (a) nor (b) is correct.

TRUE/FALSE

_____ 1. All parasites are metabolically dependent on their hosts.

_____ 2. Direct contact spread of a pathogen implies that there is actual physical contact between the current host and the new host. Therefore, transfer of a respiratory pathogen by a sneeze or a cough is not considered direct transmission.

_____ 3. It is important for endoparasites to actually penetrate the cells of the host.

_____ 4. Some pathogens have mechanisms that allow them to penetrate into host epithelial cells, while others can enter only if the epithelial cells have been damaged by other means (wounds, insect bites, etc).

_____ 5. Generally, exotoxins tend to be more heat stable than endotoxins.

_____ 6. Fever can be triggered either by an exogenous pyrogen released by the pathogen or by an endogenous pyrogen released by macrophages in response to a pathogen.

_____ 7. The net effect of inflammation is the walling off of the infected area by fibrin action and the elimination of the pathogens by phagocytosis.

_____ 8. One of the symptoms of inflammation is a generalized fever.

_____ 9. Endotoxin effects can be mediated by cytokines released from monocytes and macrophages to which the endotoxin has bound.

_____ 10. Colonization specifically refers to the multiplication of a pathogen on or within a host.

CRITICAL THINKING

1. Compare and contrast exotoxins and endotoxins. Discuss the chemical and physiological characteristics of the molecules, as well as the mechanisms of pathogenesis.

2. The skin is constantly being exposed to pathogenic, parasitic organisms. However, it is a very effective barrier against infection: it is not easily colonized and it is not readily penetrated. Discuss the various properties of the skin that make it such an effective barrier against colonization and penetration.

ANSWER KEY

Terms and Definitions

1. s, 2. k, 3. m, 4. n, 5. t, 6. aa, 7. c, 8. l, 9. z, 10. i, 11. v, 12. p, 13. x, 14. w, 15. h, 16. f, 17. q, 18. j, 19. g, 20. r, 21. y, 22. d, 23. e, 24. b, 25. u, 26. o, 27. a

Types of Nonspecific Defense Mechanisms

1. P(C), 2. C, 3. G, 4. C, 5. P, 6. B, 7. C(B), 8. G, 9. C, 10. B(C), 11. C, 12. B, 13. G, 14. C(P), 15. C, 16. B(C)

Fill in the Blank

1. parasite; ectoparasite; endoparasite 2. final; intermediate; transfer; reservoir 3. leukocidin
4. infectious disease; intoxication; toxemia 5. exotoxins; neurotoxins; cytotoxins; enterotoxins

Multiple Choice

1. d, 2. b, 3. c, 4. a, 5. d, 6. d, 7. b, 8. c, 9. c

True/False

1. T, 2. F, 3. F, 4. T, 5. F, 6. T, 7. T, 8. F, 9. T, 10. T

30   The Immune Response: Antigens and Antibodies

CHAPTER OVERVIEW

This chapter introduces specific immunity, the defense mechanism that involves the recognition of infectious agents, their products, some tumor cells, and certain macromolecules as foreign materials. The body must produce specific defensive responses that will destroy or neutralize these materials. This complex system involves several different types of responses and may lead to several different problems of various system malfunctions. In this chapter, some aspects of the immune system are introduced; other aspects are discussed in subsequent chapters.

CHAPTER OBJECTIVES

After reading this chapter you should be able to:

- explain the differences between specific and nonspecific immunity, artificial and natural immunity, and active and passive immunity
- describe the types of lymphocytes involved in immune responses and the different ways they respond to foreign substances
- describe the basic structure of antibody molecules and the different classes (isotypes) of these molecules
- describe the specificity and diversity of antibody molecules
- discuss the use of hybridomas to produce highly specific antibody molecules
- discuss the production and potential uses of catalytic antibodies

CHAPTER OUTLINE

    I.   Introduction
        A.   Immunity—the general ability of a host to resist a particular disease
        B.   Immune response—a specific and complex series of events throughout the animal's body that helps it defend against disease-causing organisms or substances
        C.   Immunology—the science that deals with these immune responses
        D.   Immunobiology—a more current term used to describe the biological basis for host defenses, growth and development, hypersensitivity, heredity, aging, cancer and transplantation
   II.   Nonspecific Resistance—general, physical, chemical, and biological barriers against disease (discussed in chapter 29); also referred to as innate or natural immunity

III. Specific Immunity—certain cells (lymphocytes) recognize the presence of foreign substances (antigens or immunogens) and act to eliminate them either by direct action or by producing specialized proteins (antibodies) that destroy the antigen or target it for destruction by other cells; also referred to as acquired or adaptive immunity
   A. Naturally acquired active immunity—an individual comes in contact with an antigen and produces sensitized lymphocytes and/or antibodies that inactivate the antigen
   B. Naturally acquired passive immunity—transfer (e.g. transplacentally or in breast milk) of antibodies from one individual (where they were actively produced) to another (where they are passively received)
   C. Artificially acquired active immunity—deliberate exposure of an individual to a vaccine (an inactivated or weakened form of the antigen) with subsequent development of an immune response
   D. Artificially acquired passive immunity—deliberate introduction of antibodies (produced elsewhere) into an individual

IV. Origin of Lymphocytes
   A. Derived from bone marrow stem cells
   B. May differentiate in thymus to become T lymphocytes (occurs during early childhood)
   C. May differentiate in fetal liver or adult bone marrow to become B lymphocytes (continues throughout life)
   D. Natural killer (NK) cells are probably derived from prethymic lymphocytes but do not have the characteristics of either B-cells or T-cells

V. Function of Lymphocytes
   A. B cells
      1. When exposed to antigen, they become plasma cells and secrete antibodies
      2. This is referred to as humoral immunity or antibody-mediated immunity
      3. Humoral immunity defends against bacteria, toxins, and viruses in body fluids
   B. T cells
      1. When exposed to antigens, they attack infected cells and release chemical mediators (cytokines) that augment the body's defenses
      2. Must be in close contact with foreign or infected cells
      3. This is referred to as cell-mediated immunity
      4. Cell-mediated immunity defends against
         a. Host cells parasitized by a microorganism
         b. Tissue cells transplanted from one host to another
         c. Cancer cells
   C. Natural killer (NK) cells
      1. Nonspecifically kill tumor cells, virus-infected cells, and other parasite-infected cells
      2. Play a role in regulating the immune response
      3. Exhibit antibody dependent cytotoxicity

VI. Antigens
   A. Prior to birth, the immune system removes most T cells specific for self-recognition determinants
   B. Antigens (immunogens) are foreign substances, such as proteins, nucleoproteins, polysaccharides, and some glycolipids, to which lymphocytes may respond
   C. Epitopes (antigenic determinant sites) are areas of an antigen that can stimulate production of specific antibodies and that can combine with them
   D. Valence—the number of epitopes on an antigen; determines number of antibody molecules an antigen can combine with at one time
   E. Hapten—a small organic molecule that is not itself antigenic but that may become antigenic when bound to a larger carrier molecule

VII. Antibodies (Immunoglobulins)—a group of glycoproteins in the blood, serum, and tissue fluids of mammals; they are produced in response to an antigen and can combine specifically with that antigen
  A. Immunoglobulin structure
    1. Multiple antigen-combining sites (usually two; some can form multimeric antibodies with up to ten combining sites)
    2. Composed of four polypeptide chains (two heavy or long, and two light or short) that form a flexible Y with a hinge region
    3. The stalk of the Y (called the Fc) is constant in amino acid sequence (i.e., the amino acid sequences of antibodies of the same subclass do not vary significantly)
    4. The arms of the Y have variable regions and constitute the antigen-binding domains (Fab)
    5. Within the Fab segment are hypervariable (or complementarity determining) regions; these regions are responsible for the diversity of antibodies
    6. There are five types of heavy chains that determine the five classes (isotypes) of immunoglobulins (IgG, IgA, IgM, IgD, and IgE)
    7. In IgG there are four subclasses, and in IgA there are two subclasses
    8. Categories of immunoglobulin types
      a. Isotypes—variations in the constant regions of heavy chains that are associated with different classes and subclasses
      b. Allotypes—genetically controlled allelic forms of the immunoglobulin molecule
      c. Idiotypes—individual-specific immunoglobulin molecules that differ in the hypervariable regions of the Fab segments
  B. Immunoglobulin function
    1. Fab region binds to antigen
    2. Fc region mediates binding to host tissue, to various cells of the immune system, to some phagocytic cells, or to the first component of the complement system
    3. Binding of antibody to an antigen does not destroy the antigen, but marks (targets) the antigen for immunological attack or attack by the nonspecific defense mechanisms that do destroy it
    4. Opsonization—coating a bacterium with antibodies to stimulate phagocytosis
  C. Immunoglobulin classes (Isotypes)
    1. IgG—monomeric protein, 70% to 75% of Ig pool
      a. Antibacterial and antiviral
      b. Enhances opsonization; neutralizes toxins
      c. Only IgG is able to cross placenta (naturally acquired passive immunity for newborn)
      d. Activates the complement system by the classical pathway
      e. Four subclasses with some differences in function
    2. IgM—pentameric protein, 10% of Ig pool
      a. First antibody made during B-cell maturation
      b. First antibody secreted into serum during primary antibody response
      c. Never leaves the bloodstream
      d. Activates complement by classical pathway
      e. Enhances phagocytosis of target cells
      f. Agglutinates bacteria and foreign red blood cells
      g. Up to 5% may be hexameric which is better able to activate the complement system than pentameric IgM
    3. IgA—dimeric protein, 15% of Ig pool
      a. Associated with secretory mucosal surfaces
      b. Protects gastrointestinal tract, respiratory tract, and genitourinary tract
      c. Also found in saliva, tears, and breast milk (protects nursing newborns)
      d. Secretory form (sIgA) helps rid the body of antibody-antigen complexes by excretion into the gut lumen and subsequent excretion from the body

4. IgD—monomeric protein, trace amounts in serum
   a. Abundant on surface of B cells
   b. May play a role in B-cell recognition of antigens
   c. Does not activate the complement system
   d. Cannot cross the placenta
5. IgE—monomeric protein, less than 1% of Ig pool
   a. Skin-sensitizing and anaphylactic antibodies
   b. When an antigen cross-links two molecules of IgE on the surface of a mast cell or basophil, it triggers release of histamine, and it increases intestinal motility, which helps to eliminate helminthic parasites

VIII. Diversity of Antibodies—a number of mechanisms contribute to the generation of antibody diversity
   A. Combinatorial joining
      1. Ig genes are interrupted or split genes for segments of the variable regions of heavy and light chains
      2. During differentiation of B cells, these genes are rearranged on the chromosome to form various combinations
      3. The number of different antibodies possible is the product of the number of light chains possible and the number of heavy chains possible
   B. Imprecise joining—during combinations, the same segments can be joined at different nucleotides, thus increasing the number of codons and the possible diversity
   C. Somatic mutations—the V regions of germ-line DNA are susceptible to a high rate of somatic mutation during B-cell development
   D. The total diversity produces more than 2×10^8 different antibody molecules

IX. Specificity of Antibodies—clonal selection theory
   A. Because of combinatorial joining and somatic mutation, there are a small number of B cells capable of responding to any given antigen; each group of cells is derived asexually from a parent cell and is referred to as a clone; there is a large, diverse population of B-cell clones that collectively are capable of responding to many possible antigens
   B. The surface receptor molecules of the B cells bind to the appropriate antigen
   C. The cell is then stimulated to divide and differentiate into two populations of cells: plasma cells and memory cells
   D. Plasma cells are protein factories that produce about 2,000 antibodies per second for their brief life span (5–7 days)
   E. Memory cells, like the original B cells, can differentiate into plasma cells if they are stimulated by being bound to the antigen; because there are more memory cells than original B cells, the secondary (anamnestic) response can be (and usually is) faster and larger than the primary response; they have long life spans (years or decades)

X. Sources of Antibodies (pure homogeneous preparations)
   A. Immunization
      1. Procedure
         a. Inject animal with antigen to stimulate primary immune response
            (1) Initial lag phase of several days
            (2) Log phase; antibody titer rises logarithmically
            (3) Plateau phase; antibody titer stabilizes
            (4) Decline phase; antibody titer decreases because the antibodies are metabolized or cleared from the circulation
            (5) Mostly IgM; low-affinity antibodies
         b. After a period of time, give animal a series of booster injections with same antigen to stimulate secondary immune response, or anamnestic response
            (1) Shorter lag phase, higher antibody titer
            (2) Mostly IgG; high-affinity antibodies (affinity maturation)

  c. Withdraw blood from animal; allow it to clot; and remove fluid (serum), which is referred to as antiserum since it is from a specifically immunized host

  2. Limitations

   a. This method results in polyclonal antibodies which have different epitope specificities; thus sensitivity is lower, and the antibodies often cross-react with closely related antigens

   b. Repeated injections with antiserum from one species into another can cause serious allergic reactions

   c. Antiserum contains a mixture of antibodies, not all of which are of interest

 B. Hybridomas—overcome some of the limitations of antisera

  1. Inject animals with antigen

  2. Separate spleen cells (which contain plasma cells)

  3. Fuse spleen cells with myeloma cells (tumor cells of the immune system that produce large quantities of antibodies and that are easy to culture)

  4. Culture fused cells (hybridomas) so that each grows into a separate colony

  5. Some plasma cells that fused with a myeloma cell will produce the desired antibody

  6. Screen colonies for those producing desired antibody

  7. Can grow many desired colonies to obtain large amounts of antibody

  8. Antibodies produced by this method are monoclonal (react with only one epitope) since they come from the fusion of a single plasma cell with the tumor cell

  9. Have a variety of uses in which high specificity is required

   a. Tissue typing for transplants

   b. Identification and epidemiological study of infectious microorganisms

   c. Identification of tumor and other surface antigens

   d. Classification of leukemias and T-cell populations

   e. Sensitive diagnostic procedures

   f. Immunotoxins—used in the targeted delivery of a toxic substance to a particular cell type; chemotherapeutic agents

XI. Catalytic Antibodies

 A. They are monoclonal antibodies made by subjecting an animal to an enzyme-substrate transition-state analogue and then producing a hybridoma

 B. The binding pocket on the antibody lowers the energy of activation of a reaction by ensuring the proper orientation of the reactant(s)

 C. Currently, catalytic antibodies have been produced that can transform relatively simple compounds, but the potential is great if antibodies that act on proteins and nucleic acids can be produced

TERMS AND DEFINITIONS

Place the letter of each term in the space next to the definition or description that best matches it.

_____ 1. Immunity that develops when a person comes in contact with an antigen during the normal course of activities

_____ 2. Immunity that develops when a person is deliberately exposed to an antigen through a vaccination procedure

_____ 3. Immunity that involves the transfer of antibodies from one host to another either across the placenta or through breast-feeding

_____ 4. Immunity that involves the transfer of antibodies from one host to another, usually by deliberate injection

_____ 5. A preparation of killed microorganisms; living, weakened microorganisms; or inactivated toxins that is administered to stimulate the development of immunity

_____ 6. Immunity that is mediated by soluble antibody proteins in the blood, and that does not require the direct participation of the cells that produced the antibodies once these antibodies have been synthesized and released

_____ 7. Immunity that involves the direct participation of T cells

_____ 8. Foreign substances, such as proteins, nucleoproteins, polysaccharides, and some glycolipids, to which lymphocytes respond

_____ 9. Areas of a molecule that stimulate production of specific antibodies and that combine with specific antibodies

_____ 10. Small organic molecules that are not themselves antigenic, but that will stimulate an immune response when coupled to a larger carrier molecule

_____ 11. Proteins that are produced in response to a foreign substance and combine with that foreign substance

_____ 12. An antibody made against a transition-state complex that can be used to catalyze a reaction

_____ 13. An antibody coupled to a chemotherapeutic agent that can be used to target the agent's delivery to a particular cell type

a. antibodies
b. antigens
c. artificially acquired active immunity
d. artificially acquired passive immunity
e. catalytic antibodies
f. cell-mediated immunity
g. epitopes
h. haptens
i. humoral immunity
j. immunotoxin
k. naturally acquired active immunity
l. naturally acquired passive immunity
m. vaccine

FILL IN THE BLANK

1. Foreign substances to which lymphocytes respond are called _____. Each of these molecules can have a number of different areas that stimulate the production of antibodies and that combine with antibodies. These areas are called _____, and the number of these areas on the surface of the molecule is the _____ of the molecule.

2. Variations in the constant regions of the heavy chains of immunoglobulins associated with the different classes and subclasses are referred to as _____; genetically controlled allelic forms of the immunoglobulin molecule are called _____; and individual specific immunoglobulin molecules differing in the hypervariable region of the Fab region are referred to as _____.

3. IgA is normally found as a _____ molecule, and consists of two monomers held together by a special polypeptide called a _____. During transport of the IgA from the mucosa-associated lymphoid tissue to the mucosal surfaces, it acquires a protein termed the _____ component. This modified molecule is found in saliva, tears, and breast milk. In breast milk it serves to protect nursing newborns by binding with antigens in the intestine and preventing absorption. This protective mechanism is referred to as _____.

4. Antibody diversity can be generated in a variety of ways. One involves chromosomal rearrangement of the gene segments that code for various regions of the antibody molecule to form a single gene combination from among many possible arrangements. This is called _____ and is responsible for the creation of more than 10 million possible antibody molecules. This is further augmented by the hypervariable regions' high rate of _____, which can lead to different amino acid sequences and, therefore, to different antibody molecules.

5. The currently accepted theory of antibody diversity suggests that a few B cells in the body can respond to a particular antigen *before* the organism is ever exposed to that antigen. Exposure to that antigen stimulates the B cells to divide and differentiate into two types of cells: _____ cells, which are short lived and produce large quantities of antibody specific for the stimulating antigen, and _____ cells, which are long lived and are responsible for the faster and better response to subsequent antigen exposures. This augmented secondary response is referred to as the _____ response. This theory of antibody formation is called the _____ theory.

MULTIPLE CHOICE

For each of the questions below select the *one best* answer.

1. A woman who is now pregnant was previously exposed to a particular disease-causing organism. As a result, she developed antibodies against that pathogen that now cross the placenta and enter the bloodstream of her unborn child. When the child is born, it will be protected against that pathogen for a short time. The protection afforded the infant is called
   a. artificially acquired active immunity.
   b. artificially acquired passive immunity.
   c. naturally acquired active immunity.
   d. naturally acquired passive immunity.

2. A child is given an attenuated strain of poliovirus, which will protect him or her from ever contracting the disease poliomyelitis. This is an example of
   a. artificially acquired active immunity.
   b. artificially acquired passive immunity.
   c. naturally acquired active immunity.
   d. naturally acquired passive immunity.

3. A child develops a case of measles that was acquired from a classmate. Upon recovery from this infection, the child will be permanently protected from contracting this disease again. This is an example of
   a. artificially acquired active immunity.
   b. artificially acquired passive immunity.
   c. naturally acquired active immunity.
   d. naturally acquired passive immunity.

4. A farmer steps on a dirty nail and sustains a deep puncture wound. His physician is concerned about the possibility of tetanus and administers tetanus antitoxin, which contains antibodies produced in other humans against the inactivated tetanus toxin. This is an example of
   a. artificially acquired active immunity.
   b. artificially acquired passive immunity.
   c. naturally acquired active immunity.
   d. naturally acquired passive immunity.

5. Which of the following are not targets for T cells?
   a. host cells that have been parasitized by microorganisms
   b. tissue cells that have been transplanted from one host to another
   c. cancer cells
   d. All of the above are targets for T cells.
6. T cells produce and secrete proteins that do not directly interact with invading microorganisms, but that augment the body's defense mechanisms. These molecules are called
   a. antibodies.
   b. cytokines.
   c. immunogens.
   d. augmenters.
7. Which of the following is not a function of natural killer cells?
   a. nonspecific destruction of tumor cells
   b. nonspecific destruction of cells infected by viruses and other parasites
   c. regulation of the immune response
   d. All of the above are functions of natural killer cells.
8. Which of the following is not a function of the Fc portion of the immunoglobulin molecule?
   a. point of attachment of immunoglobulin to host cells, including some cells of the immune system and some phagocytic cells
   b. point of attachment of immunoglobulin to the first component of the complement system
   c. point of attachment of immunoglobulin to the antigen
   d. All of the above are functions of the Fc portion of the immunoglobulin molecule.

9. The only immunoglobulin able to cross the placenta is
   a. IgA.
   b. IgM.
   c. IgG.
   d. IgD.
10. Which of the following is not considered a possible benefit of the use of monoclonal antibodies?
    a. tissue typing for transplants
    b. classification of T-cell populations
    c. targeted delivery of a chemotherapeutic agent
    d. All of the above are possible benefits of monoclonal antibodies.
11. The specific regions within the variable region on an antibody molecule that are responsible for antibody diversity and antigen specificity are referred to as _____ regions.
    a. hypervariable
    b. complementarity determining
    c. Both (a) and (b) are correct.
    d. Neither (a) nor (b) is correct.
12. Which of the following is a characteristic associated with secondary antibody responses?
    a. shorter lag phase
    b. higher antibody titer
    c. higher antibody affinity
    d. All of the above are characteristics associated with secondary antibody responses.
13. The change from low affinity antibody to high affinity antibody during secondary antibody response is referred to as
    a. isotype switching.
    b. affinity maturation.
    c. anamnestic response.
    d. idiotype diversity.

TRUE/FALSE

_____ 1. One of the major advantages of active immunity over passive immunity is that active immunity lasts a long time (years or decades), while passive immunity is very temporary (weeks or months at most).
_____ 2. An inactivated preparation of a toxin is called a toxoid.
_____ 3. T cells have been so named because they mature in the thyroid.
_____ 4. B cells have been so named because in birds they mature in an appendage of the cloaca called the bursa of Fabricius.
_____ 5. Small molecules that are not antigenic but that can stimulate an immune response when coupled to a large carrier molecule are called haptens.

_____ 6. Opsonization refers to the ability of an antibody to stimulate phagocytosis.
_____ 7. The secondary immune response is faster and results in a higher antibody titer. It is also called the anamnestic response.
_____ 8. A small percentage of IgM exists in a hexameric form rather than a pentameric form. The hexameric form is better able to activate the complement system than is the pentameric form.

CRITICAL THINKING

1. For vaccines, live, weakened (attenuated) strains of an organism are more advantageous than killed preparations. Discuss the reasons for this.

2. Using diagrams, discuss how combinatorial joining leads to the variety of possible antibody molecules that the human immune system is capable of generating.

ANSWER KEY

Terms and Definitions

1. k, 2. c, 3. l, 4. d, 5. m, 6. i, 7. f, 8. b, 9. g, 10. h, 11. a, 12. e, 13. j

Fill in the Blank

1. antigens; epitopes; valence 2. isotypes; allotypes; idiotypes 3. dimeric; J chain; secretory; immune exclusion 4. combinatorial joining; somatic mutation 5. plasma; memory; anamnestic; clonal selection

Multiple Choice

1. d, 2. a, 3. c, 4. b, 5. d, 6. b, 7. d, 8. c, 9. c, 10. d, 11. c, 12. d, 13. b

True/False

1. T, 2. T, 3. F, 4. T, 5. T, 6. T, 7. T, 8. T

31 The Immune Response: Chemical Mediators, B- and T-Cell Biology, and Immune Disorders

CHAPTER OVERVIEW

This chapter continues the discussion of the immune system with a description of the chemical mediators involved in the immune response and of the biology of T and B cells. The chapter also describes disorders of the immune system, including hypersensitivities as well as genetic and induced disorders. Finally, problems associated with the immune system, such as transplant rejection and blood transfusion reactions, are discussed.

CHAPTER OBJECTIVES

After reading this chapter you should be able to:

- discuss the various cytokines involved in the immune response and the roles that they play
- discuss the origins, development, and activities of B and T cells
- discuss superantigens
- discuss the four types of hypersensitivities (allergies) and the roles of the various immune system components in mediating these hypersensitivities
- discuss autoimmune diseases and immune deficiencies
- describe the role of the immune system in transplant rejection and blood transfusion reactions
- discuss cell-associated differentiation antigens (CDs) and their functions as cell surface receptors

CHAPTER OUTLINE

I.   Cytokines
A.   Cytokines are glycoproteins released by one cell population that acts as an intercellular mediator
1.   Monokines—released from mononuclear phagocytes
2.   Lymphokines—relapsed from T lymphocytes
3.   Interleukins—released from a leukocyte and act on another leukocyte
4.   Colony stimulating factors (CSFs)—effect is to stimulate growth and differentiation of immature leukocytes in the bone marrow
B.   Cytokines can affect various cell populations
1.   Autocrine function—affect the same cell responsible for its production
2.   Paracrine function—affect nearby cells
3.   Endocrine function—distributed by circulatory system to target cells
C.   Exert their effects by binding to cell-surface receptors called cell association differentiation antigens (CDs); possible effects include
1.   Stimulation of cell division
2.   Stimulation of cell differentiation
3.   Inhibition of cell division
4.   Apoptosis—programmed cell death

         5.   Stimulation of chemotaxis

         6.   Chemokinesis—direct cell movement

   D.   Mediators of Natural or Nonspecific Immunity

         1.   Interferon-α (IFN-α)—when secreted by infected macrophages and monocytes, it protects neighboring cells that have not yet been infected; also increases Fc receptors on macrophages and activates the lytic activity of NK cells

         2.   Interferon-β (IFN-β)—produced by fibroblasts and other cells; activates an antiviral state, modulates antibody production, induces NK activity, modulates cytotoxic T-cell activity and regulates production of cytokines and other molecules in certain cells

         3.   Tumor Necrosis Factor-α (TNF-α)—mediates host response to gram-negative bacteria and modulates the inflammatory response; also acts as an endogenous pyrogen and is cytotoxic for tumor cells

         4.   Interleukin-1 (IL-1)—produced by a variety of cells; modulates $CD4^+$ cell proliferation in the immune response and mediates production of other cytokines involved in the inflammatory response

         5.   Chemokines—low molecular weight, proinflammatory, paracrine cytokines

   E.   Activators of Effector Cells

         1.   Interferon-γ (IFN-γ)—produced by T_H1, T_C, and NK cells and functions in amplifying the activation of T-helper cells and other activities

         2.   Migration inhibition factor (MIF)—inhibits the migration of macrophages away from the site of infection

   F.   Mediators of Mature Cell Activation, Growth, and Differentiation—Interleukin-2 (IL-2)—produced by T_4 cells and stimulate T and B cell proliferation; successfully used as therapeutic agent against certain cancers

   G.   Mediators of Immature Cell Growth and Differentiation

         1.   Interleukin-3 (IL-3)—stimulates hematopoiesis

         2.   CSFs—stimulate differentiation of various leukocytes

   H.   Interleukins 8 through 16—mediate various body responses to infections (see Table 31.5)

II.   B-Cell Biology

   A.   Cells differentiate in bone marrow and lymphoid tissue

   B.   Have IgM and IgD on surface—both function as transmembrane B cell antigen receptors (BCRs)

   C.   Activated by antigen binding (clonally specific) or by mitogens (nonspecific)

   D.   T-dependent antigen triggering—causes production of IgG, IgA, and IgE

         1.   A macrophage ingests the antigen or antigen-bearing organism

         2.   The antigen is processed and appears on the surface of the antigen-presenting macrophage

         3.   A T-helper cell is signaled by this complex and by IL-1

         4.   The T cell divides and produces IL-2, which stimulates more T-cell division

         5   These T cells bind to B cells presenting the appropriate antigen on their surface

         6.   These T cells secrete B-cell growth factor (BCGF), which causes B cells to divide

         7.   These T cells also secrete B-cell differentiation factor (BCDF), which causes the B cells to differentiate into plasma cells and produce antibodies

         8.   B cells recognize antigen through BCRs; those that are stimulated differentiate into plasma cells and secrete antibody

   E.   T-independent antigen triggering causes production of IgM; used for polymeric antigens with a large number of identical epitopes; antibodies have low affinity and never switch to high-affinity IgG or other isotypes

III.   T-Cell Biology

   A.   T-cell antigen receptors—closely related to immunoglobulins, but cannot bind to free antigens; antigen is presented by an antigen-presenting cell (usually a macrophage) or a dendritic cell

   B.   Major histocompatibility complex (MHC) molecules—proteins encoded by a group of genes called the major histocompatibility complex (MHC) genes; comprise three classes (also known as human leukocyte antigens (HLA))

1. Class I—made by all cells except red blood cells; function to identify cells as "self"; primary basis of HLA typing for organ transplant; also function to present endogenous antigens to T cells
2. Class II—produced only by activated macrophages, mature B cells, some T cells, and certain cells of other tissues; function in T-cell recognition of presented exogenous antigens
3. Class III—involved in the classical and alternate complement pathways
C. Regulator T cells—control the development of effector T cells
    1. T-helper cells (T_H) are needed for T-dependent antigen triggering; recognize antigen-HLA combination on antigen-presenting cells; secrete interleukin-2 (IL-2) and activate cytotoxic T (T_C) cells; there are three types, each of which secretes a different mixture of cytokines
    2. T-suppressor cells (T_S) can be stimulated by a specific antigen to proliferate (mediated by IL-2), and to suppress immune responses; provide a negative feedback control
D. Acquired immune tolerance—nonresponse to self
    1. Clonal deletion—early in development, lymphocytes with ability to interact with self-antigens are destroyed in the thymus; mechanism is unknown
    2. Functional inactivation (clonal anergy)—lymphocytes that can interact with self-antigens are present but are inhibited either by T_S cells, or by their entry into an unresponsive state known as clonal anergy
E. Effector cells—directly attack specific target cells
    1. Cytotoxic T cells (T_C) recognize infected cells with appropriate class I MHC proteins, and are stimulated by IL-2 secreted by T_H cells to divide; they produce lymphokines that mediate destruction of infected cells either by cytolysis or by apoptosis (a programmed cell death); two possible mechanisms are:
        a. CD95 pathway—transmembrane signal transductions
        b. Perforin pathway—release of perforins that damage the target cell
    2. Delayed-hypersensitivity T cells (T_{DTH}) are responsible for type IV delayed-type hypersensitivity via secretin of lymphokines; they also activate macrophages and inhibit their migration away from the site of an inflammation
    3. Natural killer (NK) cells are nonphagocytic and granular (they may not be T cells); they do not require prior exposure to relevant antigens; they recognize cell-surface changes on virus-infected and tumor cells; they also recognize fungi, bacteria, protozoa, and helminth parasites (immune surveillance); they kill by cell-mediated lysis as follows:
        a. The NK cells are activated by interferon and/or IL-2
        b. They recognize class I MHC proteins on the target cell and bind to the cell
        c. They insert perforin I (a pore-forming protein) into the plasma membrane of the target cell
        d. They also secrete lysosomal hydrolases
        e. The attacked cell thereby undergoes cytolysis
        f. They also have Fc receptors that function by antibody-dependent cell-mediated cytotoxicity
IV. Superantigens
    A. Bacterial proteins that stimulate the immune system more extensively than do normal antigens
    B. Stimulate T cells to proliferate nonspecifically through interaction with class II MHC molecules and T-cell receptors in the absence of an antigenic peptide for which the T-cell receptor is specific
    C. Cause symptoms by way of release of massive quantities of cytokines
    D. Associated with various chronic diseases including rheumatic fever, arthritis, and others
V. Hypersensitivities—exaggerated or inappropriate immune responses that result in tissue damage to the individual
    A. Type I hypersensitivity—includes allergic reactions
        1. Occurs immediately following second contact with responsible antigen (allergen)
        2. On first exposure, B cells form plasma cells that produce IgE instead of IgG
        3. IgE binds to mast cells or basophils via Fc receptors and sensitizes them

266

4. Upon subsequent exposure, the allergen binds to these IgE-bearing cells; physiological mediators released by this binding cause anaphylaxis—smooth muscle contraction, vasodilation, increased vascular permeability, and mucus secretion
5. Systemic anaphylaxis results from a massive release of these mediators, which cause respiratory impairment, lowered blood pressure, and serious circulatory shock; death can occur within a few minutes
6. Localized anaphylaxis
   a. Hayfever if in upper respiratory tract
   b. Bronchial asthma if in lower respiratory tract
   c. Hives (skin eruptions) if food allergy
7. Desensitization to allergens involves controlled exposure to the allergen in order to stimulate IgG production; IgG molecules serve as blocking antibodies that intercept and neutralize the allergen before it can bind to the IgE-bound mast cells

B. Type II hypersensitivity—generally cytolytic or cytotoxic
1. IgG or IgM antibodies are directed against cell surface or tissue antigens
2. Cell destruction is mediated by lysis (complement activation) or by toxic mediators
3. Interferes with normal mechanisms for eliminating toxic materials
4. An example is a blood transfusion reaction in which donated blood cells are attacked by the recipient's antibodies

C. Type III hypersensitivity
1. Immune complexes are usually removed by monocytes
2. Sometimes if immune complexes are not efficiently removed, they trigger complement-mediated inflammation such as
   a. Vasculitis—inflammation of the blood vessels
   b. Glomerulonephritis—inflammation of the kidney glomerular basement membranes
   c. Arthritis—inflammation of the joints
3. Groups of type III reactions
   a. Persistent viral, bacterial, or protozoan infection, combined with a weak antibody response, leads to chronic immune complex formation and deposition in the tissues of the host
   b. Autoimmune diseases, such as systemic lupus erythematosus, lead to the prolonged production of autoantibodies and, therefore, to chronic immune complex formation; this overloads the reticuloendothelial system, which leads to immune complex deposition in the tissues
   c. Repeated inhalation of allergens can cause immune complex deposition at body surfaces (e.g., in the lungs in farmer's lung disease)

D. Type IV hypersensitivity
1. Involves T_{DTH} lymphocytes, which proliferate at the site of antigen processing by macrophages
2. The cytokines released by the T_{DTH} cells attract macrophages and basophils to the area, leading to inflammatory reactions
3. Can be used diagnostically, as in the tuberculin skin test
4. Examples of type IV hypersensitivities
   a. Allergic contact dermatitis (poison ivy, cosmetic allergies)
   b. Some chronic diseases (leprosy, tuberculosis, leishmaniasis, candidiasis, herpes simplex lesions) involve chronic T_{DTH} stimulation

VI. Autoimmune Diseases—autoimmunity is characterized by the presence of autoantibodies and is a natural consequence of aging; autoimmune disease results from activation of self-reactive T and B cells that lead to tissue damage
A. More common in older individuals

B. May involve viral or bacterial infections that cause tissue damage and that thereby release abnormally large quantities of antigen and/or cause some self-proteins to alter their form so that they are no longer recognized as self

VII. Transplantation (Tissue) Rejection

A. Autograft—donor and recipient are same individual (e.g., skin grafts)

B. Isograft—donor and recipient are identical twins

C. Allograft—donor and recipient are genetically different individuals of the same species

D. Xenograft—donor and recipient are different species

E. Mechanisms of tissue rejection reaction

1. Foreign MHC class II antigens may trigger T_H cells to help T_C cells destroy the graft; the T_C cells recognize the graft as foreign by recognizing the MHC class I antigens

2. The cells may react directly with the graft, releasing cytokines that stimulate macrophages to enter the graft and destroy it

3. Graft vs. Host Reaction—immunocompetent cells in donor tissue (e.g., bone marrow) reject the host

F. Ways to minimize rejection

1. Tissue matching—look for the best match of MHC (particularly class I) antigens

2. Immunosuppressive drugs—these however increase susceptibility to infections and cancer

G. Immunologically privileged site (e.g., cornea)—not subject to graft rejection because lymphocytes do not circulate there

VIII. Immunodeficiencies—failure to recognize and/or respond properly to antigens

A. Primary (congenital) immunodeficiencies result from a genetic disorder

B. Secondary (acquired) immunodeficiencies result from infection by immunosuppressive microorganisms (e.g., AIDS, chronic mucocutaneous candidiasis)

IX. Human Blood Types

A. The surface of red blood cells contains genetically determined sets of molecules

B. ABO grouping—based on differences in glycosphingolipid molecules

1. Type A—have type A glycoprotein molecules

2. Type B—have type B glycoprotein molecules

3. Type AB—have both types

4. Type O—have neither type on their surface

C. Antibodies exist in the bloodstream to those molecules that an individual does not have, even though no prior exposure has occurred (this is different from other antibody productions); the reasons for their existence are not entirely clear, but may be the result of exposure to similar antigens that occur in the normal intestinal microbiota

D. Transfusion of the wrong type of blood will lead to agglutination, which can block blood vessels and cause serious circulation problems

E. Rh system—based on presence (Rh^+) or absence (Rh^-) of the Rh or D antigen on the surface of the erythrocytes

1. Requires prior exposure for production of antibodies in Rh^- individuals

2. Important in transfusions, and in pregnancies if child is Rh^+ and mother is Rh^-

a. Mother produces antibodies after birth of first Rh^+ child

b. The circulating antibodies can cross placenta and agglutinate red blood cells of the next Rh^+ infant, causing hemolytic disease of the newborn

c. This can be prevented by injecting the mother with anti-Rh antibodies (RhoGAM) within 72 hours of delivery of an Rh^+ baby; these antibodies bind any Rh antigens in the mother's bloodstream and thereby prevent the formation of antibodies that could harm subsequent offspring

X. Cell-Associated Differentiation Antigens (CDs)

A. Cell surface proteins that function as receptors for various cytokines

B. Specific cell types can be identified by their complement of CDs

C. Can be secreted from cells and levels in serum can be used in disease management

TERMS AND DEFINITIONS

Place the letter of each term in the space next to the definition or description that best matches it.

_____ 1. Small molecules released by sensitized T lymphocytes that act at a short distance, that are unrelated to immunoglobulins, and that are not immunologically specific in their actions

_____ 2. Substances that nonspecifically and polyclonally trigger a large number of B and T cells into cell division

_____ 3. The ability to produce antibodies against nonself-antigens, while not producing antibodies against self-antigens

_____ 4. An exaggerated or inappropriate immune response that results in tissue damage to the individual

_____ 5. A location in the body into which lymphocytes do not circulate, which can therefore receive transplants with minimal chance for rejection, even though the tissues of the donor and the recipient may not be a close match

_____ 6. A defect in the immune system that is genetically determined

_____ 7. A defect in the immune system that is the result of infection by an immunosuppressive pathogen

_____ 8. Protein or polypeptide mediators that act on or signal between leukocytes

_____ 9. Cell surface proteins that act as receptors for various cytokines

a. acquired immune tolerance
b. acquired immunodeficiency
c. allergy
d. cell associated differentiation antigens (CDs)
e. congenital immunodeficiency
f. cytokines
g. immunologically privileged site
h. interleukins
i. mitogens

CYTOKINES AND LYMPHOKINES

Describe the role played by each of the following cytokines in the immune response.

Cytokine/Lymphokine	Role in Immune Response
CSFs	
IL-1	
IL-2	
IL-3	
Interferon-α	
Interferon-β	
Interferon-γ	
MIF	

TISSUE TRANSPLANTATION

Place the letter of each term in the space next to the situation it describes. (Terms may be used more than once.)

_____ 1. A woman donates a kidney to her aunt
_____ 2. A mechanical heart is put into a patient
_____ 3. One identical twin donates a kidney to the other
_____ 4. One fraternal twin donates a kidney to the other
_____ 5. Skin from a patient's thigh is used to cover a burn on the patient's face
_____ 6. Skin from a pig is used to cover a burn on a patient's face
_____ 7. A man donates a cornea to his brother
_____ 8. A man dies and his corneas are donated to two unrelated individuals that he never met
_____ 9. A baboon's heart is implanted into a newborn with a major heart defect
_____ 10. A woman donates a kidney to another person she has never met and is unrelated to
_____ 11. Least likely to be rejected by the immune system
_____ 12. Most likely to be rejected by the immune system

a. autograft
b. isograft
c. allograft
d. xenograft
e. none of the above

FILL IN THE BLANK

1. Small molecules that are unrelated to immunoglobulins, that are immunologically nonspecific in their action, and that are released from a variety of cells are generally called _____; if they are specifically released from sensitized T cells, they are called _____; and if they are released from monocytes, they are referred to as _____.

2. Even though B cells have receptors on their surfaces that are specific for one antigenic determinant, or _____, they usually cannot develop into _____ cells and secrete antibody without the help of _____ cells, which initiate B-cell function. This is called _____.

3. T cells that control the development and/or function of effector T cells are called _____ T cells. There are two subsets of these types of T cells. T cells that are needed for effective presentation of antigens to B cells are called _____ cells, while those that provide negative feedback control are called _____ cells.

4. The ability to produce antibodies against nonself-antigens while not producing antibodies against self-antigens is called _____. There are two hypotheses that have been proposed to explain this. One is that early in development, those lymphocytes capable of responding to self-antigens are destroyed; this is called the _____ _____ theory. The other hypothesis, called _____ _____, is that the lymphocytes that interact with self-antigens either by T_S cells that keep the T_H cells in check, or by their entry into a nonresponsive state called _____ _____.

5. Defects in one or more components of the immune system can result in its failure to recognize and respond properly to antigens; this is called an immunodeficiency and can make a person more susceptible to infection and to cancer. If it is caused by genetic errors it is termed a(n) _____ or _____ immunodeficiency. If it is caused by an infectious agent, such as a virus, it is called a(n) _____ or _____ immunodeficiency.

6. People with allergies sometimes undergo a process called _____ in which a series of allergen doses are injected beneath the skin; this results in the formation of _____ instead of IgE. The circulating _____ act as _____ _____ that intercept and neutralize the allergens before they have time to interact with IgE-bound mast cells.

7. Cytokines can affect various cell populations. If the affected cell is the same cell responsible for the production of the cytokine it is referred to as an _____ function; if they affect nearby cells it is referred to as a _____ function; if they are distributed by the circulatory system to these target cells it is referred to as an _____ function.

MULTIPLE CHOICE

For each of the questions below select the *one best* answer.

1. A substance that nonspecifically and polyclonally stimulates B and T cells to reproduce is best described as a
    a. growth hormone.
    b. mitogen.
    c. reagin.
    d. cytokine.
2. Which of the following is not true of T-independent antigen triggering?
    a. The antigens that can directly stimulate B cells usually present a large array of identical epitopes to a B cell specific for that determinant.
    b. The antibody produced is usually IgM with no switching to IgG.
    c. The antibody produced usually has a low binding affinity.
    d. All of the above are true of T-independent antigen triggering.
3. Class I MHC molecules are made by
    a. all cells of the body except red blood cells.
    b. all cells of the body.
    c. only red blood cells.
    d. activated macrophages, mature B cells, some T cells, and certain cells of other tissues.
4. Which of the following is not true about class II MHC molecules?
    a. They do not stimulate antibody production.
    b. They are required for T cells to recognize macrophages and B cells that are presenting exogenous antigens.
    c. They are made by all cells of the body except red blood cells.
    d. All of the above are true about class II MHC molecules.

5. Hives (skin eruptions associated with food allergies) are an example of which of the following types of hypersensitivities?
    a. Type I
    b. Type II
    c. Type III
    d. Type IV
6. Allergic contact dermatitis, such as the rash associated with poison ivy, is an example of which of the following types of hypersensitivities?
    a. Type I
    b. Type II
    c. Type III
    d. Type IV
7. Tissue typing for organ transplantation usually involves finding a donor and recipient that most closely match with respect to which of the following?
    a. Class I MHC proteins
    b. Class II MHC proteins
    c. Class III MHC proteins
    d. All three classes must match as closely as possible.
8. T_{DTH} cells are involved in which of the following activities?
    a. Type IV hypersensitivities
    b. Macrophage-mediated damage at the site of an inflammation
    c. Both (a) and (b) are correct.
    d. Neither (a) nor (b) is correct
9. Low-molecular weight inflammation-stimulating paracrine cytokines are referred to as _____
    a. inflammokines.
    b. chemokines.
    c. Both (a) and (b) are correct.
    d. Neither (a) nor (b) is correct.

10. Which of the following types of cells can present antigens to T cells?
    a. macrophages
    b. dendritic cells
    c. Both (a) and (b) are correct.
    d. Neither (a) nor (b) is correct.
11. Which of the following is not a function associated with cytokines?
    a. stimulation of cell division
    b. apoptosis
    c. inhibition of cell division
    d. All of the above are functions associated with cytokines.
12. Which of the following is a mechanism used by Tc cells to produce toxic effects on their target cells?
    a. CD95 transmembrane signal transduction pathway
    b. Perforin pathway
    c. Both (a) and (b) are used by Tc cells.
    d. Neither (a) nor (b) are used by Tc cells.

TRUE/FALSE

____ 1. A person's total B-cell population carries receptors that are specific for a large number of antigens; however, each mature B cell possesses receptors specific for only one epitope.

____ 2. IgM (transmembrane form) serves as the receptor for B cells.

____ 3. IgD serves as the receptor for T cells.

____ 4. T-cell receptors can only recognize antigens on the surfaces of antigen-presenting cells; they cannot bind free antigen.

____ 5. Cytotoxic T cells are capable of recognizing and destroying virus-infected cells. This can lead to tissue damage and is responsible for much of the hepatic necrosis seen in hepatitis B virus infections.

____ 6. Unlike other antibody-antigen interactions, ABO blood-transfusion reactions do not require prior exposure to the antigen.

____ 7. Like the ABO system, Rh-factor reactions do not require prior exposure to the antigen.

____ 8. Once an Rh⁻ mother has produced anti-Rh antibodies in her bloodstream, all subsequent offspring are at risk of hemolytic disease of the newborn regardless of their blood types.

____ 9. Cytotoxic T cells can kill their target cells either by cytolysis or by apoptosis (a programmed cell death).

____10. Super antigens stimulate T cells to proliferate nonspecifically causing the release of massive quantities of cytokines, which in time, can lead to tissue damage.

____11. The presence of serum autoantibodies (autoimmunity) is a natural consequence of aging and does not necessarily lead to autoimmune disease.

____12. The CDs on all cell types are nearly identical; therefore the complement of CD's on a cell cannot be used to indicate any specific cell lineage.

CRITICAL THINKING

1. Macrophages and other phagocytic cells have relatively nonspecific targets. Explain how lymphokines increase the apparent specificity of these effector cells. In your discussion indicate why this is an *apparent* and not an *actual* change in specificity, even though the result is indeed a more specific destruction of the invading organism.

2. The immune surveillance mechanisms are capable of recognizing and destroying tumor cells, which are constantly being formed in the body. This being the case, why do people ever develop cancers? How can we be sure (what evidence is there) that the immune surveillance system significantly reduces the number of cancers in the population?

ANSWER KEY

Terms and Definitions

1. f 2. c 3. a 4. c 5. g 6. e 7. b 8. h 9. d

Tissue Transplantion

1. c 2. e 3. b 4. c 5. a 6. d 7. c 8. c 9. d 10. c 11. a 12. d

Fill in the Blank

1. cytokines; lymphokines; monokines 2. epitope; plasma; T-helper (T_H); T-dependent antigen triggering 3. regulator; helper (T_H); suppressor (T_S); 4. acquired immune tolerance; clonal deletion; functional inactivation; clonal anergy 5. primary; congenital; secondary; acquired 6. desensitization; IgG; IgG; blocking antibodies; paracrine; endocrine 7. autocrine; paracrine; endocrine

Multiple Choice

1. b 2. d 3. a 4. c 5. a 6. d 7. a 8. c 9. b 10. c, 11. d, 12. c

True/False

1. T 2. T 3. F 4. T 5. T 6. T 7. F 8. F 9. T 10. T 11. T 12.F

32   The Immune Response: Antigen-Antibody Reactions

CHAPTER OVERVIEW

This chapter continues the discussion of the immune system with a description of the reactions between antigens and antibodies and the ways in which these reactions protect higher animals against infectious agents, their products, and certain macromolecules. The chapter concludes with a discussion of antigen-antibody reactions in vitro and their uses in disease diagnosis, microorganism identification, and immunological monitoring.

CHAPTER OBJECTIVES

After reading this chapter you should be able to:

- discuss the various ways in which antigen-antibody binding initiates the participation of other defense mechanisms, such as the complement system, phagocytosis, chemotaxis, and inflammation
- discuss antibody-dependent, cell-mediated cytotoxicity; opsonization; neutralization; and immune complex formation
- describe various in vitro diagnostic/identification procedures that employ antigen-antibody reactions

CHAPTER OUTLINE

I.   Introduction
    A.   Antibodies are bifunctional—they bind to the target antigen they recognize as foreign, and they enable other defense components to react with it
    B.   Variable domain (Fab)—binds to target antigen
    C.   Constant domain (Fc)—interacts with cells of the immune system and other host defense mechanisms
II. Antigen-Antibody Binding—occurs within the pocket formed by folding the V_H and V_L regions of the Fab domain
    A.   Binding is due to weak, noncovalent bonds
    B.   Shapes of epitope and binding site must be complementary for efficient binding
    C.   The high complementarity provides for the high specificity associated with antigen-antibody binding
III. Antigen-Antibody Reactions in the Animal Body (In Vivo)
    A.   The complement system is a series of protein components that must be activated in a cascade fashion (i.e., the activation of one component results in the activation of the next, and so on)
        1.   Results in lysis of antibody-coated bacteria and eucaryotic cells
        2.   Mediates inflammation
        3.   Attracts and activates phagocytic cells
        4.   There are three activation pathways; each results in destruction of the target cell, but their triggering mechanisms are different
            a.   The classical pathway is dependent on antigen-antibody interactions to trigger it; it is fast and efficient

b. The lectin pathway is activated by mannose-binding lectins (MBLs) that have been secreted by liver; activation leads to opsonization
c. The alternative pathway does not require antigen-antibody binding; it is nonspecific and inefficient, but contributes to innate resistance
5. The final step in the pathway is the formation of a membrane attack complex that creates a pore in the membrane of the target cell
6. The pore allows entry of destructive enzymes or leads to osmotic rupture of the target cell
B. Toxin neutralization—antibody (antitoxin) binding to toxin renders the toxin incapable of attachment or entry into target cells
C. Viral neutralization—binding prevents viral attachment to target cells
D. Adherence inhibiting antibodies—sIgA prevents bacterial adherence to mucosal surfaces
E. Antibody-dependent cell-mediated cytotoxicity—involves the complement system, NK cells, or release of cytotoxic mediators from effector cells that attach to the target cell by means of F_c receptors
F. IgE and parasitic infection—in the presence of elevated IgE levels, eosinophils bind parasites and release lysosomal enzymes
G. Opsonization
1. Prepares the microorganism for phagocytosis; phagocytes recognize the Fc portion of IgG or IgM antibodies coating the surface of the foreign microorganism
2. Phagocytosis can also be stimulated by components of the complement system, whether initiated by the classical or alternative pathways
H. Inflammation can be mediated by IgE attachment to mast cells and basophils, or by the binding of one of the complement components to mast cells and platelets; this complement component is also a powerful chemoattractant for macrophages, neutrophils, and basophils
I. Immune complex formation—two or more antigen-binding sites per antibody molecule lead to cross-linking, forming precipitins (molecular aggregates) or agglutinins (cellular aggregates); agglutination that specifically involves red blood cells is called hemagglutination
J. In vivo testing involves immediate or delayed skin testing for the presence of antibodies to various antigens
IV. Antigen-Antibody Reactions In Vitro (serology)
A. Agglutination—visible clumps or aggregates of cells or of coated latex microspheres; if red blood cells are agglutinated, this is called hemagglutination
B. Agglutination inhibition can be used to detect serum antibodies or to detect the presence of specific substances (e.g., illegal drugs) in urine samples by a competition assay
C. Complement fixation—irreversible alterations to complement components that are initiated by the binding of antibody to antigen; used to detect the presence of serum antibodies, thereby indicating prior exposure to a pathogen
1. If immune complexes are formed, then complement is used up and lysis will not occur when sensitive indicator cells are added
2. If immune complexes are not formed, then complement is not used up and lysis will occur when sensitive indicator cells are added
D. Enzyme-linked immunosorbent assay (ELISA)
1. Indirect immunosorbent assay—detects serum antibody
a. Antigen is coated on test wells and serum is added
b. If antibodies are present, they will bind antigen; if not, they will wash off
c. Add to the plate an enzyme that is covalently coupled to a second antibody against first immunoglobulin
d. If antigen was present, the second antibody will bind; if not, it will wash off
e. Add colorless substrate (chromogen) for the enzyme and measure colored product formation spectrophotometrically; no colored product will form if everything washed off
2. Double antibody sandwich assay—detects antigens in a sample

a. Antibody is coated onto test wells and sample is added
b. If antigen is present in sample it will bind; if not it will wash off
c. React with antibody against the antigen; if antigen was present in the sample, this antibody will bind; if not, it will wash off
d. Continue with steps (c), (d), and (e) as in the indirect assay

E. Immunodiffusion—involves the precipitation of immune complexes in an agar gel after diffusion of one or both components
1. Single radial immunodiffusion (RID) assay—quantitative
2. Double diffusion assay (Ouchterlony technique)—lines of precipitation form where antibodies and antigens have diffused and met; determines whether antigens share identical determinants

F. Immunoelectrophoresis—antigens are first separated by electrophoresis according to charge, and are then visualized by the precipitation reaction; greater resolution than diffusion assay

G. Immunofluorescence—dyes coupled to antibody molecules will fluoresce (emit visible light) when irradiated with ultraviolet light
1. Direct—used to detect antigen-bearing organisms fixed on a microscope slide
2. Indirect—used to detect the presence of serum antibodies

H. Immunoprecipitation—soluble antigens form insoluble immune complexes that can be detected by formation of a precipitin

I. Neutralization—an antibody that is mixed with a toxin or a virus will neutralize the effects of the toxin or the infectivity of the virus; this is determined by subsequent assay

J. Radioimmunoassay (RIA)—purified antigen labeled with a radioisotope competes with unlabeled sample for antibody binding

K. Serotyping—antigen-antibody specificity is used to differentiate among various strains (serovars) of an organism

TERMS AND DEFINITIONS

Place the letter of each term in the space next to the definition or description that best matches it.

_____ 1. A group of circulating plasma proteins that acts in a cascade fashion and plays a major role in an animal's defensive immune response

_____ 2. The action of antibodies that makes a toxin or a virus unable to attach to its target cell and therefore unable to damage it

_____ 3. The phenomenon in which target cells are coated with antibody to prepare them for ingestion by phagocytic cells

_____ 4. Cross-linking of antibody and antigen molecules into large aggregates, which is due to multivalency of both antibody and antigen molecules

_____ 5. Cross-linking of antibody and antigen-bearing cells into large aggregates, which is due to multivalency of both antibody and target cells

_____ 6. The branch of immunology that deals with antibody-antigen reactions that occur outside an animal's body (in vitro)

_____ 7. A colorless substrate that is acted on by the enzyme portion of an enzyme-antibody ligand to produce a colored end product

_____ 8. In vitro antibody-antigen reactions that are used to differentiate strains of microorganisms

a. agglutinin reaction
b. chromogen
c. complement system
d. neutralization
e. opsonization
f. precipitation reaction
g. serology
h. serotyping

FILL IN THE BLANK

1. Activation of the complement system can proceed by two different mechanisms. One is dependent on the reaction between antibodies and antigens to initiate it and is called the _____ pathway. The other does not require the binding of antibodies to antigens for its activation; instead, it plays an important role in defense against _____ invasion by bacteria and some fungi. This is called the _____ pathway.

2. One end point of the classical complement pathway is the formation of a _____ complex, which creates a _____ in the plasma membrane of the target cell. If the cell is eucaryotic, Na^+ and H_2O enter, which leads to _____ _____. On the other hand, if the target cell is a gram-_____ bacterium, the enzyme _____ enters through this structure and digests the peptidoglycan. Gram-_____ bacteria are resistant to complement because they lack a(n) _____ membrane.

3. When antibodies bind to certain types of antigens, those antigens lose the ability to bind to and therefore cannot enter their target cells. This is called _____ and is most effective against _____ and _____.

4. The phagocytic process can be greatly enhanced by _____, which is the coating of microorganisms with _____ and/or _____ in order to facilitate recognition and ingestion by phagocytic cells.

5. Since antibodies have at least two antigen-binding sites, and most antigens have at least two determinants, cross-linking can occur, producing large aggregates called _____. If this involves molecular antigens, the antibody is called a(n) _____; if it involves cross-linking of cells or particles, the antibody is called a(n) _____; if it specifically involves red blood cells, the antibody is called a(n) _____. These large aggregates are more readily phagocytized.

6. One of the most widely used serological tests is the _____, and it involves the linkage of an antibody molecule to an enzyme whose activity can be measured calorimetrically. One variety of this test is primarily used to detect the presence of antigen and is called the _____ assay; another variety of this test is primarily used to detect the presence of antibody in a sample and is called the _____ assay.

MULTIPLE CHOICE

For each of the questions below select the *one best* answer.

1. Which of the following describes how the complement system aids in the defensive responses of an organism?
    a. lysis of antibody-coated eucaryotic cells or bacteria
    b. attraction of phagocytic cells
    c. activation of phagocytic cells
    d. All of the above describe how the complement system aids in the defensive responses of an organism.

2. The complement proteins act sequentially, and this results in an amplification of the activity of the response. This type of response is referred to as a(n)
    a. synergism.
    b. cascade.
    c. avalanche.
    d. amplified response.

3. Which of the following is not normally capable of viral neutralization?
    a. IgA
    b. IgM
    c. IgD
    d. IgG

4. Which of the following is normally involved in inhibiting the adherence of bacteria to mucosal surfaces?
   a. IgA
   b. IgM
   c. IgD
   d. IgG

5. The release of lysosomal vesicles from eosinophils in the presence of IgE is called
   a. opsonization.
   b. degranulation.
   c. exocytotic lysis.
   d. inflammation.

6. When complement binds to an antibody-antigen complex, it becomes used up and is no longer available to lyse sensitized red blood cells. This can be used diagnostically in an assay referred to as
   a. complement utilization.
   b. complement titration.
   c. complement fixation.
   d. complement complexation.

7. Reactions between antibodies and antigens that are detected by precipitation formation in an agar gel are referred to as
   a. immunoprecipitation assays.
   b. immunodiffusion assays.
   c. immunoaggregation assays.
   d. immunofixation assays.

8. Immunological assays that are used to differentiate strains of an organism are referred to as
   a. serotyping.
   b. immunotyping.
   c. isotyping.
   d. antigen typing.

9. Neutralizing antibody against a toxin is referred to as a(n)
   a. toxoid.
   b. antitoxin.
   c. antitoxoid.
   d. neutrotox.

TRUE/FALSE

_____ 1. Once activated by antibodies binding to antigens, the actions of the other components of the immune system are not specific (i.e., those components will interact with anything in the vicinity).

_____ 2. Agglutination inhibition reactions are currently used as rapid assays for the presence of illegal drugs in urine samples.

_____ 3. Opsonizing antibodies must be against surface components, not internal components, if they are to effectively stimulate phagocytosis.

_____ 4. Respiratory allergies are normally diagnosed by immediate skin tests, while food allergies are normally diagnosed by delayed skin tests.

_____ 5. The titer of antibody in serum is the reciprocal of the highest dilution of serum showing an agglutination reaction.

_____ 6. Immunoprecipitation reactions will occur as long as the antibody is present in equal or greater amounts than the antigen.

CRITICAL THINKING

1. Describe the indirect immunosorbent variation of ELISA procedures. Diagram what will happen at each step for a positive serum sample, and for a negative serum sample. How would you modify this procedure if you wanted to detect antigen rather than antibody?

2. Describe the Ouchterlony double diffusion assay. In your description, include a discussion of the three major types of results (identity, partial identity, and nonidentity) that can be obtained.

ANSWER KEY

Terms and Definitions

1. c, 2. d, 3. e, 4. f, 5. a, 6. g, 7. b, 8. h

Fill in the Blank

1. classical; intravascular; alternative 2. membrane attack; pore; osmotic; lysis; negative; lysozyme; positive; outer 3. neutralization; toxins; viruses 4. opsonization; antibody; complement 5. immune complexes; precipitin; agglutinin; hemagglutinin 6. enzyme-linked immunosorbent assay (ELISA); double antibody sandwich; indirect immunosorbent

Multiple Choice

1. d, 2. b, 3. c, 4. a, 5. b, 6. c, 7. b, 8. a, 9. b

True/False

1. F, 2. T, 3. T, 4. T, 5. T, 6. F

Part IX MICROBIAL DISEASES AND THEIR CONTROL

33 Antimicrobial Chemotherapy

CHAPTER OVERVIEW

The control or the destruction of microorganisms that reside within the bodies of humans and other animals is of tremendous importance. This chapter introduces the principles of chemotherapy, and discusses the ideal characteristics for successful chemotherapeutic agents (including the concept of selectively damaging the target microorganism while minimizing damage to the host). The chapter also presents characteristics of some commonly used antibacterial, antifungal, and antiviral drugs.

CHAPTER OBJECTIVES

After reading this chapter you should be able to:

- discuss the various ways in which antimicrobial agents can damage pathogens while causing minimal damage to the host
- discuss the various factors that influence the effectiveness of a chemotherapeutic agent
- discuss the increasingly serious problem of drug-resistant pathogens
- discuss the increasing demand for and availability of antifungal and antiviral agents

CHAPTER OUTLINE

I. The Development of Chemotherapy
    A. Ancient herbal remedies were actually primitive forms of chemotherapy
    B. Paul Ehrlich (1904-1909)—aniline dyes and arsenic compounds
    C. Gerhard Domagk, and Jacques and Therese Trefouel (1939)—sulfanilamide
    D. Ernest Duchesne (1896) discovered penicillin; however, this discovery was not followed up and was lost for 50 years until rediscovered by a librarian
    E. Alexander Fleming (1928) accidentally discovered the antimicrobial activity of penicillin on a contaminated plate; however, follow-up studies convinced him that penicillin would not remain active in the body long enough to be effective
    F. Howard Florey and Ernst Chain (1939) aided by the biochemist, Norman Heatley, worked from Fleming's published observations, obtained a culture from him, and demonstrated the effectiveness of penicillin
    G. Selman Waksman (1944)—streptomycin; this success led to a worldwide search for additional antibiotics, and the field has progressed rapidly since then

II. General Characteristics of Antimicrobial Drugs
  A. Selective toxicity with minimal side effects
    1. Therapeutic dose—the drug level required for clinical treatment of a particular infection
    2. Toxic dose—the drug level at which the agent becomes too toxic for the host (produces undesirable side effects)
    3. Therapeutic index—the ratio of toxic dose to therapeutic dose: the larger the better, all other things being equal
  B. Broad spectrum activity (activity against a wide variety of pathogens) is more desirable than narrow spectrum activity, but this is not crucial
  C. Drug can be cidal (able to kill) or static (able to reversibly inhibit growth)
  D. Chemotherapeutic agents can occur naturally or be synthetic or semisynthetic (chemical modifications of naturally occurring antibiotics)
III. Determining the Level of Antimicrobial Activity
  A. Dilution susceptibility tests
    1. The lowest concentration of the antibiotic resulting in no microbial growth is the minimal inhibitory concentration (MIC)
    2. Tubes showing no growth are subcultured into tubes of fresh medium that contain no antibiotic to determine the lowest concentration of the drug from which the organism does not recover—the minimal lethal concentration (MLC)
  B. Disk diffusion tests—disks impregnated with specific drugs are placed on agar plates inoculated with the test organism; clear zones (no growth) will be observed if the organism is sensitive to the drug; the size of the clear zone is used to determine the relative sensitivity according to tables prepared for the various available drugs; zone width is a function of initial concentration, solubility and diffusion rate of the antibiotic
  C. Measurement of drug concentrations in the blood can be done using microbiological, chemical, immunological, enzymatic, and/or chromatographic assays
IV. Mechanism of Action of Antimicrobial Agents
  A. Inhibition of cell wall synthesis
  B. Inhibition of protein synthesis
  C. Inhibition of nucleic acid synthesis
  D. Disruption of cell membranes
  E. Inhibition of metabolic activities (antimetabolites)
V. Factors Influencing the Effectiveness of Antimicrobial Drugs
  A. Factors influencing a drug's ability to reach the site of infection
    1. Mode of administration
       a. Oral
       b. Topical
       c. Parenteral (injection)
    2. Susceptibility to various bodily defense mechanisms (e.g., penicillin is rapidly degraded in the stomach, but the penicillin derivative ampicillin is more acid stable)
  B. Factors influencing drug concentration in the body, which must exceed the pathogen's MIC for the drug to be effective; this will depend on
    1. Amount of drug administered
    2. Route of administration
    3. Speed of uptake
    4. Rate of clearance (elimination) from the body
  C. The nature of the pathogen, including its inherent susceptibility and the presence of its active growth, may have a profound effect on susceptibility to particular drugs
  D. Drug resistance has become an increasing problem
VI. Antibacterial Drugs
  A. Sulfonamides, or sulfa drugs, are structural analogues of metabolic intermediates (also called antimetabolites); they inhibit folic acid synthesis

B. Quinolones inhibit bacterial DNA gyrase, thereby disrupting replication, repair, and other processes involving DNA

C. Penicillins inhibit cell wall synthesis

D. Cephalosporins also inhibit cell wall synthesis, but have a broader spectrum of activity; they can be given to some patients with penicillin allergies

E. Tetracyclines inhibit protein synthesis

F. Aminoglycosides (streptomycin, kanamycin, neomycin, tobramycin, gentamicin) inhibit protein synthesis

G. Erythromycin and other macrolides inhibit protein synthesis

H. Chloramphenicol inhibits protein synthesis; it has a broad spectrum but is toxic

VII. Drug Resistance

    A. Mechanisms of drug resistance

        1. Exclusion of drug from cell—changes in ability of drug to bind to and/or penetrate the cell or the ability to pump the drug out of the cell once it has entered

        2. Enzymatic inactivation of the drug—chemical modification of the drug by cellular enzymes can render it inactive before it has a chance to damage the cell

        3. Alteration of target enzyme or organelle—modification of the target so that it is no longer susceptible to the action of the drug

        4. Use of alternative pathways and increased production of the target metabolite have been used by some organisms to minimize the effects of the drug

    B. The origin and transmission of drug resistance involve chromosomal or plasmid genes for drug resistance

        1. Drug resistance has become an increasing problem

        2. Superinfection by drug-resistant pathogens may result because of the lack of competition from drug-sensitive strains

VIII. Antifungal Drugs—fungal infections are more difficult to treat than bacterial infections, because the greater similarity between fungi and host limits the ability of a drug to have a selective point of attack

    A. Superficial mycoses are infections of superficial tissues and can often be treated by topical application of antifungal drugs such as miconazole, nystatin, and griseofulvin, thereby minimizing toxic systemic side effects

    B. Systemic mycoses are more difficult to treat and can be fatal; however, amphotericin B and flucytosine have been used with limited success; amphotericin B is highly toxic and must be used with care; flucytosine must be converted by the fungus to an active form, and animal cells are incapable of this; some selectivity is possible, but severe side effects have been observed with both drugs

    C. Drug resistant fungal strains are also beginning to emerge

IX. Antiviral Drugs—selectivity has been a problem because viruses use the metabolic machinery of the host; recently, however, drugs that inhibit virus-specific enzymes (such as amantadine, vidarabine, acyclovir, and azidothymidine [AZT]) have been found and offer hope for this area in the future, another promising area is the possibility of using human interferon (a naturally produced antiviral substance) as an effective treatment; develoment of drug resistance is minimized by the use of a cocktail of several drugs at high doses

TERMS AND DEFINITIONS

Place the letter of each term in the space next to the definition or description that best matches it.

_____ 1. The ratio of therapeutic dose to toxic dose

_____ 2. Compounds used in the treatment of disease that kill or prevent the growth of microorganisms at concentrations low enough to avoid undesirable damage to the host

_____ 3. Chemotherapeutic antimicrobial agents that are natural products of microorganisms

_____ 4. Activities of a chemotherapeutic agent that damage the host either by inhibiting the same process in the host as in the target organism or by damaging other processes

_____ 5. Describes an antibiotic that attacks many different pathogens

_____ 6. Describes an antibiotic that is effective against only a limited variety of pathogens

_____ 7. Antimicrobial drugs that are totally synthesized by one or more microorganisms

_____ 8. Antimicrobial drugs that are artificially synthesized by chemical processes

_____ 9. Natural antibiotics that have been chemically modified

_____ 10. Lowest concentration of a drug necessary to prevent the growth of a particular microorganism

_____ 11. Lowest concentration of a drug necessary to kill a particular microorganism

_____ 12. Drugs that block the function of metabolic pathways

_____ 13. Small circular DNA molecules that can exist separately from the chromosome or be integrated into it

_____ 14. Plasmids that bear one or more resistance genes

a. antibiotics
b. antimetabolites
c. broad spectrum drug
d. chemotherapeutic agent
e. minimal inhibitory concentration
f. minimal lethal concentration
g. narrow spectrum drug
h. natural antibiotics
i. plasmids
j. R factors
k. semisynthetic chemotherapeutic agents
l. side effects
m. synthetic chemotherapeutic agents
n. therapeutic index

DRUGS AND THEIR DISCOVERERS

Match the following scientists with their discoveries.

_____ 1. Use of the arsenic compound Salvarsan as a treatment for syphilis

_____ 2. Use of Prontosil Red (sulfanilamide) as a treatment for streptococcal and staphylococcal infections

_____ 3. First discovered penicillin, but discovery was lost

_____ 4. Rediscovered penicillin, but did not pursue the significance of it

_____ 5. Co-discoverer of the therapeutic value of penicillin

_____ 6. Co-discoverer of the therapeutic value of penicillin

_____ 7. Discovered streptomycin and stimulated intense search for other antibiotics

_____ 8. Biochemist who helped with work demonstrating the therapeutic value of penicillin

a. Chain
b. Domagk
c. Duchesne
d. Ehrlich
e. Fleming
f. Florey
g. Heatley
h. Waksman

DRUG ACTION

For each of the drugs listed below indicate: (1) its mechanism of action (e.g., cell wall synthesis inhibitor, cell membrane damage); (2) whether it is primarily cidal or static; and (3) what type of organism it is used against (bacteria, fungi, virus).

Drug	Mechanism	Cidal/Static	Organism
1. Penicillin			
2. Polymyxin B			
3. Gentamicin			
4. Acyclovir			
5. Flucytosine			
6. Sulfonamides			
7. Miconazole			
8. Rifampin			
9. Cephalosporins			
10. Ciprofloxacin			
11. Tetracyclines			
12. Isoniazid			

FILL IN THE BLANK

1. The _____ is the ratio of the therapeutic dose to the _____ dose. The larger this ratio, the _____ the chemotherapeutic agent (all other things being equal).
2. A number of useful drugs act as _____; they block the functioning of metabolic pathways by acting as _____ inhibitors of key enzymes.
3. Many forms of penicillin are not suitable for a(n) _____ route of administration because they are relatively unstable in _____.
4. Antibiotics that are taken by mouth are said to have a(n) _____ route of administration; those given by injection have a(n) _____ route of administration; while those applied to the skin are said to have a(n) _____ route of administration.
5. Sulfonamide is a competitive inhibitor of _____ synthesis and thereby causes a reduction in concentration of this metabolite. Ultimately, this decreases the supply of _____ and _____, which are needed for nucleic acid synthesis. Sulfonamides do not affect humans because humans do not synthesize _____ and must ingest it in their diet.
6. One of the most serious threats to the successful treatment of disease is the spread of _____ pathogens. This problem has been aggravated by _____ use of antibiotics.

MULTIPLE CHOICE

For each of the questions below select the *one best* answer.

1. Which of the following is not true about drug side effects?
   a. Side effects may result from drug inhibition of the same process in the host as in the target cell.
   b. Side effects may result from drug damage to a different process in the host cell than in the target cell.
   c. Side effects usually result from damage to the host's liver when it attempts to eliminate the drug from the system.
   d. All of the above are true about drug side effects.

2. In which of the following cases would the action of the drug be considered primarily static?
   a. The MLC is equal to the MIC.
   b. The MLC is lower than the MIC.
   c. The MLC is 2 to 4 times higher than the MIC.
   d. The MLC is 10 to 20 times higher than the MIC.

3. The most selective antibiotics are those that interfere with bacterial cell wall synthesis. This is because
   a. bacterial cell walls have a unique structure not found in eucaryotic cells.
   b. bacterial cell wall synthesis is easy to inhibit, while animal cell wall synthesis is more resistant to the actions of the drugs.
   c. animal cells do not take up the drugs.
   d. animal cells inactivate the drugs before they can do any damage.

4. Which of the following will not have an effect on the concentration achieved in the blood by a particular antibiotic?
   a. the route of administration
   b. the speed of uptake from the site of administration
   c. the rate at which the drug is eliminated from the body
   d. All of the above will affect the concentration of the drug in the blood.

5. Combinations of drugs are often more effective because each drug enhances the effect of the other. This is called
   a. multiplicity.
   b. synergism.
   c. complementation.
   d. additivity.

6. Which of the following is not a common mechanism by which microorganisms develop drug resistance?
   a. enzymatic inactivation of the drug
   b. exclusion of the drug from the cell
   c. use of an alternative pathway to bypass the drug-sensitive pathway
   d. All of the above are common mechanisms by which microorganisms develop drug resistance.

7. Which of the following is(are) used to discourage the development of drug resistance?
   a. sufficiently high drug doses to destroy any resistant mutants that may arise spontaneously
   b. use of two drugs simultaneously with the hope that each will prevent the emergence of resistance to the other
   c. avoidance of indiscriminate use of drugs
   d. All of the above are used to discourage development of drug resistance.

8. Which of the following is not a reason that treatment of fungal infections generally have been less successful than treatment of bacterial infections?
   a. Fungi use the metabolic machinery of the host and therefore cannot be selectively attacked.
   b. Fungi are more similar to human cells than are bacteria, and many drugs that inhibit or kill fungi are toxic for humans.
   c. Fungi have detoxifying systems that rapidly inactivate many drugs.
   d. All of the above are reasons that treatment of fungal infections has been less successful than treatment of bacterial infections.

9. The drug level required for the clinical treatment of a particular infection is called the
   a. therapeutic dose.
   b. toxic dose.
   c. therapeutic index.
   d. None of the above are correct.

10. Which of the following affects the size of the clear zone in a disk diffusion test of antimicrobial susceptibility?
   a. the initial concentration of the drug
   b. the solubility of the drug
   c. the diffusion rate of the drug
   d. All of the above are correct.

11. Antibiotic resistance is becoming a major problem in
   a. bacterial diseases
   b. fungal diseases
   c. both (a) and (b)
   d. neither (a) nor (b)

TRUE/FALSE

____ 1. A drug that disrupts a microbial function not found in animal cells usually has a lower therapeutic index.

____ 2. Static agents do not kill infectious organisms and therefore are not useful as chemotherapeutic agents.

____ 3. Protein synthesis inhibitors have a high therapeutic index because they can usually discriminate between procaryotic and eucaryotic ribosomes.

____ 4. Sulfonamides and other drugs that inhibit folic acid synthesis have a high therapeutic index because humans must obtain folic acid in their diets while microorganisms synthesize their own.

____ 5. Drugs can usually penetrate the blood-brain barrier, and therefore, it is easy to treat nervous system infections, such as meningitis.

____ 6. The fungus *Candida albicans* is normally present in various parts of the body and causes a problem (superinfection) only when bacterial competition is eliminated by antibiotic treatment.

____ 7. Isoniazid is a narrow-spectrum antibiotic. However, it is considered useful because it is one of the few drugs that are effective against tuberculosis.

____ 8. Drugs with highly toxic side effects are usually used only in life-threatening situations where suitable alternatives are not available.

____ 9. There are no effective antiviral drugs because viruses use the metabolic machinery of their hosts and therefore have no selective point of attack.

____ 10. One way in which organisms may exhibit resistance to a drug is to be able to pump the drug out of the cell immediately after it has entered.

____ 11. Drug resistance can be reduced by the use of multiple drugs simultaneously at high doses.

CRITICAL THINKING

1. Antibiotics are natural products of certain microorganisms. What advantages do these antibiotics provide for the organisms that produce them?

2. Viruses generally use the metabolic machinery of their hosts. Therefore, they should present no selective point of attack for potential antiviral drugs. Yet, recently there have been several antiviral drugs developed that have a reasonably high therapeutic index. Explain.

ANSWER KEY

Terms and Definitions

1. n, 2. d, 3. a, 4. l, 5. c, 6. g, 7. h, 8. m, 9. k, 10. e, 11. f, 12. b, 13. i, 14. j

Drugs and Their Discoverers

1. d, 2. b, 3. c, 4. e, 5. f, 6. a, 7. h, 8. g

Drug Action

1. cell wall synthesis inhibitor; cidal; bacteria 2. disrupts cell membranes; cidal; bacteria 3. protein synthesis inhibitor; cidal; bacteria 4. DNA synthesis inhibitor; N/A; viruses 5. disrupts RNA function; cidal; fungi 6. antimetabolite; static; bacteria 7. disrupts membrane permeability; cidal; fungi 8. RNA synthesis inhibitor; static; bacteria 9. cell wall synthesis inhibitor; cidal; bacteria 10. DNA synthesis inhibitor; cidal; bacteria 11. protein synthesis inhibitor; static; bacteria 12. antimetabolite; static or cidal; bacteria

Fill in the Blank

1. therapeutic index; toxic; better 2. antimetabolites; metabolic antagonists; competitive 3. oral; stomach acid 4. oral; parenteral; topical 5. folic acid; purines; pyrimidines; folic acid 6. drug resistant; indiscriminate

Multiple Choice

1. c, 2. d, 3. a, 4. d, 5. b, 6. d, 7. d, 8. a, 9. a, 10. d, 11. c

True/False

1. F, 2. F, 3. T, 4. T, 5. F, 6. T, 7. T, 8. T, 9. F, 10. T, 11.T

34 Clinical Microbiology

CHAPTER OVERVIEW

This chapter presents important information for understanding the field of clinical microbiology, which involves the detection and identification of pathogens that are the etiological agents of infectious disease. Identification may be based on the results of some combination of morphological, physiological, biochemical, and immunological procedures. Time may be critical in life-threatening situations; therefore, rapid identification systems and computers can be used to greatly speed up the process. Procedures that are useful for determining microorganism sensitivity to various antimicrobial agents are also discussed.

CHAPTER OBJECTIVES

After reading this chapter you should be able to:

- describe the functions and/or services performed by clinical microbiology laboratories
- discuss the need for proper specimen selection, collection, handling, and processing
- discuss the various procedures used to identify microorganisms in specimens
- explain the methods used for testing the sensitivity of microorganisms to antimicrobial agents
- discuss the use and advantage of computers in clinical microbiology

CHAPTER OUTLINE

I.   Specimens—portions of human material that are tested, examined, or studied to determine the presence or absence of specific microorganisms; universal safety precautions have been recommended by the CDC to address safety issues in specimen handling
   A.   Specimens should be:
      1.   Representative of the diseased area
      2.   Adequate in quantity for a variety of diagnostic tests
      3.   Devoid of contamination, particularly by microorganisms indigenous to the skin and mucous membranes
      4.   Forwarded promptly to the clinical laboratory
      5.   Obtained prior to the administration of any antimicrobials
   B.   Collection
      1.   Sterile swab—skin and mucous membranes
      2.   Needle aspiration—blood and cerebrospinal fluid—skin surface microorganisms must be excluded by the use of stringent antiseptic techniques
      3.   Intubation—stomach
      4.   Catheterization—urine
      5.   Clean-catch midstream urine—first urine voided is not collected, because it is likely to be contaminated with surface organisms
      6.   Sputum—mucous secretion expectorated from the lungs, bronchi, and/or trachea
   C.   Handling—includes any special additives (e.g., anticoagulants) and proper labeling

    D.   Transport—should be timely; temperature control may be needed; special treatment may be needed for anaerobes

II.   Identification of Microorganisms from Specimens
    A.   Microscopy—direct examination of specimen, or examination of specimen after various staining procedures
    B.   Growth and Biochemical characteristics
        1.   Viruses—identified by isolation in living cells, immunoassays, or nucleic acid technology
           a.   Host organisms for culture
              (1)   Embryonated hen's eggs
              (2)   Laboratory animals
              (3)   Cell or tissue cultures
              (4)   Transgenic cell lines from transgenic animals
           b.   Detection of virus growth
              (1)   Cytopathic effects, lesions, plaques
              (2)   Hemadsorption—the binding of red blood cells to the surface of infected cells
              (3)   Interference with subsequent infection by another virus
        2.   Fungi—identified by direct microscopic examination; immunofluorescence; yeast can be identified by the use of rapid ID methods (e.g., API 20C; Microring YT)
        3.   Parasites—identified by examining specimens for eggs, cysts, larvae, or vegetative cells; immunofluorescence is often used
        4.   Bacteria—identified by growth on selective and differential media; hemolytic, metabolic, and fermentative properties are also used
        5.   Rickettsias—identified by isolation (hazardous) or by immunoassays
        6.   Chlamydiae—identified by immunofluorescence and growth in cell culture
        7.   Mycoplasmas—identified immunologically or by the use of DNA probes
    C.   Rapid methods of identification
        1.   Manual biochemical techniques—API 20E system for enterobacteria
           a.   Consists of 20 microtube inoculation tests
           b.   Results are converted to a seven- or nine-digit profile number
           c.   The number is compared to the *API Profile Index* to determine the name of the bacterium
        2.   Mechanized/automated techniques (e.g., BIOLOG)—based on substrate catabolism
        3.   Immunological techniques—quick test kits that usually do not require growing the organism
    D.   Immunological techniques—detection of antigens or serum antibodies in specimens by the various procedures discussed in chapter 32
    E.   Bacteriophage typing—the host range specificities of bacteriophages are dependent upon surface receptors on the particular bacteria; therefore, this can be a reliable method of identification
    F.   Molecular methods (some have been previously discussed)
        1.   Enzyme kinetic comparisons
        2.   Nucleic Acid-based detection methods—ssDNA molecules that have been cloned from organism or prepare by PCR technology can be used in hybridization procedures; rRNA genes can be used to identify bacterial strains (ribotyping)
        3.   Gas-liquid chromatography (GLC)—used to identify specific microbial metabolites, cellular fatty acids, and products of pyrolysis; usually for nonpolar substances that are extractable in ether
        4.   Plasmid fingerprinting—separation and detection of the number and molecular weight of different plasmids, which are often consistently present in a strain of bacteria

III.   Susceptibility Testing—used to help physician decide which drug(s) and which dosage(s) to use

IV.   Computers in Clinical Microbiology
    A.   Test ordering—specific requests, patient data, and accession number
    B.   Result entry
    C.   Report printing—flexible format to meet the needs of physician

D. Laboratory management
E. Interfaced with automated instruments

TERMS AND DEFINITIONS

Place the letter of each term in the space next to the definition or description that best matches it.

_____ 1. An item or a part typical of a group or whole
_____ 2. A portion of human material that is tested, examined, or studied to determine the presence or absence of specific microorganisms
_____ 3. A dacron-tipped polystyrene applicator used for obtaining specimens from skin or mucous membranes
_____ 4. Secretion from glands near the oral cavity
_____ 5. Secretion from lungs, bronchi, and/or trachea
_____ 6. An observable change that occurs in cells as a result of viral infection
_____ 7. Phenomenon that occurs because of an alteration of the membrane of a virus-infected cell so that red blood cells will adhere to it

a. clinical specimen
b. cytopathic effect
c. hemadsorption
d. saliva
e. specimen
f. sputum
g. swab

FILL IN THE BLANK

1. The major concern of the clinical microbiologist is to rapidly _____ and _____ microorganisms from clinical specimens.
2. There are three types of catheters that are used to collect urine samples. When the urethra is narrow or has strictures, a _____ catheter is used; if a single sample is needed, and the urethra has no stricture and is reasonably wide, then a _____ catheter is used; and if multiple samples are required over a prolonged period of time, a _____ catheter is used.
3. In the _____ method of urine collection, the first urine voided is not used since it is likely to be contaminated with microorganisms that normally colonize the lower portion of the urethra. The urine behind this is more likely to contain those microorganisms found in the _____.
4. Cell cultures are divided into three general classes: those derived directly from tissues are called _____ cultures; when these are subcultured and consist of diploid fibroblasts that undergo a finite number of divisions, they are called _____ cultures; when they are derived from transformed, heteroploid cells that can be subpassaged indefinitely, they are referred to as _____ cultures.
5. Viral replication in cell cultures is detected in three ways: virus infection resulting in observable cell morphology changes is referred to as a _____; when it results in the plasma membrane alteration so that red blood cells adhere to it firmly, it is called _____; if there are no visible effects but the cells become resistant to infection with certain other viruses, this is referred to as _____.

MULTIPLE CHOICE

For each of the questions below select the *one best* answer.

1. Which of the following is *not* normally used to collect a urine sample?
   a. clean-catch midstream
   b. needle aspiration
   c. catheterization
   d. All of the above are normally used to collect urine samples.
2. Which of the following should be written or imprinted on the culture request form?
   a. patient information
   b. type or source of sample
   c. physician's choice of tests
   d. All of the above should be on the culture request form.
3. Which of the following is/are *not* normally used in the identification of microorganisms?
   a. microscopic examination
   b. growth or biochemical characteristics
   c. immunological techniques that detect antibodies or microbial antigens
   d. All of the above are used in the identification of microorganisms.
4. Which of the following is normally used to detect spirochetes in skin lesions in early syphilis?
   a. bright-field microscopy
   b. phase-contrast microscopy
   c. dark-field microscopy
   d. immunofluorescence microscopy
5. Which of the following is *not* normally used to culture viruses?
   a. growth on artificial media
   b. growth in cell cultures
   c. growth in embryonated hen's eggs
   d. growth in whole animals
6. Which of the following is *not* likely to be useful in the identification of bacteria?
   a. source of the culture specimen
   b. growth patterns on selective and differential media
   c. hemolytic, metabolic, and fermentative properties
   d. All of the above are useful in the identification of bacteria.

7. Which of the following is *not* a reason that immunological tests for antibodies against a particular infectious organism might yield negative results?
   a. The person might not be infected with the particular organism.
   b. The organism might be poorly immunogenic and may not stimulate sufficient antibody production to be detectable.
   c. There might not have been sufficient time since the onset of infection for an antibody response to develop.
   d. All of the above might yield negative results.
8. Which of the following is *not* true about the use of DNA:rRNA hybrids for identification as compared to the use of DNA:DNA hybrids?
   a. DNA:rRNA hybrids are more sensitive, and therefore, fewer microorganisms are required.
   b. DNA:rRNA hybrids are more specific, and they show less cross-hybridization to other species.
   c. DNA:rRNA hybrids are formed more rapidly; test requires two hours or less for results.
   d. All of the above are true about the use of DNA:rRNA hybrids as compared to the use of DNA:DNA hybrids.
9. The basic principle of plasmid fingerprinting is that
   a. microbial isolates of the same strain will contain the same number of plasmids with the same molecular weights.
   b. microbial isolates of different strains will have different plasmids (either in number, molecular weight, or both).
   c. Both (a) and (b) are correct.
   d. Neither (a) nor (b) is correct.

TRUE/FALSE

_____ 1. For best results, specimens should be obtained prior to the administration of any antimicrobial agents.

_____ 2. Needle aspirations of blood samples usually employ an anticoagulant in order to prevent entrapment of microorganisms in a clot, which would then make isolation difficult.

_____ 3. The Gram stain is used for bacteria that have cell walls, while the acid-fast stain is used primarily for wall-less bacteria.

_____ 4. Rickettsias are routinely isolated and identified by culture methods because these are relatively inexpensive and safe to use.

_____ 5. Bacteria can usually be identified by morphological examination, and biochemical tests are only needed for confirmation of identification.

_____ 6. Detection of an elevated antibody titer is used to indicate an active, ongoing infection.

_____ 7. Strains of bacteria that are infected by different phage isolates are referred to as phagovars.

_____ 8. The combination of gas-liquid chromatography (for separation of molecules) with mass spectrometry, nuclear magnetic resonance spectroscopy, and other analytical techniques for identification of separated components makes the identification of possible etiological agents of infectious diseases directly from body fluid specimens without first isolating the infectious organism.

_____ 9. Rapid ID systems such as BIOLOG identify bacteria based on substrate utilization characteristics.

_____ 10. The use of probes to rRNA genes that are scattered throughout the bacterial genomes to determine bacterial identity after endonuclease digestion of the DNA is called ribotyping.

CRITICAL THINKING

1. Compare and contrast the use of cell cultures, embryonated hen's eggs, and whole animals for growing viruses. What are the advantages and the disadvantages of each method? Why are all three sometimes used for the same clinical specimen and for different clinical specimens?

2. Describe the various ways that computers can be used in the clinical microbiology laboratory. How does each use contribute to the more efficient running of the laboratory? Are there any disadvantages to the use of computers? If so, what are they?

ANSWER KEY

Terms and Definitions

1. e, 2. a, 3. g, 4. d, 5. f, 6 b, 7. c

Fill in the Blank

1. isolate; identify 2. hard; soft; Fowley 3. clean-catch midstream; urinary bladder 4. primary; semicontinuous cell; continuous cell 5. cytopathic effect; hemadsorption; interference

Multiple Choice

1. b, 2. d, 3. d, 4. c, 5. a, 6. d, 7. d, 8. b, 9. c

True/False

1. T, 2. T, 3. F, 4. F, 5. F, 6. F, 7. T, 8. T, 9. T, 10. T

35 The Epidemiology of Infectious Disease

CHAPTER OVERVIEW

This chapter discusses the epidemiological parameters used to institute effective control, prevention, and eradication measures within an affected or potentially affected population. This chapter also discusses the epidemiology of hospital-acquired (nosocomial) infections, which have been of increasing concern in recent years.

CHAPTER OBJECTIVES

After reading this chapter you should be able to:

- define *epidemiology* and explain how it relates to infectious diseases
- discuss the statistical parameters used to define and describe various infectious diseases
- discuss the need to identify the etiologic agent in order to trace the origin and manner of spread of an infectious disease outbreak
- describe the five epidemiological links (characteristics of the infectious organism, source and/or reservoir, mode of transmission, susceptibility of the host, and exit mechanisms) in the infectious disease cycle of any epidemic
- discuss the increase in nosocomial infections in recent years and the consequences of these infections

CHAPTER OUTLINE

I.   Introduction
    A.   Epidemiology—the science that evaluates the occurrence, determinants, distribution, and control of health and disease in a defined human population
    B.   Epidemiologist—one who practices epidemiology (a disease detective)
    C.   Disease—an impairment of the normal state of an organism or any of its components, which hinders the performance of vital functions
II. Epidemiological Terminology
    A.   Sporadic disease—occurs occasionally at irregular intervals in a human population
    B.   Endemic disease—maintains a relatively steady low-level frequency at a moderately regular interval
    C.   Hyperendemic—a gradual increase in frequency above the endemic level, but not to the epidemic level
    D.   Epidemic—sudden increase in frequency above the endemic level
    E.   Index case—the first case in an epidemic
    F.   Outbreak—an epidemic-like increase in frequency, but in a very limited (focal) segment of the population
    G.   Pandemic—a long-term increase in frequency in a large (usually worldwide) population
    H.   Zoonoses—diseases of animals that can be transmitted to humans

III.	Measuring Frequency: The Tools of Epidemiologists
   A.	Statistics—the mathematics of collection, organization, and interpretation of numerical data
   B.	Morbidity—the number of new cases in a specific time period per unit of population
   C.	Prevalence—number of individuals infected at any one time per unit of population
   D.	Mortality—number of deaths from a disease per number of cases of the disease
IV.	Infectious Disease Epidemiology—Concerns of an Epidemiologist
   A.	What organism caused the disease?
   B.	What is the source and/or reservoir of the disease?
   C.	How is the disease transmitted?
   D.	What host and environmental factors facilitated development of the disease within a defined population?
   E.	How can the disease best be controlled or eliminated?
V.	Recognition of an Infectious Disease in a Population
   A.	Recognition involves various surveillance methods to monitor the population for disease occurrence and for demographic analysis
   B.	Surveillance involves identification of the signs and/or symptoms of a disease
      1.	Signs—objective changes in the body (e.g. fever, rash, etc.) that can be directly observed
      2.	Symptoms—subjective changes (e.g. pain, appetite loss, etc.) experienced by the patient
      3.	Disease syndrome—a set of signs and symptoms that is characteristic of a disease
   C.	Characteristic Patterns of an Infectious Disease
      1.	Incubation period—the period after pathogen entry but before signs and symptoms appear
      2.	Prodromal stage—onset of signs and symptoms but not yet clear enough for diagnosis
      3.	Period of illness—disease is most severe and has characteristic signs and symptoms
      4.	Period of decline (convalescence)—signs and symptoms begin to disappear
   D.	Correlation with a single causative organism uses methods described in chapter 34 for the organism's isolation and identification
VI.	Recognition of an Epidemic
   A.	Common-source epidemic—characterized by a sharp rise to a peak, and then a rapid but not as pronounced decline in the number of cases; usually results from exposure of all infected individuals to a single common, contaminated source, such as food or water
   B.	Propagated epidemic—characterized by a gradual increase and then a gradual decline in the number of cases; usually results from the introduction of one infected individual into a population, who then infects others; these in turn infect more, until an unusually large number of individuals within the population are infected
   C.	Herd immunity—the resistance of a population to infection and to the spread of an infectious organism because of the immunity of a large percentage of the population; this limits the effective contact between infective and susceptible individuals
   D.	Antigenic shifts—genetically determined changes in the antigenic character of a pathogen so that it is no longer recognized by the host's immune system (e.g., new flu strains); smaller changes are called antigenic drift
VII.	The Infectious Disease Cycle: Story of a Disease—Links in the infectious disease chain
   A.	What pathogen caused the disease? Epidemiologists must determine the etiology (cause) of a disease
      1.	Koch's postulates (or modifications of them) are used if possible
      2.	The clinical microbiology laboratory plays an important role in the isolation and identification of the pathogen
      3.	Communicable disease—one that can be transmitted from one host to another
   B.	What was the source and/or reservoir of the pathogen?
      1.	Source—location from which organisms are immediately transmitted to the host
      2.	Period of infectivity—the time during which the source is infectious or is disseminating the organism
      3.	Reservoir—site or natural environmental location where organism is normally found

4. Carrier—an infected individual who is a potential source of infection for others
   a. Active carrier—a carrier with an overt clinical case of the disease
   b. Convalescent carrier—an individual who has recovered from the disease but continues to harbor large numbers of the pathogen
   c. Healthy carrier—an individual who harbors the pathogen but is not ill
   d. Incubatory carrier—an individual who harbors the pathogen but is not *yet* ill
   e. Casual (acute, transient) carriers—any of the above carriers who harbor the pathogen for a brief period (hours, days, or weeks)
   f. Chronic carriers—any of the above carriers who harbor the pathogen for long periods (months, years, or life)
C. How was the pathogen transmitted?
   1. Airborne—suspended in air; travels a meter or more
      a. Droplet nuclei—may come from sneezing, coughing, or vocalization
      b. Dust particles—may be important in airborne transmission because microorganisms adhere readily to dust
   2. Contact—touching between source and host
      a. Direct (person-to-person)—physical interaction between infected person and host
      b. Indirect—involves an intermediate, such as eating utensils, thermometers, dishes, glasses, and bedding
      c. Droplets—large particles that travel less than one meter through the air
   3. Vehicle (fomite)—food and water, as well as those intermediates described for indirect contact
   4. Vector-borne—living transmitters, such as arthropods or vertebrates
      a. External (mechanical) transmission—passive carriage of the pathogen on the body of the vector with no growth of the organism during transmission
      b. Internal transmission—carried within the vector
         (1) Harborage—organism does not undergo morphological or physiological changes within the vector
         (2) Biologic—organism undergoes morphological or physiological changes within the vector
D. Why was the host susceptible to the pathogen? Depends on defense mechanisms of the host and the pathogenicity of the organism
E. How did the pathogen leave the host?
   1. Active escape—movement of organism to portal of exit
   2. Passive escape—excretion in feces, urine, droplets, saliva, or desquamated cells
VIII. Virulence and the Mode of Transmission
A. A virus that is spread by direct contact (e.g., rhinoviruses) cannot afford to make the host so ill it cannot be spread effectively
B. A virus that is vector-borne can afford to be highly virulent
C. Pathogens that do not survive well outside the host and that do not use a vector are likely to be less virulent while pathogens that can survive for long periods of time outside the host tend to be more virulent
IX. The Emergence of New Diseases
A. New diseases have emerged in the past few decades such as AIDS, Hepatitis C and E, hantavirus, Lyme disease, Legionnaire's disease, toxic shock *E. coli* 0157:H7, cryptosporidiosis, hepatitis G and others
B. Systematic epidemiology focuses on the ecological and social factors that influence the development and emergence of disease
C. Rapid transportation systems aid in the spread of disease out of areas where they are endemic
D. Ecological disruption can favor the spread of disease (e.g., loss of deer predators lead to an increase in the deer and deer tick population, thereby spreading Lyme disease to human hosts)
E. Narcotic use and sexual promiscuity have stimulated the spread of disease

X. Control of Epidemics
   A. Reduce or eliminate the source of infection through:
      1. Quarantine and isolation of cases and carriers
      2. Destruction of an animal reservoir, if one exists
      3. Treatment of sewage to reduce water contamination
      4. Therapy that reduces or eliminates infectivity of individuals
   B. Break the connection between the source and susceptible individuals through sanitization, disinfection, vector control, etc.; examples include:
      1. Chlorination of water supplies
      2. Pasteurization of milk
      3. Supervision and inspection of food and food handlers
      4. Destruction of insect vectors with pesticides
   C. Reduce the number of susceptible individuals—increase herd immunity
   D. Role of the public health system: epidemiological guardian—a network of health professionals involved in surveillance, diagnosis, and control of epidemics
XI. Nosocomial Infections—produced by infectious agents that develop within a hospital or other clinical care facility and that are acquired by patients while they are in the facility; infections that are incubating within the patient at the time of admission are not considered nosocomial
   A. Source
      1. Endogenous—patient's own microbiota
      2. Exogenous—microbiota other than the patient's
      3. Autogenous—cannot be determined to be endogenous or exogenous
   B. Control, prevention, and surveillance should include proper handling of the patient and the materials provided to the patient, as well as monitoring of the patient for signs of infection
   C. The hospital epidemiologist (other terms are also used) is an individual (usually a registered nurse) responsible for developing and implementing policies to monitor and control infections and communicable disease; usually reports to an infection control committee or other similar group

TERMS AND DEFINITIONS

Place the letter of each term in the space next to the definition or description that best matches it.

____ 1. The science that evaluates the determinants, occurrence, distribution, and control of health and disease in a defined population

____ 2. Term that describes a disease that occurs occasionally at regular intervals

____ 3. Term that describes a disease that maintains a relatively steady, low-level frequency

____ 4. Term that describes a disease that gradually increases in frequency

____ 5. Term that describes a disease that rapidly increases in frequency

____ 6. Term that describes a disease that shows a long-term increase in a large (usually worldwide) population

____ 7. The first case in an epidemic

____ 8. A disease of animals that can be transmitted to humans

____ 9. The mathematics of the collection, organization, and interpretation of numerical data

____ 10. The number of new cases of a disease during a specific time period per unit of population

____ 11. The total number of cases of a disease at any one time per unit of population

____ 12. The number of deaths from a given disease per number of infected individuals

____ 13. Term that describes a disease capable of being transmitted from one individual to another

____ 14. The location from which the causative organism is immediately transmitted to a host

____ 15. The location at which the causative organism is normally found

____ 16. An infected individual who is a potential source of infection for others

____ 17. Inanimate objects involved in the transmission of an infectious organism

____ 18. Living transmitters of an infectious organism

____ 19. Infections that are acquired in a hospital or other health care facility

____ 20. An impairment of the normal state of an organism or any of its components that hinders the performance of vital functions

____ 21. The cause of a disease

____ 22. The period of time in which the disease is most severe and has characteristic signs and symptoms

____ 23. Objective changes in the body

____ 24. The study of ecological and social factors that influence the development and emergence of disease

____ 25. Subjective changes experienced by the patient

a. carrier
b. communicable
c. convalescence
d. disease
e. disease syndrome
f. endemic
g. epidemic
h. epidemiology
i. etiology
j. hyperendemic
k. incubation period
l. index case
m. morbidity
n. mortality
o. nosocomial
p. pandemic
q. period of illness
r. prevalence
s. prodromal stage
t. reservoir
u. signs
v. source
w. sporadic
x. statistics
y. symptoms
z. systemic epidemiology
aa. vectors
bb. vehicles
cc. zoonosis

_____ 26. The period in which the signs and symptoms of a disease begin to disappear

_____ 27. The period after pathogen entry but before the onset of signs and symptoms

_____ 28. A set of signs and symptoms that are characteristic of a particular disease

_____ 29. The period in which signs and symptoms begin but are not yet clear enough for a diagnosis

FILL IN THE BLANK

1. When referring to diseases of animal populations, moderate prevalence is referred to as _____; a sudden outbreak is termed _____; and wide dissemination is called _____. Diseases of animals that can be transmitted to humans are termed _____.

2. Epidemics that are characterized by a sharp rise to a peak and then a rapid but not as pronounced decline in the number of individuals infected, and that usually results from exposure of all infected individuals to a single contaminant, such as food or water, is called a _____ epidemic; one that is characterized by a slow and prolonged rise and then a gradual decline in the number of infected individuals, and that usually results from the introduction of a single infected individual who spreads the disease to others, is called a _____ epidemic.

3. An organism that can be transmitted from one host to another is said to be _____. The immediate location from which the organism is transmitted to a host (either directly or through an intermediate) is called the _____, while the location where the organism is normally found is called the _____.

4. A carrier is an infected individual that can transmit the organism to other potential hosts. If the infected individual has an overt clinical case of the disease, he/she is referred to as a(n) _____ carrier; if the person has recovered from the infectious disease but can still transmit the organism, he/she is referred to as a(n) _____ carrier; if the individual can transmit the organism but is not ill, he/she is a(n) _____ carrier; if the individual is not yet ill but will become ill, he/she is referred to as a(n) _____ carrier.

5. Organisms that spread in the air and that travel a meter or more from source to host are said to be _____, while those which travel less than a meter are referred to as _____.

6. The transmission of organisms by a coming together or a touching of the source and the host is called _____ transmission. If an actual physical interaction is involved, it is called _____; if it involves touching, kissing, or sexual activity between two individuals, it is sometimes referred to as _____ transmission. If an intermediate (usually inanimate) is involved, it is called _____.

7. Living transmitters of an infectious organism are called _____. If the organism is carried on the body, this is called _____ transmission. If it is carried within the body, this is referred to as _____ transmission. In the latter situation, if the organism undergoes no morphological or physiological changes, this is called a _____ transmission phase, while if it undergoes morphological or physiological changes, this is called a _____ transmission phase.

8. In order to maintain its life cycle, the infectious organism must escape from the host. If the organism moves on its own to the portal of exit and then escapes, it is called _____ escape; if it is secreted out of the host in feces, urine, droplets, salvia, or desquamated cells, it is called _____ escape.

9. The resistance of a population to infection and spread of an infectious organism because of the immunity of a large percentage of the population is referred to as _____ immunity. It works because it limits the contact between _____ and _____ individuals.

MULTIPLE CHOICE

For each of the questions below select the *one best* answer.

1. Which of the following is(are) the major concern(s) of epidemiologists?
   a. the discovery of factors essential to disease occurrence
   b. the development of methods for disease prevention
   c. Both (a) and (b) are major concerns.
   d. Neither (a) nor (b) are major concerns.

2. A disease that gradually rises above the endemic level but is not at an epidemic level is referred to as
   a. outbreak.
   b. hyperendemic.
   c. sporadic.
   d. pandemic.

3. A disease that occurs suddenly but is a focal occurrence or in too limited a segment of the population is called a(n)
   a. outbreak.
   b. hyperendemic.
   c. sporadic.
   d. pandemic.

4. A disease that is of epidemic proportions but encompasses a broad (usually worldwide) population is referred to as
   a. outbreak.
   b. hyperendemic.
   c. sporadic.
   d. pandemic.

5. A disease that occurs occasionally at irregular intervals in a human population is referred to as
   a. outbreak.
   b. hyperendemic.
   c. sporadic.
   d. pandemic.

6. In a propagated epidemic, as the number of infected individuals increases, the number of susceptible individuals decreases. When the number of susceptible individuals decreases below a critical level, the probability of spread from an infected individual to a susceptible host becomes so small that the disease can no longer sustain propagation. This phenomenon is referred to as
   a. population resistance.
   b. susceptibility diminution.
   c. herd immunity.
   d. propagation termination.

7. Which of the following is *not* a mechanism by which new, susceptible individuals enter a population?
   a. birth of new individuals
   b. migration of individuals into the population
   c. evolution of disease-causing organisms to forms that are no longer recognized by host immune mechanisms
   d. All of the above are ways that new susceptible individuals enter a population.

8. Carriers that harbor the infectious organism for only a brief period of time (hours, days, or weeks) are referred to as
   a. casual carriers.
   b. acute carriers.
   c. transient carriers.
   d. All of the above are this type of carrier.

9. Which of the following is not normally a way that airborne organisms are introduced into the air by humans or animals?
   a. sneezing
   b. evaporation of sweat
   c. coughing
   d. vocalization (talking)

10. Which of the following does not refer to inanimate objects involved in the transmission of disease-causing organisms?
    a. vectors
    b. vehicles
    c. fomites
    d. All of the above refer to inanimate objects involved in the transmission of disease-causing organisms.

11. Which of the following is not used to reduce or eliminate the source of infection?
    a.   treatment of sewage to reduce water contamination
    b.   destruction of vectors by spraying insecticides
    c.   destruction of an animal reservoir of infection
    d.   All of the above eliminate the source of an infection.

12. Which of the following contributes to the emergence of new diseases?
    a.   rapid transportation systems and the mobility of the population
    b.   ecological disruption such as loss of predators and/or rainforest destruction
    c.   increased drug usage and sexual promiscuity
    d.   All of the above contribute to the emergence of new diseases.

TRUE/FALSE

_____ 1. A measured increase in the morbidity rate of a disease may forecast the need for health care professionals to prepare for an increase in mortality.

_____ 2. Epidemiologists monitor usage of specific antibiotics, antitoxins, vaccines, and other prophylactic measures in order to recognize the existence of a particular disease in the population.

_____ 3. Demographic data, such as movements of specific populations during various times of the year, is not a particularly useful predictor of disease occurrence.

_____ 4. The evolution of disease-causing organisms to forms that are no longer recognized by host immune mechanisms, but that still cause the same disease is called antigenic shift.

_____ 5. Generally the goal of public health officials is to make sure that at least 70% of the population is immunized against a particular disease-causing organism. This provides enough herd immunity to offer at least partial protection to those not immunized.

_____ 6. Most zoonoses are caused by wild animal populations because there are so many of them.

_____ 7. Nosocomial infections are those that become apparent while a patient is hospitalized.

_____ 8. Pathogens that are spread by direct contact tend to be more virulent than those that are vector-borne.

CRITICAL THINKING

1. Explain the concept of herd immunity. How can it develop naturally? How can it be stimulated artificially? Cite examples of both methods.

2. Explain antigenic drift and how it overcomes herd immunity and leads to new outbreaks of the same disease.

3. The authors present a number of surveillance methods used by epidemiologists. Discuss these methods in the context of a particular disease of past or current importance (e.g., smallpox, AIDS) to show how the data collected by these methods increase our knowledge of the disease process.

ANSWER KEY

Terms and Definitions

1. h, 2. w, 3. f, 4. j, 5. g, 6. p, 7. l, 8. cc, 9. x, 10. m, 11. r, 12. n, 13. b, 14. v, 15. t, 16. a, 17. bb, 18. aa, 19. o, 20. d, 21. i, 22. q, 23. v, 24. z, 25. y, 26. c, 27. k, 28. e, 29. s

Fill in the Blank

1. enzootic; epizootic; panzootic; zoonoses 2. common-source; propagated 3. communicable; source; reservoir 4. active; convalescent; healthy; incubatory 5. air-borne; droplet spread 6. contact; direct contact; person-to-person; indirect contact 7. vectors; external (mechanical); internal; harborage; biologic 8. active; passive 9. herd; infective; susceptible

Multiple Choice

1. c, 2. b, 3. a, 4. d, 5. c, 6. c, 7. d, 8. d, 9. b, 10. a, 11. b, 12. d

True/False

1. T, 2. T, 3. F, 4. T, 5. T, 6. F, 7. F, 8. F

36 Human Diseases Caused by Viruses

CHAPTER OVERVIEW

This chapter discusses viruses that are pathogenic to humans, with emphasis on those viral diseases occurring in the United States.

CHAPTER OBJECTIVES

After reading this chapter you should be able to:

- describe those viral diseases that are transmitted through the air and that directly or indirectly involve the respiratory system
- discuss viral diseases transmitted by arthropod vectors
- discuss viruses requiring direct contact for transmission, because they are so sensitive to environmental conditions that they are unable to survive for significant periods of time outside their hosts
- discuss viral diseases that are food- or waterborne
- discuss slow virus diseases that may be caused by viruses or prions
- discuss viral diseases that do not readily fit into any of the above categories

CHAPTER OUTLINE

I. Airborne Diseases
    A. Chickenpox (varicella) and shingles (zoster)
        1. Chickenpox is acquired by inhaling virus-laden droplets into the respiratory system (epidemic)
        2. Incubation period is 10 to 23 days
        3. Confers permanent immunity, but the virus enters a latent stage in the nuclei of sensory nerve roots
        4. When the person who harbors the virus is under stress, the virus can emerge and cause sensory nerve damage and painful vesicle formation, a condition known as shingles (zoster)
        5. Treated with acyclovir or famciclovir in immunocompromised patients
        6. Infection can be prevented or shortened by a live attenuated vaccine
    B. Influenza (flu)—undergoes frequent antigenic variation
        1. Antigenic drift—small variation
        2. Antigenic shift—large variation
        3. Treatment is symptomatic
        4. Enters by receptor-mediated endocytosis
        5. Can be rapidly identified in clinical specimens by an enzyme immunoassay (EIA)
        6. Epidemics vary greatly in severity and mortality rate
    C. Measles (rubeola)—skin disease with respiratory spread; infection confers permanent immunity; MMR vaccine is used for prevention; serious outbreaks are still reported in North America and Europe, especially among college students

D. Mumps
   1. Spread in saliva and respiratory droplets
   2. Can be serious in postpubescent male (infects testes and causes sterility)
   3. MMR vaccine is used for prevention
E. Respiratory syndromes and viral pneumonia—acute respiratory viruses
   1. A *syndrome* is a set of signs and symptoms that occur together and characterize a particular disease
   2. Associated with rhinitis, tonsillitis, laryngitis, and bronchitis
   3. Immunity is incomplete, and reinfection is common
   4. Viral pneumonia is clinically nonspecific, and symptoms may be mild or severe (death is possible)
   5. Treatment is supportive, not curative
   6. Respiratory syncytial virus (RSV) is the most dangerous cause of respiratory infection in young children
F. Rubella (German measles)
   1. Infection spread by respiratory droplets
   2. Mild in children
   3. Disastrous for pregnant women in first trimester—causes congenital rubella syndrome, which leads to fetal death, premature delivery, and congenital defects
   4. A vaccine (MMR: measles, mumps, rubella) is available
G. Smallpox (variola)—in the past was devastating but has been eradicated
   1. It has obvious (easily identifiable) clinical features
   2. There are virtually no asymptomatic carriers
   3. It infects only humans (there are no animal or environmental reservoirs)
   4. It has a short period of infectivity
   5. A vigorous worldwide vaccination program was undertaken
   6. The last case from a natural infection occurred in Somali in 1977

II. Arthropod-Borne Diseases
A. Viruses multiply in tissues of insect vectors without producing disease
B. Vector acquires a lifelong infection
C. Three clinical syndromes are common
   1. Undifferentiated fevers, with or without a rash
   2. Encephalitis—often with a high case fatality rate
   3. Hemorrhagic fevers—frequently severe and fatal
D. Permanent immunity; no vaccines available; treatment is supportive
E. Some examples are yellow fever, Ebola viral hemorrhagic fever, various forms of encephalitis, and Colorado tick fever

III. Direct Contact Diseases
A. Acquired immune deficiency syndrome (AIDS)
   1. Caused by human immunodeficiency virus (HIV), a lentivirus within the family *Retroviridae,* which is believed to have evolved in Africa from a virus that infects African green monkeys
   2. Requires direct exposure of the person's bloodstream to body fluids containing the virus; groups most at risk are:
      a. Homosexual/bisexual men
      b. Intravenous drug users
      c. Transfusion patients and hemophiliacs
      d. Prostitutes
      e. Newborn children of infected mothers
   3. The virus targets $CD4^+$ cells such as certain T cells, macrophages, dendritic cells, and monocytes

4. Four types of pathological changes may ensue
   a. AIDS-related complex (ARC)—mild fever, weight loss, lymph node enlargement, and presence of antibodies to HIV; can develop to full-blown AIDS
   b. AIDS—antibodies not sufficient to prevent infection; virus establishes itself in $CD4^+$ immunocompetent cells, which then proliferate in the lymph nodes and cause the lymph nodes to collapse; leads to depletion of T-cell progenitors, which cripples the immune system; this leaves the person open to opportunistic infections
   c. AIDS dementia and other evidence of central nervous system damage; the virus can cross the blood-brain barrier
   d. AIDS-related cancer—Kaposi's sarcoma (caused by human herpesvirus 8; HVV-8), carcinoma of the mouth and rectum, B-cell lymphomas
4. Co-infection with human herpesvirus-6 (HHV-6) can greatly enhance the development of AIDS
5. Diagnosis is by viral antigen detection or by viral antibody detection (seroconversion)
6. Drugs such as AZT, ddI and ddC target the reverse transcriptase and slow progress but do not reverse or cure the disease, nor do they affect transmission of the virus by infected individuals
7. New drugs being used are protease inhibitors; they also slow progress of the disease but do not cure it
8. Vaccines to stimulate production of neutralizing antibodies are currently under investigation
9. Prevention and control involves screening of blood and blood products, education, and safe sexual practices (use of condoms)
B. Cold sores (fever blisters)
   1. Caused by herpes simplex type 1 (HSV-1)
   2. Blister at site of infection is due to viral- and host-mediated tissue destruction
   3. Lifetime latency with periodic reactivation in times of physical or emotional stress
   4. Herpetic keratitis—recurring infections of the cornea; leads to blindness
   5. Drugs are available that are effective against cold sores
C. Common cold—caused by many different rhinoviruses, many of which do not confer durable immunity; at one time it was thought to be spread by explosive sneezing, but now it is believed to be primarily spread by hand-to-hand contact; treatment is supportive and symptomatic
D. Cytomegalovirus inclusion disease
   1. Most infections are asymptomatic but infection can be serious in immunologically compromised individuals
   2. Infected cells have intranuclear inclusion bodies
   3. Spread by contaminated saliva and urine
E. Genital herpes
   1. Caused by herpes simplex type 2 (HSV-2)
   2. Most frequently transmitted by sexual contact
   3. Symptoms include fever, burning sensation, genital soreness, and blisters in the infected area
   4. Blisters heal spontaneously, but the virus remains latent and is periodically reactivated
   5. Congenital (neonatal) herpes is spread to an infant during vaginal delivery; therefore, infected females should deliver children by caesarean section
   6. There is no cure, but acyclovir decreases healing time, duration of viral shedding, and duration of pain
F. Human Herpesvirus 6 Infections
   1. Etiologic agent of *exanthem subitum* (rash) in infants, a short-lived disease characterized by a high fever of 3 to 4 days duration, followed by a macular rash
   2. $CD4^+$ cells are the main sites of viral replication
   3. Tropism is wide including $CD8^+$ T cells, natural killer cells, and probably epithelial cells
   4. Transmission is probably by way of saliva
   5. It can lead to pneumonitis in immunocompromised individuals

6. It has been implicated in chronic fatigue syndrome and lymphadenitis in immunocompetent adults
7. Diagnosis is by immunofluorescence or enzyme immunoassay
8. There is neither treatment nor prevention currently available

G. Human Parvovirus B19 Infections
   1. Cause mild symptoms (fever, headaches, chills, malaise) in most normal adults
   2. Cause *erythema infectiosum* in children
   3. Can cause a joint disease in some adults
   4. Can cause serious aplastic crisis in immunocompromised individuals or those with sickle-cell disease or autoimmune hemolytic anemia
   5. Can cause anemia and fetal hydrops (the accumulation of fluid in the tissues) in infected fetuses
   6. Spread by a respiratory route
   7. Antiviral antibodies are the principal means of defense, and treatment is by means of commercial immunoglobulins containing anti-B19 and human monoclonal antibodies to B19
   8. Infection is usually followed by lifelong immunity

H. Leukemia—certain leukemias (adult T-cell leukemia and hairy-cell leukemia) are caused by retroviruses (HTLV-1 and HTLV-2, respectively) and are spread similarly to AIDS; they are often fatal and there is no effective treatment although interferon (INF-α) has shown some promise

I. Mononucleosis (Infectious)
   1. Caused by the Epstein-Barr virus (EBV), a herpesvirus
   2. Spread by mouth-to-mouth contact ("kissing disease")
   3. Occurs frequently in late adolescence
   4. Infects B cells
   5. Disease is self-limited and requires supportive treatment with plenty of rest
   6. EBV is also associated with Burkitt's lymphoma and nasopharyngeal carcinoma in certain parts of the world

J. Rabies—a disease of carnivorous and insectivorous animals
   1. Transmitted by bites or infected animals, aerosols in caves where bats roost, or contaminated scratches, abrasions, open wounds, or mucous membranes with saliva of infected animals; most are caused by dog bites in countries where canine rabies is still endemic
   2. Multiplies in skeletal muscle and connective tissue
   3. Migrates to central nervous system; produces characteristic Negri bodies (masses of virus particles or unassembled viral subunits)
   4. Symptoms progress and death results from destruction of the part of the brain that regulates breathing
   5. Vaccines conferring short-term immunity are available and must be given soon after exposure (postexposure vaccination is effective because of the long incubation period of the virus)
   6. Prevention and control involves annual preexposure vaccination of dogs and cats, postexposure vaccination of humans, and frequent preexposure vaccination of humans at special risk

K. Viral Hepatitides
   1. Hepatitis is any inflammation of the liver; currently nine viruses are recognized as causing hepatitis; some have not been well characterized
   2. Hepatitis B (serum hepatitis) is transmitted by blood transfusions, contaminated equipment, unsterile needles, or any body secretion; also transplacental transmission to fetus occurs
      a. Most cases are asymptomatic
      b. Symptomatic cases involve fever, appetite loss, and fatigue
      c. Death can result from liver cirrhosis or HBV-related liver cancer

       d.   Control measures include
           (1)   Excluding contact with contaminated materials
           (2)   Passive immunotherapy within seven days of exposure
           (3)   Vaccination of high-risk groups

    3.   Hepatitis C is spread by intimate contact with virus-contaminated blood, *in utero* from mother to fetus, by the fecal-oral route, or through organ transplants; disease is milder than hepatitis B with no jaundice

    4.   Hepatitis D is caused by hepatitis D virus (HDV) formally called the delta agent, which only causes disease if the individual is coinfected with hepatitis B virus; coinfection may lead to a more serious acute or chronic infection than that normally seen with HBV alone

    5.   Treatment is supportive, to allow the liver damage to resolve and the liver to repair itself, and the use of the hepatitis B vaccine

    6.   Recently, hepatitis F and hepatitis G have been identified and are currently being investigated

IV.   Food-Borne and Waterborne Diseases

   A.   Gastroenteritis (viral)—acute viral gastroenteritis
     1.   Caused by Norwalk and Norwalk-like viruses, rotaviruses, caliciviruses, and astroviruses
     2.   Disease severity may range from asymptomatic infection, to mild diarrhea, to severe and occasionally fatal dehydration
     3.   Leading cause of childhood death in developing countries where malnutrition is common
     4.   Seen most frequently in infants
     5.   Treatment is supportive and symptomatic

   B.   Hepatitis A—caused by the hepatitis A virus (HAV)
     1.   Spread by fecal-oral contamination of food or drink, or by infected shellfish that live in contaminated water
     2.   Mild intestinal infections sometimes progress to liver involvement
     3.   Low mortality rate (less than 1%); resolves in four to six weeks; infections in children
     4.   Produces strong immunity
     5.   A killed vaccine (Havrix) is now available

   C.   Hepatitis E
     1.   Implicated in many epidemics in developing countries in Asia, Africa, and Central and South America
     2.   Infection is associated with fecal-contaminated drinking water
     3.   HEV enters the blood from the gastrointestinal tract, replicates in the liver, is released from hepatocytes into the bile, and is subsequently excreted in the feces
     4.   HEV, like HAV, usually runs a benign course and is self-limiting
     5.   Can be fatal (10%) in pregnant women in their last trimester
     6.   There are no specific measures for prevention other than those aimed at improving the level of health and sanitation in affected areas

   D.   Poliomyelitis (polio, infantile paralysis)
     1.   Virus is stable and remains infectious in food and water
     2.   Multiplies in throat and intestinal mucosa
     3.   Enters bloodstream and causes viremia (99% of viremia cases are transient with no clinical disease)
     4.   Enters central nervous system (less than 1% of cases), leading to paralysis
     5.   Treatment is supportive, but vaccines have been extremely effective (less than 10 cases per year; no endogenous reservoir)

V.   Slow Virus Diseases
   A.   Progressive pathological process caused by a virus or a prion
   B.   Clinical silence for months or years, followed by progressive clinical disease, ending in profound disability or death

VI. Other Diseases
   A. Diabetes mellitus (juvenile onset) is suspected to have a viral etiology
   B. Viral arthritis does not cause permanent joint damage; associated with various viruses that primarily cause other diseases
   C. Warts are caused by papillomaviruses; treatment involves removal of warts, physical destruction, or injection of interferon

TERMS AND DEFINITIONS

Place the letter of each term in the space next to the definition or description that best matches it.

_____ 1. Small genetic variations in influenza virus that reduce herd immunity

_____ 2. Large genetic variations in influenza virus that reduce herd immunity

_____ 3. Lesions in the oral cavity that are red with a bluish-white center and that are diagnostic of measles

_____ 4. Deposition of bile in the skin and mucous membranes as a result of liver damage

_____ 5. Masses of viruses or unassembled viral subunits found in neurons infected with rabies virus

_____ 6. Progressive pathological processes caused by viruses or prions that remain clinically silent for a prolonged period of months or years, eventually leading to profound disability or death

a. antigenic drift
b. antigenic shift
c. jaundice
d. Koplik spots
e. Negri bodies
f. slow virus diseases

FILL IN THE BLANK

1. Changes in the antigenicity of influenza virus are referred to as antigenic _____; if such changes are small, they are referred to as antigenic _____; if the changes are relatively large, they are referred to as antigenic _____.

2. The arthropod-borne viruses can cause three different clinical syndromes: undifferentiated _____ with or without a _____; an inflammation of the brain, referred to as _____, which often has a high case fatality rate; and _____ fevers, which are also frequently severe and fatal.

3. Yellow fever is so named because of the jaundice it causes in severe cases. This jaundice is due to the deposition of bile pigments in the _____ and _____ as a result of damage to the _____.

4. Human immunodeficiency virus (HIV) is the causative agent of _____. In addition to the classical symptoms recognized as being associated with this virus, HIV can also damage the central nervous system. Virus-infected _____ cross the blood-brain barrier and cause an abnormal proliferation of _____ cells that surround neurons. This gives rise to a loss of memory, which is referred to as _____.

5. Herpes simplex type 2 virus is the causative agent of _____. It is most frequently spread by _____ but can be passed to a newborn during vaginal delivery, resulting in _____ herpes.

VIRAL DISEASES OF HUMANS

Complete the following table by providing the mode of transmission, the treatment, and the availability of an effective vaccine for each disease. Select your answers from among the following possibilities:

Mode of Transmission	Type of Treatment	Effective Vaccine
Airborne	Arrest progress	Yes
Arthropod-borne	Reduce severity	No
Direct contact	Symptomatic relief	
Food- or waterborne	Supportive	
Slow virus disease	None/prevention of bacterial infection	
Other/unknown		

Disease/Virus	Mode of Transmission	Type of Treatment	Effective Vaccine
1. Shingles 2. Eastern equine encephalitis 3. Infectious mononucleosis 4. Kuru 5. Hepatitis A 6. Hepatitis B 7. Measles 8. Common cold 9. AIDS 10. Poliomyelitis 11. Warts 12. Smallpox 13. Human herpesvirus 6 14. Human parvovirus B19			

MULTIPLE CHOICE

For each of the questions below select the *one best* answer.

1. The rash associated with German measles is most likely caused by
   a. the virus infecting the cells of the skin.
   b. the virus infecting the cells of the dermal layer beneath the skin.
   c. an immunological reaction to the virus.
   d. All of the above can cause this rash.

2. The vaccine against rubella (German measles) is recommended for
   a. all individuals not previously exposed to the virus.
   b. all children not previously exposed to the virus.
   c. all women of childbearing age not previously exposed to the virus.
   d. all children and all women of childbearing age not previously exposed to the virus.

3. Which of the following is(are) potential complications of mumps?
   a. meningitis
   b. inflammation of the testes in the postpubescent male
   c. Both (a) and (b) are potential complications.
   d. Neither (a) nor (b) is a potential complication.

4. Which of the following contributed to the eradication of smallpox from the world?
   a. There are no hosts, reservoirs, or vectors, other than humans.
   b. There are no asymptomatic carriers.
   c. Spread of the virus has been prevented by a vigorous worldwide vaccination program.
   d. All of the above contributed to the eradication of smallpox.

5. Which of the following can stimulate a reactivation of the virus that causes cold sores?
   a. physical stress, such as fever, cold wind, trauma, and hormonal changes
   b. emotional stress
   c. Both (a) and (b) can stimulate reactivation.
   d. Neither (a) nor (b) can stimulate reactivation.

6. Which of the following is not true about the use of acyclovir to treat genital herpes?
   a. It decreases the duration of virus shedding.
   b. It kills the virus and thereby cures the disease if it is applied for a long enough time.
   c. It decreases the duration of pain.
   d. All of the above are true about the use of acyclovir to treat genital herpes.

7. Which of the following is not true of retroviruses in humans?
   a. They cause all known leukemias in humans.
   b. They cause adult T-cell leukemia in humans.
   c. They cause hairy-cell leukemia in humans.
   d. None of the above are true of retroviruses in humans.

8. Which of the following is not true about rabies?
   a. It can be transmitted to humans by the bite of an infected animal.
   b. It can be caused by aerosol in caves where bats roost.
   c. It can be caused by contamination of abrasions, wounds, scratches, or mucous membranes with saliva from an infected animal.
   d. All of the above are true about rabies.

9. Which of the following is the mode of transmission for hepatitis C?
   a. contaminated blood and blood products
   b. fecal-oral route
   c. Both (a) and (b) are correct.
   d. Neither (a) nor (b) is correct.

10. Acute viral gastroenteritis is the leading cause of death in
    a. children worldwide.
    b. children in the United States.
    c. children in developing countries where malnutrition is a major problem.
    d. children in day-care centers.

11. Prevention and control of rabies involves
    a. preexposure vaccination of dogs and cats.
    b. postexposure vaccination of humans.
    c. preexposure vaccination of humans at high risk.
    d. All of the above are used in the prevention and control of rabies.

12. Drugs used to slow the progress of AIDS fall into which of the following categories?
    a. reverse transcriptase inhibitors
    b. protease inhibitors
    c. Both (a) and (b) are correct.
    d. Neither (a) nor (b) is correct.

TRUE/FALSE

_____ 1. Chickenpox and shingles are caused by the same virus.

_____ 2. Treatment of flu is symptomatic only, since there are no drugs that will affect the duration or severity of the disease.

_____ 3. Viral pneumonia is not specifically diagnosed, but is assumed when bacterial and mycoplasmal pneumonia have been eliminated as possibilities.

_____ 4. Control of yellow fever depends on control of the insect vector.

_____ 5. Because HIV (the causative agent of AIDS) infects cells of the immune system, no antibodies are ever produced against this virus.

_____ 6. The drug azidothymidine (AZT) has been successfully used in the treatment of AIDS. It does not cure the disease, but it does arrest its progress and inhibits its transmission.

_____ 7. The most likely mode of transmission of the common cold is by hand-to-hand contact between a rhinovirus donor and a susceptible recipient.

_____ 8. Infectious mononucleosis is called the kissing disease because that is the primary mode of transmission.

_____ 9. There was a tremendous interest in developing an effective vaccine against poliovirus because a high percentage of infections leads to paralytic central nervous system damage.

_____ 10. Slow virus diseases are so named because the viruses that cause them have extremely slow replication cycles.

_____ 11. Herpetic keratitis is a herpes simplex virus (HSV)-related infection of the eye that can cause blindness.

_____ 12. Respiratory syncytial virus is the most dangerous cause of respiratory infection in young children.

_____ 13. The last case of natural infection of smallpox occurred in Somali in 1977.

_____ 14. Coinfection with HIV and HHV-6 can greatly enhance the development of AIDS.

_____ 15. Flu epidemics vary greatly in severity and mortality.

CRITICAL THINKING

1. Discuss the pathology of AIDS. Why do people develop antibodies against HIV, the causative agent of AIDS, and yet are not protected against the disease by those antibodies? Does this mean that an effective vaccine against the virus is not possible? Why or why not?

2. There are at least two reasons why people continue to get the common cold throughout their lives. Explain what they are and why permanent immunity against this disease is not possible.

ANSWER KEY

Terms and Definitions

1. a, 2. b, 3. d, 4. c, 5. e, 6. f

Fill in the Blank

1. variation; drift, shift 2. fevers; rash; encephalitis; hemorrhagic 3. skin; mucous membranes; liver 4. AIDS; macrophages; glial; dementia 5. genital herpes; sexual contact; congenital (neonatal)

Viral Diseases of Humans

1. airborne; none/prevention; no 2. arthropod-borne; supportive; no 3. direct contact; supportive; no 4. slow virus disease; none; no 5. food-borne; symptomatic relief; no 6. direct contact; supportive; yes 7. airborne; none; yes 8. direct contact; symptomatic relief; no 9. direct contact; arrest progress; no 10. waterborne; supportive; yes 11. other/unknown; reduce severity; no 12. airborne; none; yes 13. direct contact; none; no
14. direct contact; arrest progress; no

Multiple Choice

1. c, 2. d, 3. c, 4. d, 5. c, 6. b, 7. a, 8. d, 9. c, 10. c, 11. d, 12. c

True/False

1. T, 2. F, 3. T, 4. T, 5. F, 6. F, 7. T, 8. T, 9. F, 10. F, 11. T, 12. T, 13. T, 14. T, 15. T

37 Human Diseases Caused Primarily by Gram-Positive and Gram-Negative Bacteria

CHAPTER OVERVIEW

This chapter discusses some of the more important gram-positive and gram-negative bacteria that are pathogenic to humans.

CHAPTER OBJECTIVES

After reading this chapter you should be able to:

- describe bacterial diseases that are transmitted through the air and that involve the respiratory system
- describe bacterial diseases that are transmitted through the air and that cause diseases of the skin or systemic disease
- discuss arthropod-borne bacterial diseases
- discuss bacterial diseases that require direct contact and that usually involve the skin and underlying tissues
- discuss food-borne and waterborne bacterial infections and bacterial intoxications
- discuss diseases and their sequelae such as sepsis and septic shock that cannot be categorized under a specific mode of transmission

CHAPTER OUTLINE

I.   Airborne Diseases
  A.   Diphtheria—*Corynebacterium diphtheriae*
    1.   Usually affects poor people living in crowded conditions
    2.   Caused by an exotoxin (diphtheria toxin) produced by lysogenized bacteria
    3.   Symptoms include nasal discharge, fever, cough, and the formation of a pseudomembrane in the throat
    4.   Treatment—neutralization of toxin with antitoxin, and penicillin or erythromycin to eliminate the bacteria
    5.   Vaccine—diphtheria-pertussis-tetanus vaccine (DPT)
    6.   Eradication may be possible with a vigorous worldwide vaccination campaign, because only humans serve as a reservoir for this organism
  B.   Legionnaires' disease and Pontiac fever—*Legionella pneumophila*
    1.   Legionnaires' disease (legionellosis)
      a.   Bacteria normally found in soil and aquatic ecosystems
      b.   Bacteria also found in air-conditioning systems and shower stalls
      c.   Infection causes cytotoxic damage to lung alveoli
      d.   Symptoms include fever, cough, headache, neuralgia, and bronchopneumonia
      e.   Common-source spread
      f.   Treatment is supportive; administration of erythromycin or rifampin

2. Pontiac fever
    a. Resembles an allergic disease more than an infection
    b. Symptoms are the same as for legionellosis, but pneumonia does not occur
    c. It usually spontaneously resolves in two to five days
C. Meningitis—caused by a variety of organisms and conditions
    1. Bacterial (septic) meningitis is diagnosed by the presence of bacteria in the cerebrospinal fluid
    2. Bacterial meningitis is treated with various antibiotics, depending on the specific bacterium involved
    3. Aseptic (nonbacterial) meningitis syndrome is more difficult to treat but the mortality is generally low
D. *Mycobacterium avium—M. intracellulaire* pneumonia
    1. Organisms are normal soil and water inhabitants
    2. Both the respiratory and the gastrointestinal tracts have been proposed as portals of entry; the gastrointestinal tract is thought to be the most common site of colonization and dissemination
    3. Pulmonary infection is similar to tuberculosis; most often seen in elderly patients with preexisting pulmonary disease
    4. Occurs in 15 to 40% of AIDS patients; is becoming a severe problem
    5. Symptoms include fever, malaise, weight loss, and diarrhea
    6. Treatment is usually multiple drug therapy
E. Pertussis (whooping cough)—*Bordetella pertussis*
    1. Highly contagious disease that primarily affects children
    2. Transmission is by droplet inhalation
    3. Toxins are responsible for most of the symptoms
    4. The disease progresses in stages
        a. Catarrhal stage—inflamed mucous membranes; resembles a cold
        b. Paroxysmal stage—prolonged coughing sieges with inspiratory whoop
        c. Convalescent stage—may take months (some fatalities)
    5. Permanent or long-lasting immunity
    6. Treatment with erythromycin, tetracycline, or chloramphenicol
    7. Vaccine—DPT
F. Streptococcal diseases—"strep throat," scarlet fever, pneumonia, etc.; treated with penicillin, cephalosporin, sulfonamide combination, or erythromycin; no available vaccines except one for streptococcal pneumonia; most streptococcal diseases are caused by varieties of *Streptococcus pyogenes*
    1. Cellulitis, impetigo, and erysipelas
        a. Cellulitis—diffuse, spreading infection of subcutaneous tissue characterized by redness and swelling
        b. Impetigo—superficial cutaneous infection commonly seen in children
        c. Erysipelas—acute infection of the dermis characterized by reddish patches
    2. Invasive Streptococcus A infections
        a. Dependent on specific strains and predisposing factors
        b. Causes necrotizing fasciitis that destroys the sheath covering skeletal muscle
        c. Also causes myositis, the inflammation and destruction of skeletal muscle and fat tissue
        d. Pyogenic exotoxins A and B are produced by 85% of the bacterial isolates
        e. Tissues die from lack of oxygen
        f. Can also trigger a toxic shocklike syndrome (TSLS) with a mortality rate over 30%
        g. Rapid treatment with penicillin G reduces the risk of death

315

3. Poststreptococcal diseases—onset is one to four weeks after an acute streptococcal infection
   a. Glomerulonephritis (Bright's disease)—antibody-mediated inflammatory reaction (type III hypersensitivity); may spontaneously heal or may become chronic; possibly a kidney transplant or lifelong renal dialysis may eventually be necessary
   b. Rheumatic fever—autoimmune disease involving the heart valves, other parts of the heart joints, subcutaneous tissues and central nervous system; mechanism is unknown; occurs primarily in children ages 6 to 15 years old; therapy is directed at decreasing inflammation and fever, as well as controlling cardiac symptoms and damage
4. Scarlet fever (scarlatina)—skin-shedding rash mediated by an erythrogenic toxin; caused by a lysogenic bacteriophage; treatment is with penicillin
5. Streptococcal sore throat (strep throat)—inflammatory response with lysis of erythrocytes and leucocytes; treatment is with penicillin, primarily to minimize the possibility of subsequent rheumatic fever and glomerulonephritis
6. Streptococcal pneumonia
   a. Opportunistic (endogenous) infection (caused by *S. pneumoniae*) from normal microbiota
   b. Individuals usually have one of the following predisposing factors:
      (1) Viral infection of the respiratory tract
      (2) Physical injury to the respiratory tract
      (3) Alcoholism
      (4) Diabetes
   c. Treatment is with penicillin, or erythromycin
   d. Vaccine (Pneumovax) is available
G. Tuberculosis—*Mycobacterium tuberculosis*
   1. Forms nodules (tubercles) in the lungs
   2. Diagnosis by isolation of organism, chest X ray, skin test, or DNA probes
   3. Chemotherapeutic and prophylactic treatment is isoniazid and rifampin, and streptomycin and/or ethambutol
   4. New cases are appearing either by recent transmission (25 to 33%) or by reactivation of old dormant infections (67 to 75%)
   5. Multidrug resistant strains are appearing in the population due to spontaneous mutations
II. Arthropod-Borne Diseases
A. Lyme disease (LD, Lyme borreliosis)—*Borrelia burgdorferi, B. garinii* and *B. afzelii*
   1. Tick-borne, with deer, mice, or the woodrat as the natural reservoir
   2. Symptoms include localized rash, flulike symptoms
   3. Disease can progress to include heart inflammation, arthritis, and neurological symptoms
   4. Years later, it can cause symptoms resembling Alzheimer's disease and multiple sclerosis with behavioral changes as well
   5. Laboratory diagnosis is by isolation of the spirochete, PCR to detect DNA in the urine, or serological testing (ELISA or Western Blot)
   6. Treatment with penicillin or tetracycline is effective if administered early
B. Plague—*Yersinia pestis*
   1. Flea-borne, from rodents
   2. Sporadic in the U.S. (about 25 cases per year)
   3. Bacteria survive and proliferate inside phagocytic cells
   4. Symptoms include subcutaneous hemorrhages, fever, and enlarged lymph nodes (buboes)
   5. Mortality rate is 50 to 70% if untreated
   6. Treatment—streptomycin or tetracycline
   7. If untreated it may invade lungs (pneumonic plague), resulting in 100% mortality if unrecognized within 12 to 24 hours
   8. A vaccine is available for people at high risk

III. Direct Contact Diseases
    A. Anthrax—*Bacillus anthracis*
        1. Causes ulcerated skin lesions or influenza-like symptoms; headache, fever, and nausea are major symptoms
        2. Treatment—penicillin, usually in combination with streptomycin; other treatment possibilities are erythromycin or tetracycline; cephalosporius or chloramphenicol
        3. Vaccine—available for animals and persons with high occupational risk
    B. Bacterial Vaginosis
        1. Disease is sexually transmitted with polymicrobic etiology
        2. Autoinfection in women from the rectum, which is inhabited by these organisms
        3. Disease is mild but is a risk factor for obstetric infections, various adverse outcomes of pregnancy, and pelvic inflammatory disease
        4. Diagnosis is based on fishy odor and microscopic observation of clue cells (sloughed-off vaginal epithelial cells covered with bacteria) in the discharge
        5. Treatment is with metronidazole to kill the anaerobes necessary for continuation of the disease
    C. Cat Scratch Disease (CSD)
        1. Probably caused by *Bartonella henselae*
        2. Diagnosis is based on the clinical history of a cat scratch or bite and subsequent swelling of the regional lymph nodes and by PRR amplification of appropriate gene sequences
        3. It is typically self-limiting with abatement of symptoms over a period of days to weeks
    D. Chancroid-genital ulcer disease
        1. It is sexually transmitted, caused by the gram-negative bacillus, *Haemophilus ducreyi*
        2. The bacterium enters the skin through a break in the epithelium
        3. After 4 to 7 days a papular lesion develops with swelling and white blood cell infiltration
        4. A pustule forms and ruptures leading to a painful ulcer on the penis or vagina
        5. It is a cofactor in the transmission of AIDS
        6. Diagnosis is by isolating the bacterium
        7. Treatment is with erythromycin or ceftriaxone
        8. Prevention is by use of condoms or abstinence
    E. Gas gangrene or clostridial myonecrosis—*Clostridium perfringens*
        1. Found in soil and intestinal tract microbiota
        2. Obligate anaerobe; problem in deep puncture wound where organism can grow readily and produce toxin
        3. Tissue necrosis (gangrene) occurs because of the toxin; growth of the organism leads to production of carbon dioxide and hydrogen gas
        4. Treatment—antitoxins, extensive surgical wound debridement, antibiotics, and hyperbaric oxygen
        5. Amputation may be necessary to prevent spread
    F. Gonorrhea—*Neisseria gonorrhoeae*
        1. Sexually transmitted disease of the genitourinary tract, eye, rectum, and throat
        2. Invades mucosal cells, causes inflammation and formation of pus
        3. Males—urethral discharge; painful, burning urination
        4. Females—frequently asymptomatic; some vaginal discharge; may lead to pelvic inflammatory disease (PID)
        5. Infection, if disseminated through the bloodstream can involve other organs
        6. Birth through infected vagina can result in neonatal eye infections (ophthalmia neonatorum, or conjunctivitis of the newborn) leading to possible blindness
        7. Diagnosis is by culture of the organism and/or use of a DNA probe
        8. Treatment—several combination antibiotic treatment regimens have been found to be effective; silver nitrate is often used in the eyes of newborns to prevent infection

9. Control by public education, diagnosis, treatment of symptomatic and asymptomatic individuals, and use of condoms

G. Leprosy—*Mycobacterium leprae*
   1. Severely disfiguring skin disease
   2. Usually requires prolonged exposure to nasal secretion of heavy bacteria shedders
   3. The incubation period may be three to five years, or even longer
   4. Starts as skin lesion and progresses slowly; most lesions heal spontaneously, those that don't are of two types:
      a. Tuberculoid (neural)—mild, nonprogressive, delayed-type hypersensitivity
      b. Lepromatous (progressive)—relentlessly progressive disfigurement
   5. Diagnosis is by observation in biopsy specimens and by serodiagnostic tests
   6. Treatment—long-term use of sulfa drugs (diacetyl/dapsone), rifampin, and clofazimine
   7. Control by identification and treatment of patients
   8. Children of contagious parents should be given prophylactic drug therapy until their parents are treated and have become noninfectious

H. Peptic ulcer disease and gastritis—*Helicobacter pylori*
   1. The evidence of pathogenicity of this organism is now very strong, if not overwhelming
   2. Alters gastric pH to favor its own growth near the mucosal cells
   3. Colonizes gastric mucus-secreting cells
   4. Releases toxins that damage epithelial mucosal cells
   5. Transmission is probably propagated, but common source has not been definitively ruled out
   6. Diagnosis is by culture of gastric biopsy specimens and serological testing
   7. Treatment includes antibiotics and bismuth subsalicylate (Pepto-Bismol)

I. Staphylococcal diseases
   1. Staphylococci are facultative anaerobes and are usually catalase positive
   2. *S. aureus*—coagulase positive, pathogenic; causes severe chronic infections
   3. *S. epidermidis*—coagulase negative, less invasive, opportunistic pathogens associated with nosocomial infections
   4. Many of the pathogenic strains are slime producers (SP); slime is a viscous extracellular glycoconjugate that allows the bacteria to adhere to smooth surfaces such as medical prostheses and catheters
   5. Slime producers form biofilms on prosthetic devices
   6. Slime also inhibits neutrophil chemotaxis, phagocytosis and the antimicrobial agents vancomycin and teicoplanin
   7. Disease can be produced in any organ of the body but is most likely to occur in individuals whose defenses have been compromised
   8. They produce exotoxins and substances that promote invasiveness
   9. Can cause food poisoning, localized abscesses, impetigo contagiosum, toxic shock syndrome, and staphylococcal scalded skin syndrome
   10. Diagnosis is by culture identification, catalase and coagulase tests, serology, DNA fingerprinting, and phage typing
   11. Several antibiotics can be used for treatment but isolates should be tested for sensitivity because of the existence of many drug-resistant strains

J. Syphilis—*Treponema pallidum*
   1. Sexually transmitted or congenitally acquired in utero
   2. Disease progresses in stages
      a. Primary stage—lesion (chancre) at infection site that can transmit organism during sexual intercourse
      b. Secondary stage—skin rash and other, more general symptoms
      c. Latent stage—not communicable after two to four years except possibly congenitally
      d. Tertiary stage—degenerative lesions (gummas) in the skin, bone, and nervous system
   3. Diagnosed by clinical history, microscopic examination, and serology

4. Treatment—penicillin in early stages, tertiary stage is highly resistant to treatment
5. Immunity is incomplete
6. Control is by public education, treatment, follow-up on sources and contacts, sexual hygiene, and prophylaxis (use of condoms)

K. Tetanus—*Clostridium tetani*
1. Found in soil, dust, hospital environments, and mammalian feces
2. Low invasiveness, but in deep tissues with low oxygen tension, the spores germinate; when the vegetative cells lyse, they release tetanospasmin (an exotoxin)
3. Toxin causes prolonged muscle spasms
4. A hemolysin (tetanolysin) is also produced
5. Prevention is important and involves:
   a. Active immunization with toxoid (DPT)
   b. Debridement of wounds
   c. Prophylactic use of antitoxin
   d. Administration of penicillin

L. Tularemia—*Francisella tularensis* is spread from animal reservoirs by a variety of mechanisms; a vaccine is available for high-risk laboratory workers

IV. Food-Borne and Waterborne Diseases

A. Cholera—*Vibrio cholerae*
1. Bacteria adhere to the intestinal mucosa of the small intestine
2. They are not invasive, but secrete cholera enterotoxin (choleragen)
3. Choleragen stimulates hypersecretion of water and chloride ions, while inhibiting adsorption of sodium ions; leads to fluid loss
4. Death may result from volume depletion-induced increased protein concentrations in blood, causing circulatory shock and collapse
5. Diagnosis is by culture of the bacterium from feces and serotyping
6. Treatment is rehydration therapy (fluid and electrolyte replacement) plus tetracycline or chloramphenicol
7. With treatment, mortality rate is reduced from higher than 50% (untreated) to less than 1% (treated)
8. Control is based on proper sanitation

B. Botulism—*Clostridium botulinum*
1. Food toxication (poisoning)
2. Frequently caused by canned foods that contain endospores, which germinate and produce an exotoxin (neurotoxin)
3. Can cause death by respiratory or cardiac failure
4. Diagnosis is by hemagglutination testing or toxigenicity testing in animals using the patient's serum, stools, or vomitus
5. Treatment is supportive; also antitoxin administration
6. Control involves safe food processing practices in the food industry and in home canning; also not feeding honey to babies under one year of age helps prevents infant botulism

C. *Campylobacter jejuni* gastroenteritis (campylobacteriosis)
1. Transmitted by contaminated food or water, contact with infected animals, or anal-oral sexual activity
2. Causes diarrhea, fever, intestinal inflammation and ulceration, and bloody stools
3. Treatment is supportive, with fluid and electrolyte replacement
4. Erythromycin is used in severe cases, but disease is usually self-limited

D. Salmonellosis—*Salmonella typhimurium* and others
1. Food-borne, particularly in poultry, eggs, and egg products; also in contaminated water
2. Food infection (bacteria must multiply and invade the intestinal mucosa)
3. Enterotoxin and cytotoxin destroy intestinal epithelial cells
4. Causes abdominal pain, cramps, diarrhea, and fever

5. Fluid loss can be a problem, particularly for children and elderly people
6. Treatment is fluid and electrolyte replacement
7. Controlled with good food preparation practices

E. Shigellosis—*Shigella* spp
1. Shigellosis or bacterial dysentery is most prevalent in children 1 to 4 years old
2. Small infectious dose (10 to 100 bacteria)
3. It is a particular problem in day care centers and custodial institutions where there is crowding
4. They do not usually spread beyond the colon epithelium
5. Identification is based on biochemical characteristics and serology
6. Self-limiting in adults but may be fatal in children
7. Treatment is fluid and electrolyte replacement
8. Prevention is a matter of personal hygiene and a clean water supply

F. Staphylococcal food poisoning—*Staphylococcus aureus*
1. Caused by ingestion of improperly stored or prepared food in which the organism has grown
2. Organism produces several enterotoxins that cause severe abdominal pain, diarrhea, vomiting, and nausea
3. Symptoms come quickly (one to six hours) and leave quickly (24 hour)
4. Treatment is fluid and electrolyte replacement

G. Traveler's Diarrhea and *Escherichia coli* Infections
1. Traveler's diarrhea is caused by certain viruses, bacteria or protozoa normally absent from the traveler's environment
2. *E. coli* is one of the major causative agents and may cause disease by a variety of mechanisms
3. Six categories or strains of diarrheagenic *E. coli* are now recognized
4. Enterotoxigenic *E. coli* (ETEC) produces two enterotoxins that are responsible for symptoms including hypersecretion of electrolytes and water into the intestinal lumen
5. Enteroinvasive *E. coli* (EIEC) multiplies within the intestinal epithelial cells; may also produce a cytotoxin and an enterotoxin
6. Enteropathogenic *E. coli* (EPEC) causes effacing lesions, destruction of brush border microvilli on intestinal epithelial cells
7. Enterohemorrhagic *E. coli* (EHEC) causes attaching-effacing lesions leading to hemorrhagic colitis; it also releases toxins that kill vascular epithelial cells; E. coli 0517:H7 is a major form of BHEC that has caused many outbreaks of hemorrhagic colitis in the U.S.
8. Enteroaggregative *E. coli* (EAggEC) forms clumps adhering to epithelial cells, toxins have not been identified but are suspected from the type of damage done
9. Diffusely adhering *E. coli* (DAEC) adheres in a uniform pattern to epithelial cells and is particularly problematic in immunologically naive or malnourished children
10. Diagnosis is based on past travel history and symptoms
11. Lab diagnosis is by isolation of the specific type of *E. coli* from feces and identification using DNA probes, determination of virulence factors, and the polymerase chain reaction
12. Treatment is electrolyte replacement plus antibiotics
13. Prevention and control involve avoiding contaminated food and water

H. Typhoid fever—*Salmonella typhi*
1. Caused by ingestion of food or water contaminated with human or animal feces
2. Symptoms are fever, headache, abdominal pain, and malaise that last several weeks
3. Treatment is with chloramphenicol, cephalosporin, or ampicillin
4. Control involves purification of drinking water, pasteurization of milk, preventing carriers from handling food, and complete patient isolation
5. A vaccine is available for high-risk individuals

V. Sepsis and Septic Shock
    A. Cannot be categorized under a specific mode of transmission
    B. Sepsis
        1. Systemic response to a microbial infection
        2. Manifested by fever or retrograde fever, heart rate > 90 beats per minute, respiratory rate > 20 breaths per minute, a pCO_2 < 32 mmHg, a leukocyte count > 12,000 cells per ml or < 4,000 cells per ml
    C. Septic shock
        1. Sepsis associated with severe hypotension (low blood pressure)
        2. Gram-positive bacteria, fungi, and endotoxin-containing gram-negative bacteria can initiate the pathogenic cascade of sepsis leading to septic shock
        3. Lipopolysaccharide (LPS), an integral component of the outer membrane of gram-negative bacteria has been implicated
    D. Pathogenesis begins with localized proliferation of the microorganism
        1. Bacteria may invade the bloodstream or may proliferate locally and release various products into the bloodstream
        2. Products include structural components (endotoxins) and secreted exotoxins
        3. These products stimulate the release of endogenous mediators of shock from plasma cells, monocytes, macrophages, endothelial cells, neutrophils, and their precursors
        4. The endogenous mediators have profound effects on the heart, vasculature, and other body organs
        5. Death ensues if one or more organ systems fail completely
    E. Diagnosis is based primarily on symptoms
    F. Early administration of antimicrobial therapy is important for effective management; broad spectrum antibiotics are used prior to culture results; this is followed by specifically tailored treatment
    G. Monoclonal antibodies to endotoxin can attenuate the adverse effects in shock patients

TERMS AND DEFINITIONS

Place the letter of each term in the space next to the definition or description that best matches it.

_____ 1. Inflammation of the membranes around the brain or spinal cord

_____ 2. A diffuse, spreading infection of subcutaneous skin tissue

_____ 3. An acute inflammation of the dermal layer of the skin in people with a history of streptococcal sore throat

_____ 4. An inflammatory disease of the membranous sites within the kidney where blood is filtered

_____ 5. A hard nodule of bacteria in the lungs surrounded by lymphocytes, macrophages, and connective tissue

_____ 6. An enlarged lymph node that is associated with the plague

_____ 7. An ulcerated papule on the skin associated with anthrax

_____ 8. Tissue necrosis caused by the anaerobic growth of *Clostridium perfringens* in puncture wounds

_____ 9. A small, painless, reddened ulcer with a hard ridge that appears at the infection site in the early stages of syphilis

_____ 10. A degenerative lesion in the skin, bone, and nervous system that is associated with tertiary syphilis

_____ 11. Sepsis accompanied by severe hypotension (low blood pressure) that is often fatal

_____ 12. A viscous extracellular glycoconjugate that allows bacteria to adhere to smooth surfaces

_____ 13. An infection caused the person's own microbiota

a. bubo
b. cellulitis
c. chancre
d. endogenous infection
e. erysipelas
f. eschar
g. gangrene
h. glomerulonephritis
i. gumma
j. meningitis
k. slime
l. septic shock
m. tubercle

BACTERIAL DISEASES OF HUMANS

Complete the following table by providing the name of the organism, the mode of transmission, the treatment, and the availability of an effective vaccine for each disease. For the latter three items, select your answers from among the following possibilities:

Mode of Transmission	Type of Treatment	Effective Vaccine
Airborne	Antibiotic	Yes
Arthropod-borne	Antitoxin	No
Direct contact	Supportive	
Food- or waterborne		

Disease	Organism	Mode of Transmission	Type of Treatment	Effective Vaccine
1. Tetanus				
2. Lyme disease				
3. Botulism				
4. Whooping cough				
5. Typhoid fever				
6. Meningitis				
7. Peptic ulcer disease				
8. Plague				
9. Syphilis				
10. Leprosy				
11. Chancroid				

FILL IN THE BLANK

1. The DPT vaccine protects against _____, _____ (pertussis), and _____.
2. Bacterial meningitis is also called _____ meningitis, while meningitis caused by other types of agents, such as viruses, is called _____ meningitis.
3. The lesions caused by *Mycobacterium tuberculosis* can take various forms; if they are hard nodules surrounded by lymphocytes, macrophages, and connective tissue, they are called _____; later they may take on a cheeselike consistency, and are referred to as _____ lesions; and if they calcify, they are called _____, which show up prominently in a chest X ray.
4. Whooping cough or _____ is divided into three stages. The onset of the _____ stage is insidious and resembles the common cold; the _____ stage is characterized by prolonged sieges of coughing followed by an inspiratory "whoop" from which the disease gets its name; the final recovery period is called the _____ stage and may last for several months.
5. The plague is sometimes referred to as _____ plague because of the enlarged lymph nodes, which are called _____; if the bacteria invade the lungs, they may cause a more serious disease referred to as _____ plague.
6. Leprosy is sometimes referred to as _____ disease and can take two distinct forms: _____ (neural) leprosy is a mild, nonprogressive hypersensitivity to surface antigens of the organism; those individuals who do not develop the hypersensitivity have a relentlessly progressive form referred to as _____ leprosy.
7. Traveler's diarrhea can be caused by _____, _____, or _____ normally absent from the traveler's environment. One of the major causative agents is _____ of which six categories are now recognized.

MULTIPLE CHOICE

For each of the questions below select the *one best* answer.

1. Legionnaires' disease is so named because
   a. it only strikes members of the French Foreign Legion.
   b. it was first identified at a convention of the American Legion.
   c. it strikes predominantly old soldiers.
   d. None of the above are correct.
2. Which of the following is not a streptococcal disease?
   a. cellulitis
   b. erysipelas
   c. scarlet fever
   d. All of the above are streptococcal diseases.
3. Lyme disease was so named because
   a. it is caused by eating too many limes.
   b. it is caused by not eating enough limes.
   c. it was first observed and described among people of Old Lyme, Connecticut.
   d. it was first diagnosed in a patient named Harold Lyme.

4. Which of the following is(are) used in the treatment of gas gangrene?
   a. surgical removal of foreign material and dead tissue from the wound
   b. antitoxins against the necrosis-causing toxins released by the organism
   c. hyperbaric chambers containing pressurized oxygen to poison the anaerobic organism that causes the condition
   d. All of the above are used in the treatment of gas gangrene.
5. Diseases of the gastrointestinal tract that are caused by the ingestion of bacterial exotoxins even in the absence of the viable microorganisms that produced them are called
   a. food intoxications (poisonings).
   b. food infections.
   c. Both (a) and (b) are correct.
   d. Neither (a) nor (b) is correct.

6. Diseases of the gastrointestinal tract that require colonization by a viable microorganism to produce symptoms are called
   a. food intoxications (poisonings).
   b. food infections.
   c. Both (a) and (b) are correct.
   d. Neither (a) nor (b) is correct.
7. Which of the following is not considered a predisposing factor for streptococcal pneumonia?
   a. diabetes
   b. alcoholism
   c. viral respiratory infections
   d. All of the above are predisposing factors for streptococcal pneumonia.
8. Which of the following is a mode of transmission for *Campylobacter jejuni*?
   a. contaminated food or water
   b. contact with infected animals
   c. anal-oral sexual activity
   d. All of the above are correct.
9. Invasive Streptococcus A infections depend on
   a. specific strains.
   b. predisposing factors.
   c. Both (a) and (b) are correct.
   d. Neither (a) nor (b) is correct.
10. New cases of tuberculosis are the result of
    a. recent transmission.
    b. reactivation of old dormant infections.
    c. Both (a) and (b) are correct.
    d. Neither (a) nor (b) is correct.

11. Sloughed-off vaginal epithelial cells covered with bacteria found in the discharge associated with bacterial vaginosis are called
    a. nurse cells.
    b. clue cells.
    c. vaginotal cells.
    d. diagnostic cells.
12. Control of syphilis involves
    a. follow-up on sources and contacts.
    b. sexual hygiene.
    c. use of condoms.
    d. All of the above help control syphilis.
13. Which of the following is not involved in the control of botulism?
    a. safe food-processing practices in the food industry
    b. safe food-processing practices in home canning
    c. not feeding honey to babies less than one year of age
    d. All of the above are involved in the control of botulism.
14. Shigellosis, or bacterial dysentery, is a particular problem in
    a. daycare centers.
    b. crowded custodial institutions.
    c. Both (a) and (b) are correct.
    d. Neither (a) nor (b) is correct.
15. Lab diagnosis of traveler's diarrhea is by isolation of the specific type of *E. coli* from feces and identification by
    a. using DNA probes.
    b. determination of virulence factors.
    c. the polymerase chain reaction.
    d. All of the above are correct.

TRUE/FALSE

____ 1. *Corynebacterium diphtheriae* can only cause diphtheria when it is lysogenized by a phage responsible for the production of the toxin.

____ 2. Once the initial case has occurred, Legionnaires' disease spreads as a propagated epidemic.

____ 3. Bacterial meningitis is more difficult to treat than meningitis caused by other agents.

____ 4. Although permanent immunity is not established, outbreaks of streptococcal sore throat are less frequent among adults because of their accumulation of antibodies to many different serotypes.

____ 5. There are more male than female asymptomatic carriers of gonorrhea.

____ 6. Because of advances in education and treatment, there has recently been a dramatic decrease in the incidence of syphilis and other sexually transmitted diseases.

____ 7. Gastroenteritis caused by *Campylobacter jejuni* has a high mortality rate (30 to 50%) if left untreated.

____ 8. *Legionella pneumophila* causes both legionellosis and Pontiac fever.

_____ 9. If a vigorous worldwide vaccination program were undertaken, diphtheria, like smallpox, could be eradicated because there is no nonhuman reservoir.

_____ 10. Mycobacterial (MAC) pneumonia has been seen in 15 to 40% of AIDS patients and is becoming an increasing problem.

_____ 11. Because leprosy is not very contagious, it usually is not necessary to treat uninfected children of infected parents.

_____ 12. The evidence is very strong, if not overwhelming, that *Helicobacter pylori* is the primary cause of peptic ulcers disease.

_____ 13. Bacterial vaginosis is frequently the result of autoinfection from the rectum where the causative organisms usually reside.

_____ 14. *Staphylococcus epidermidis* are more invasive than *Staphylococcus aureus*.

_____ 15. Shigellosis is often fatal in both adults and children.

_____ 16. Sepsis and septic shock cannot be categorized under a specific mode of transmission.

_____ 17. Septic shock is actually the result of endogenous mediators released form the patient's own cells in response to bacterial endotoxins or exotoxins.

_____ 18. *E. coli* 0157:H7 is a major cause of hemorrhagic colitis in the U.S.

_____ 19. Tularemia is primarily spread from animal reservoirs.

CRITICAL THINKING

1. In 1976, an outbreak of a respiratory disease occurred at a hotel in Philadelphia where a convention of the American Legion was being held. The previously unrecognized disease was named Legionnaires' disease. Assume you were the epidemiologist involved in this. How would you identify the source and/or reservoir for the causative agent, and how would you determine if it was a common-source or propagated epidemic?

2. What is the difference between a food infection and a food intoxication (poisoning)? Give at least two examples of each and discuss how the treatments vary for the different types of food-borne diseases.

ANSWER KEY

Terms and Definitions

1. j, 2. b, 3. e, 4. h, 5. m, 6. a, 7. f, 8. g, 9. c, 10. i, 11. l, 12. k, 13. d

Bacterial Diseases of Humans

1. *Clostridium tetani;* direct contact; antibiotics, antitoxin, supportive; yes 2. *Borrelia burgdorferi;* arthropod-borne; antibiotics; no 3. *Clostridium botulinum;* food-borne; supportive; antitoxin; no 4. *Bordetella pertussis;* airborne; antibiotics; yes 5. *Salmonella typhi;* food-borne; antibiotics; no 6. *Neisseria meningitidis* and others; airborne; antibiotics; yes 7. *Helicobacter pylori;* direct contact; antibiotics, supportive; no 8. *Yersinia pestis;* arthropod-borne; antibiotics; no 9. *Treponema pallidum;* direct contact; antibiotics; no 10. *Mycobacterium leprae;* direct contact; antibiotics; no 11. *Haemophilus ducreyi;* direct contact; antibiotics; no

Fill in the Blank

1. diphtheria; whooping cough; tetanus 2. septic; aseptic 3. tubercle; caseous; Ghon complexes 4. pertussis; catarrhal; paroxysmal; convalescent 5. bubonic; buboes; pneumonic 6. Hansen's; tuberculoid; lepromatous 7. bacteria; fungi; protozoa; *Escherichia coli*

Multiple Choice

1. b, 2. d, 3. c, 4. d, 5. a, 6. b, 7. d, 8. d, 9. c, 10. c, 11. b, 12. d, 13. d, 14. c, 15. d

True/False

1. T, 2. F, 3. F, 4. T, 5. F, 6. F, 7. F, 8. T, 9. T, 10. T, 11. F, 12. T, 13. T, 14. F, 15. F, 16. T, 17. T, 18. T, 19. T

38   Human Diseases Caused by Other Bacteria (Chlamydiae, Mycoplasmas, Rickettsias); Dental and Nosocomial Infections

CHAPTER OVERVIEW

This chapter discusses some of the more important chlamydia, mycoplasmas, and rickettsia that are pathogenic to humans. Dental and nosocomial infections are also discussed.

CHAPTER OBJECTIVES

After reading this chapter you should be able to:

- describe diseases caused by *Chlamydia trachomatis, C. psittaci,* and *C. pneumoniae*
- describe the genitourinary diseases caused by *Mycoplasma hominis* and *Ureaplasma urealyticum,* and the respiratory disease caused by *M. pneumonia*
- discuss the typhus and spotted fever groups of rickettsias
- discuss the bacterial odontopathogens involved in tooth decay and periodontal disease
- discuss bacterial nosocomial infections, such as bacteremias, burn wound infections, respiratory tract infections, surgical wound infections, and urinary tract infections

CHAPTER OUTLINE

I. Chlamydial Diseases
    A. Chlamydial Pneumonia—*C. pneumoniae*
        1. Mild upper respiratory infection with some lower respiratory tract involvement
        2. Infections are common but sporadic; about 50% of adults have antibodies to *C. pneumoniae*
        3. Transmitted from human to human without a bird or animal reservoir
        4. Diagnosis is based on symptoms and a microimmunofluorescence test
        5. Treatment is with tetracycline and erythromycin
    B. Inclusion conjunctivitis—*C. trachomatis*
        1. Sensitive to sulfonamides; forms inclusions containing glycogen
        2. Causes epithelial cell infections of the conjunctiva, pharynx, respiratory tract, urethra, cervix, and uterine tubes
        3. Spreads primarily by sexual contact
        4. Resolves spontaneously over several weeks or months
        5. Diagnosed by direct immunofluorescence, Giemsa stain, or nucleic acid probes
        6. Treatment is tetracycline or erythromycin
        7. Inclusion conjunctivitis of newborns is established from contact with an infected birth canal
    C. Lymphogranuloma venereum—*C. trachomatis*
        1. Sexually transmitted disease (STD)

2. Occurs in phases
   a. Primary phase—ulcer on genitals that heals with no scar
   b. Secondary phase—enlargement of lymph nodes (buboes); fever, chills, and anorexia are common
   c. Late phase—fibrotic changes and abnormal lymphatic drainage leading to fistulas and/or urethral or rectal strictures; leads to untreatable fluid accumulation in the penis, scrotum, or vaginal area
3. Treated by aspiration of buboes and by antibiotics in early phases; by surgery in late phase
4. Controlled by education, prophylaxis, and early diagnosis and treatment

D. Nongonococcal urethritis (NGU)—inflammation of the urethra not caused by *Neisseria gonorrhoeae*
   1. Caused by a variety of agents including *C. trachomatis*
   2. Organisms are sexually transmitted—50% are caused by chlamydia; NGU is the most common STD in the U.S.
   3. May be inapparent in many males, or may cause inflammation of genital tract
   4. Asymptomatic in some females, but may cause pelvic inflammatory disease (PID), which can lead to sterility
   5. Serious disease in pregnant females: leads to miscarriages, stillbirth, inclusion conjunctivitis, and infant pneumonia
   6. Rapid diagnostic tests are now available
   7. Treatment is with various antibiotics

E. Psittacosis (ornithosis)—*C. psittaci*
   1. Resistant to sulfonamides; inclusions do not contain glycogen
   2. Infectious disease in birds, which can be transmitted to humans
   3. Spread by handling infected birds or by inhalation of dried bird excreta; occupational hazard in the poultry industry (particularly to workers in turkey processing plants)
   4. Infects respiratory tract, liver, spleen, and lungs, causing inflammation, hemorrhaging, and pneumonia
   5. Diagnosis based on isolation of *C. psittaci* from blood or sputum or on serology
   6. Treatment is with tetracycline, which decreases the mortality rate from 20 to 2%

F. Trachoma—*C. trachomatis*
   1. Greatest single cause of blindness in the world, although uncommon in the U.S.
   2. Transmitted by hand-to-hand contact or by contact with infected soaps and towels
   3. First infection usually heals spontaneously with no lasting effects
   4. Reinfection leads to vascularization of the cornea (pannus formation) and scarring of the conjunctiva
   5. Treatment is the same as for inclusion conjunctivitis
   6. Prevention and control is by health education, personal hygiene, and access to clean water for washing

II. Mycoplasmal Diseases
A. Genitourinary diseases—*Mycoplasma homonis* and *Ureaplasma urealyticum*
   1. Colonization is related to sexual activity
   2. Bacteria are difficult to recognize because they are not usually cultured in the clinical microbiology laboratory
   3. Diagnosis is usually by recognition of clinical syndromes
   4. Treatment is usually tetracycline or erythromycin

B. Primary atypical pneumonia—*M. pneumonia*
   1. Spread by close contact and/or airborne droplets
   2. Common and mild in infants; more serious in older children and young adults
   3. Symptoms vary from none to serious pneumonia
   4. Diagnosis is considered if bacteria cannot be isolated and viruses cannot be detected; rapid antigenic detection kits are now available

5. Treatment is usually tetracycline or erythromycin
III. Rickettsial Diseases (Rickettsioses)
    A. Ehrlichiosis
        1. First case was diagnosed in the United States in 1986
        2. Caused by a new species of *Rickettsiaceae, Ehrlichia chaffeensis*
        3. Transmitted from unknown animal vectors to humans by *Dermacentor andersoni* and *Amblyomma americanum* ticks
        4. Infects circulating monocytes causing a nonspecific febrile illness (human monocytic ehrlichiosis; HME) that resembles Rocky Mountain spotted fever
        5. Diagnosis is by serological tests
        6. Treatment is with tetracycline
        7. In 1994 a new form (human granulocytic ehrlichiosis; HGE) was discovered
    B. Epidemic (louse-borne) typhus—*Rickettsia prowazekii*
        1. Transmitted from person to person by the body louse (in the U.S., a reservoir is the southern flying squirrel)
        2. Organism is found in insect feces, and feces are deposited when the insect takes a blood meal; as the person scratches, the bite becomes infected
        3. Causes vasculitis (infection of blood vessel endothelial cells); leads to headache, fever, muscle aches, and a characteristic rash
        4. If untreated, recovery takes two weeks, but mortality rate is 50%
        5. Recovery gives a solid immunity that also cross-protects against endemic (murine) typhus
        6. Treatment is usually tetracycline and chloramphenicol
    C. Endemic (murine) typhus—*R. typhi*
        1. Transmitted from rats by fleas
        2. Occurs in isolated areas around the world, including southeastern and Gulf coast states, especially Texas
        3. Similar to epidemic typhus, but milder with lower mortality rate (less than 5%)
        4. Treatment is the same as for epidemic typhus
    D. Q fever—*Coxiella burnetii*
        1. Can survive outside host by forming endosporelike structures
        2. Transmitted by ticks between animals, and by contaminated dust to humans; an acute zoonotic disease
        3. Occupational hazard among slaughterhouse workers, farmers, and veterinarians
        4. Starts with mild respiratory symptoms, but an acute onset of severe headache, muscle pain, and fever
        5. Rarely fatal, but 10% of cases develop endocarditis in 5 to 10 yrs
        6. Treatment is usually tetracycline and chloramphenicol
        7. Prevention and control consists of vaccination researchers and other of high occupational risk as well as pasteurization of cow and sheep milk in areas of endemic Q fever
    D. Rocky Mountain spotted fever—*R. rickettsii*
        1. Transmitted by the wood tick or the dog tick
        2. Can destroy blood vessels in the heart, lungs, or kidneys, and lead to death; causes a characteristic rash
        3. Treatment is usually chloramphenicol and chlortetracycline
IV. Dental Infections—caused by various odontopathogens
    A. Dental plaque
        1. Enamel adsorbs acidic glycoproteins—acquired enamel pellicle
        2. Colonized by *S. gordonii, S. mitis* and *S. oralis*
        3. This environment provides binding surface for other microorganisms, such as *S. mutans* and *S. sobrinus* via intergeneric bacterial adherence (coaggregation)
        4. The latter two organisms hydrolyze sucrose to glucose and fructose, and polymerize the glucose into polymers (glucans) that act like a cement which forms a dental plaque ecosystem

330

B. Dental caries (tooth decay)
   1. Dental plaque allows growth of anaerobes
   2. Anaerobes produce lactic acid and acetic acid, which remain at that site undiluted because of the impermeability of the dental plaque
   3. Lactic and acetic acids cause demineralization of the enamel (repair or remineralization occurs, except when diet is too rich in fermentable substrates); leads to cavitation (dental caries)
   4. If enamel is breached and bacteria reach dentin and pulp, the tooth can die
   5. Drugs are not available to prevent dental caries
   6. Good dental hygiene (brushing, flossing, fluoride) and minimal ingestion of sucrose are the main strategies for the prevention of dental caries
C. Periodontal disease—initiated by formation of subgingival plaque
   1. Main species involved is *Porphyromonas gingivalis*
   2. Can be mild inflammation—gingivitis
   3. Can lead to loss of teeth if unchecked
   4. Controlled by plaque removal and good dental hygiene; oral surgery and antibiotics may be necessary in some cases
V. Nosocomial Infections
A. Bacteremia—transient presence of bacteria in the blood (6% of all nosocomial infections)
   1. Bacteria can be introduced via intravenous infusion, respiratory devices, catheters, and other procedures and devices that penetrate the body's normal defense mechanisms (primary bacteremia)
   2. Infection may also result from infections at another body site (secondary bacteremia)
   3. Occurs primarily in newborns and the elderly, whose natural defense mechanisms are compromised
   4. Organisms commonly found are coagulage negative staphylococci (CONS), *S. aureus,* and *E. coli*
B. Burn wounds are frequently infected by *Pseudomonas aeruginosa* and *Staphylococcus aureus* and account for about 8% of all nosocomial infections
C. Respiratory tract diseases (18%) involve pneumonias and other infections related to the use of respiratory devices that are used to aid breathing or administer medications; the microorganisms most often found are *S. aureus, P. aeruginosa,* and *Enterabacter* spp.
D. Surgical site infections (SSI; 17%) are a particular problem when surgery is on the gastrointestinal, respiratory, or genitourinary tracts; frequent bacterial isolates are *S. aureus* CoNS, and *Enterococcus* spp.
E. Urinary tract infection (UTI) are the most common of all nosocomial infections (39%) and generally result from catheterization; frequent isolates are *E. coli, Enterococcus* spp. and *P. aeruginosa*
F. Miscellaneous infections include skin and eye infections that are common in newborns but not in adults; infections often occur in immunosuppressed patients and cancer patients; organ transplant patients receiving immunosuppressive therapy are also highly susceptible

TERMS AND DEFINITIONS

Place the letter of each term in the space next to the definition or description that best matches it.

____ 1. An enlarged, tender lymph node
____ 2. Vascularization of the cornea
____ 3. An inflammation of the urinary tract
____ 4. An inflammation of the blood vessels
____ 5. A covering of dextrans and microorganisms that forms over the teeth
____ 6. Cavitation of the teeth caused by lactic acid and acetic acid produced by microorganisms
____ 7. Transient occurrence of bacteria in the bloodstream
____ 8. Binding of one bacterial species to another

a. bacteremia
b. bubo
c. dental caries
d. dental plaque
e. intergeneric bacterial adherence (coaggregation)
f. pannus
g. urethritis
h. vasculitis

DISEASES AND THEIR CAUSATIVE AGENTS

Select the organism which causes each disease listed below. If more than one organism is associated with a particular disease, select all of the appropriate alternatives.

____ 1. Inclusion conjunctivitis
____ 2. Epidemic typhus
____ 3. Primary atypical pneumonia
____ 4. Lymphogranuloma venereum
____ 5. Ornithosis
____ 6. Trachoma
____ 7. Rocky Mountain spotted fever
____ 8. Burn wound infections
____ 9. Endemic typhus
____ 10. Q fever

a. *Rickettsia rickettsii*
b. *Coxiella burnetii*
c. *Chlamydia psittaci*
d. *Rickettsia typhi*
e. *Mycoplasma pneumonia*
f. *Rickettsia prowazekii*
g. *Chlamydia trachomatis*
i. *Pseudomonas aeruginosa*

FILL IN THE BLANK

1. Lymphogranuloma venereum is a sexually transmitted disease that has a worldwide distribution, but it is more common in _____ climates. In the primary phase, an ulcer appears in the genital area; it heals quickly and generally leaves no scar. In the secondary phase, the regional lymph nodes become enlarged and tender, and are referred to as _____. If not treated, a late phase with fibrotic changes and abnormal lymphatic drainage occurs, leading to the formation of abnormal passages from abscesses to a hollow organ. These passages are called _____. This can lead to untreatable fluid accumulation in the penis, scrotum, or vagina.

2. Any inflammation of the urethra not caused by *Neisseria gonorrhoeae* is referred to as _____ or _____. Males have few, if any, manifestations of the disease. Females may be _____ (i.e., they have no clinical manifestations), or they may have a severe infection that is characterized as _____ or _____.

3. *Chlamydia psittaci* causes a disease that was first recognized in parrots and parakeets; it was referred to as _____. However, it is now recognized in pigeons, chickens, ducks, turkeys, and other birds, and has been given the more general name _____. When the organism invades the lungs, it can cause _____, _____, and _____.

4. In trachoma, the first infection site usually heals spontaneously. Reinfection leads to vascularization of the _____ (called _____ formation), which leads to scarring of the _____, which may result in blindness.

5. Periodontal disease is a group of diverse clinical entities affecting the periodontium, the supporting structure of a tooth that includes the _____, the periodontal _____, the bones of the jaw, and the _____. If the inflammation is restricted to the gums, it is referred to as _____; if it involves bone destruction, it is called _____.

6. Direct introduction of bacteria into the body by way of intravenous infusion, transducers, respiratory devices, prostheses, catheters, endoscopes, hemodialysis units, or parenteral nutrition is called a _____, while an infection at another site (urinary tract, respiratory tract, surgical site, etc.) that leads to the presence of bacteria in the blood is called a(n) _____.

MULTIPLE CHOICE

For each of the questions below select the *one best* answer.

1. Which of the following is(are) sensitive to sulfonamides?
   a. *C. trachomatis*
   b. *C. psittaci*
   c. Both (a) and (b) are correct.
   d. Neither (a) nor (b) is correct.

2. Which of the following form(s) glycogen negative inclusions?
   a. *C. trachomatis*
   b. *C. psittaci*
   c. Both (a) and (b) are correct.
   d. Neither (a) nor (b) is correct.

3. The use of tetracycline has lowered the mortality rate of psittacosis
   a. from 50% to 5%.
   b. from 20% to 2%.
   c. from 20% to less than 1%.
   d. from 2% to less than 0.1%.

4. Which of the following does not normally cause genitourinary disease?
   a. *Mycoplasma homonis*
   b. *Ureaplasma urealyticum*
   c. Neither (a) nor (b) causes genitourinary disease.
   d. Both (a) and (b) cause genitourinary disease.

5. Which of the following rickettsias can survive outside the host by forming endosporelike structures?
   a. *Rickettsia prowazekii*
   b. *Rickettsia typhi*
   c. *Coxiella burnetii*
   d. *Rickettsia rickettsii*

6. Which of the following diseases is not caused by rickettsias?
   a. Rocky Mountain spotted fever
   b. Q fever
   c. Typhus fever
   d. All of the above are caused by rickettsias.

7. Which of the following commonly uses an avian reservoir?
   a. *C. psittaci*
   b. *C. pneumoniae*
   c. *C. trachomatis*
   d. All of the above use an avian reservoir.

8. Psittacosis is generally considered to be an occupational hazard for which of the following groups of people?
   a. hog farmers
   b. turkey processing plant workers
   c. students working on a microbiology study guide
   d. All of the above are correct.

9. Prevention and control of Q fever consists of
   a. vaccination of high risk individuals.
   b. pasteurization of cow and sheep milk in areas of endemic Q fever.
   c. Both (a) and (b) are correct.
   d. Neither (a) nor (b) is correct.

TRUE/FALSE

_____ 1. Different serotypes of *C. trachomatis* cause different disease states.

_____ 2. Different disease states can be caused by the same serotype of *C. trachomatis* if the mode of transmission is different.

_____ 3. *Chlamydia trachomatis* is the greatest single cause of blindness throughout the world.

_____ 4. Mycoplasmas that cause genitourinary disease can be readily cultured and identified so that specific treatment can be initiated.

_____ 5. Mycoplasmal pneumonia is considered only when bacterial and viral etiologies have been eliminated as possibilities.

_____ 6. Recovery from epidemic typhus confers permanent immunity and also protects against endemic typhus.

_____ 7. Rocky Mountain spotted fever is so named because it only occurs in the Rocky Mountain area.

_____ 8. The most common nosocomial infection is associated with surgical sites.

_____ 9. Nongonococcal urethritis (NGU) is the most common STD in the U.S.

_____ 10. *C. pneumoniae* infections are common but sporadic; about 50% of adults have antibody to the organism.

_____ 11. Transmission of *C. pneumoniae* requires an avian reservoir.

_____ 12. Erhlichiosis is a newly discovered disease first diagnosed in the U.S. in 1987.

CRITICAL THINKING

1. Discuss the steps involved in the development of dental caries. At each step, name the microorganisms involved and describe how they contribute to tooth decay.

2. You are a hospital epidemiologist who is concerned about the increase in nosocomial infections. What procedures are you using or can you use to decrease their frequency? Be specific in your recommendations.

ANSWER KEY

Terms and Definitions

1. b, 2. f, 3. g, 4. h, 5. d, 6. c, 7. a, 8. e

Diseases and Their Causative Agents

1. g, 2. f, 3. e, 4. g, 5. c, 6. g, 7. a, 8. h, 9. d, 10. b

Fill in the Blank

1. tropical; buboes; fistulas 2. nongonococcal urethritis; NGU; asymptomatic; pelvic inflammatory disease; PID 3. psittacosis; ornithosis; inflammation; hemorrhaging; pneumonia 4. cornea; pannus; conjunctiva 5. cementum; membrane; gums; gingivitis; periodontosis 6. primary bacteremia; secondary bacteremia

Multiple Choice

1. a, 2. b, 3. b, 4. d, 5. c, 6. d, 7. a, 8. b, 9. c

True/False

1. T, 2. T, 3. T, 4. F, 5. T, 6. T, 7. F, 8. F, 9. T, 10. T, 11. F, 12. T

39   Human Diseases Caused by Fungi and Protozoa

CHAPTER OVERVIEW

This chapter discusses some of the more important fungal and protozoan diseases of humans. The clinical manifestations, diagnosis, epidemiology, pathogenesis, and treatment of selected diseases are presented.

CHAPTER OBJECTIVES

After reading this chapter you should be able to:

- discuss the five groups—superficial, cutaneous, subcutaneous, systemic, and opportunistic—of diseases caused by fungi (mycoses) including *Pneumocystis carinii* pneumonia
- discuss some of the more important diseases caused by protozoans (including amebiasis, giardiasis, malaria, the hemoflagellate diseases, toxoplasmosis, and trichomoniasis)

CHAPTER OUTLINE

I.   Fungal Diseases—mycoses
    A.   Superficial mycoses
        1.   Most occur in the tropics
        2.   Limited to the outer surface of the hair and the skin
        3.   Referred to as piedras or tineas because of the hard nodules that are formed
        4.   Treatment involves removal of skin scales and infected hairs
        5.   Prevention is by good personal hygiene
    B.   Cutaneous mycoses—dermatophytoses, ringworms, tineas
        1.   Occur worldwide; most common fungal diseases
        2.   Three genera, *Epidermophyton, Microsporum,* and *Trichophyton,* are involved
        3.   Diagnosed by microscopic examination of skin biopsies and by culture on Sabouraud's glucose agar
        4.   Treatment—topical ointments, oral griseofulvin, or oral itraconazole (sporanox)
        5.   Different diseases are distinguished according to the causative agent and the area of the body affected (e.g., tinea barbae—beard hair; tinea pedis—athlete's foot)
        6.   Includes such common diseases as ringworm, jock itch, and athlete's foot
    C.   Subcutaneous mycoses
        1.   Introduced in soil-contaminated puncture wounds
        2.   Develop slowly over a period of years
        3.   Nodule develops and then ulcerates; organisms spread along lymphatic channels; more nodules then develop at other locations
        4.   Diagnosis is by culture of the infected tissue
        5.   Treatment—5-flucytosine, iodides, amphotericin B, and surgical excision
        6.   Examples include chromomycosis, maduromycosis, sporotrichosis

D. Systemic mycoses
   1. Caused by dimorphic fungi, except for *Cryptococcus neoformans* which has only a yeast form
   2. Usually acquired by inhalation of spores from soil
   3. Begins as lung lesions, becomes chronic, and disseminates through the bloodstream to other organs
   4. Blastomycosis—*Blastomyces dermatitidis;* antibiotics are used; surgery may be necessary to drain large abscesses
   5. Coccidiomycosis—*Coccidioides immitis;* usually an asymptomatic or mild respiratory infection that spontaneously resolves in a few weeks; occasionally progresses to chronic pulmonary disease; antibiotics are used for treatment
   6. Cryptococcosis—*Cryptococcus neoformans;* minor transitory pulmonary infection that can disseminate and cause meningitis if it reaches the central nervous system; found in approximately 15% of AIDS patients; treatment involves amphotericin B and flucytosine or intraconazole
   7. Histoplasmosis—*Histoplasma capsulatum;* facultative intracellular parasite that causes disease of the reticuloendothelial system; symptoms are usually those of mild respiratory involvement; occasional dissemination; occupational hazard for spelunkers and bat guano miners; treatment is amphotericin B, ketoconazole, or intraconazole
E. Opportunistic mycoses—normally harmless, but can cause disease in a compromised host
   1. Aspergillosis—*Aspergillus fumigatus* or *A. flavus;* pulmonary involvement with potential systemic dissemination; in severely compromised individuals, invasive aspergillosis may occur and fill the lungs with mycelia (aspergillomas); treatment involves treating the underlying illness to strengthen resistance and the use of intraconazole
   2. Candidiasis—*Candida albicans;* usually involves the skin and mucous membranes of people whose normal microbiota has been disturbed by prolonged antibiotic therapy
      a. Oral candidiasis (thrush)—mouth; common in newborns
      b. Paronychia—subcutaneous tissues of the digits
      c. Onychomycosis—subcutaneous tissues of the nails
      d. Intertriginous candidiasis—warm, moist areas such as axillae, groin, and skin folds
      e. Diaper candidiasis—can occur if diapers are not changed frequently
      f. Candidal vaginitis—occurs when lactobacilli are depleted by diabetes, antibiotics, oral contraceptives, pregnancy, or other factors
      g. Balanitis—sexually transmitted to males; infection of the glans penis (primarily in uncircumcised males)
      h. Treatment involves topical agents; oral antibiotics are used for systemic candidiasis
   3. *Pneumocystis carinii* pneumonia is caused by a eucaryotic protist
      a. It was once considered a protozoan parasite but recent comparisons of rRNA and DNA sequences from several genes have shown it to be more closely related to fungi
      b. Disease occurs almost exclusively in immunocompromised hosts including more than 80% of AIDS patients
      c. It is transmitted by inhalation of infected material and remains localized in the lungs, even in fatal cases
      d. Definitive diagnosis involves demonstrating the presence of the organism in infected lung material or PCR analysis
      e. Treatment is by oxygen therapy and combination drug therapy
II. Protozoan Diseases
A. Amoebiasis (amebic dysentery)—*Entamoeba histolytica;* ingested cysts excyst in the intestine and proteolytically destroy the epithelial lining of the large intestine; disease severity may be asymptomatic to fulminating dysentery, exhaustive diarrhea, and abscesses of the liver, lungs, and brain; treatment with several antibiotics is possible; prevention involves avoiding contaminated water; hyperchlorination or iodination can destroy waterborne cysts

B. Cryptosporidiosis—*Cryptosporidium parvum*; found in the intestines of many birds and mammals; oocysts are shed into the environment in fecal material; they are then ingested and excyst in the small intestine; the released sporozoites parasitize intestinal epithelial cells, causing diarrhea; treatment is supportive; patients will usually recover, but the disease can be fatal in late stage AIDS patients

C. Freshwater amoebae—*Naegleria* and *Acanthamoebae;* facultative parasites that cause primary amebic meningoencephalitis and keratitis (particularly among wearers of soft contact lenses); found in fresh water and soil

D. Giardiasis—*Giardia lamblia;* most common cause of waterborne epidemic diarrheal disease; common in wilderness areas because many animal carriers shed cysts into otherwise "clean" water; varies in severity; asymptomatic carriers are common; may be chronic or acute; treatment is usually metronidazole (Flagyl); prevention involves avoiding contaminated water and the use of slow sand filters in the processing of drinking water

E. Malaria—*Plasmodium* species; transmitted by bite of an infected female *Anopheles* mosquito; reproduces in the liver and also penetrates erythrocytes; a periodic sudden release of merozoites, toxins, cell debris from the infected erythrocytes and TNF-α and interleukin-1 from macrophages trigger the characteristic attack of chills and fever; anemia can result, and the spleen and liver often hypertrophy; treatment is by chloroquine or related drugs

F. Hemoflagellate diseases
1. Leishmaniasis—caused by flagellated protozoans; transmitted by sandflies; can be mucocutaneous, cutaneous, or visceral; symptoms vary with the particular etiological organism involved; treated with pentavalent antimonial compounds; recovery usually confers permanent immunity; vector and reservoir control and epidemiological surveillance are the best options for control
2. Trypanosomiasis—transmitted by tsetse flies; causes interstitial inflammation and necrosis of the lymph nodes, brain, and heart; causes sleeping sickness (uncontrollable lethargy) and Chagas' disease, which is currently untreatable; vaccines are not useful because the parasite can change its coat to avoid the immune response

G. Toxoplasmosis—*Toxoplasma gondii,* fecal-oral transmission from infected animals; also transmitted by ingestion of undercooked meat and by congenital transfer, blood transfusion, or tissue transplant; most cases are asymptomatic or resemble mononucleosis; can be fatal in immunocompromised individuals; a major cause of death in AIDS patients; treatment is with antibiotics

H. Trichomoniasis—*Trichomonas vaginalis;* host accumulates leukocytes at the site of infection; in females, this leads to a yellow purulent discharge and itching; in males, most infections are asymptomatic; treatment is with metronidazole; one of the most common sexually transmitted diseases

TERMS AND DEFINITIONS

Place the letter of each term in the space next to the definition or description that best matches it.

_____ 1. The discipline that deals with fungi that cause human disease

_____ 2. General term for diseases caused by fungi

_____ 3. A tumorlike deformity that results from a subcutaneous fungal infection

_____ 4. An inflammation of the cornea

_____ 5. A process involving multiple asexual fission for some protozoa

_____ 6. Transmitted by the bites of infected arthropods, a protozoan that infects the blood and tissues of humans

a. hemoflagellates
b. keratitis
c. medical mycology
d. mycetoma
e. mycoses
f. schizogony

FILL IN THE BLANK

1. A fungal infection of beard hair is called tinea _____; a fungal infection of scalp hair is called tinea _____; a fungal infection of the smooth or bare parts of the skin is called tinea _____; a fungal infection of the groin is called tinea _____; and a fungal infection of the feet is called tinea _____.

2. An _____ fungal organism is one that is generally harmless in its normal environment but that can become pathogenic in a _____ host. The most important of these mycoses are systemic _____ and _____.

3. In aspergillosis, if the infected individual is hypersensitive, the disease is called _____ aspergillosis; if large fungus balls form in the lungs, it is called _____ aspergillosis; if it spreads from the lungs to other organs, it is called _____ aspergillosis; and if extensive mycelia form in the lungs of severely compromised patients, it is called _____ aspergillosis.

4. In candidiasis, if the disease involves the mouth, it is referred to as _____ candidiasis or _____; if it involves the female genital tract, it is referred to as _____ candidiasis; and if it is sexually transmitted to a male who then has an infection of the glans penis, it is referred to as _____.

5. Leishmaniasis involving the mouth, nose, and throat is called _____; if it involves the reticuloendothelial system, it is called _____; while if it involves the formation of red papules at the site of an insect bite which is usually on the face and ears, it is called _____.

MULTIPLE CHOICE

For each of the questions below select the *one best* answer.

1. Which of the following terms is(are) used to describe cutaneous mycoses?
   a. dermatophytoses
   b. ringworms
   c. tineas
   d. All of the above are used to describe cutaneous mycoses.

2. Which of the following is not a subcutaneous mycosis?
   a. chromomycosis
   b. coccidiomycosis
   c. maduromycosis
   d. sporotrichosis

3. In sporotrichosis, the lesions can remain localized or can spread throughout the body. In the latter case it is referred to as
   a. extracutaneous sporotrichosis.
   b. disseminated sporotrichosis.
   c. systemic sporotrichosis.
   d. fulminating sporotrichosis.
4. Which of the following fungi that cause systemic mycoses is not dimorphic?
   a. *Blastomyces dermatitidis*
   b. *Coccidioides immitis*
   c. *Cryptococcus neoformans*
   d. *Histoplasma capsulatum*
5. Which of the following statements about *Pneumocystis carinii* is *not* true?
   a. The disease it causes occurs almost exclusively in immunocompromised hosts.
   b. The organism and the disease remain localized in the lungs, even in fatal cases.
   c. The incidence of this disease in relatively young patients is largely attributed to its prevalence as an opportunistic infection in AIDS patients.
   d. All of the above are true about *Pneumocystis carinii*.

6. Which of the following diseases is(are) caused by trypanosomes?
   a. sleeping sickness
   b. Chagas' disease
   c. Both (a) and (b) are caused by trypanosomes.
   d. Neither (a) nor (b) is caused by trypanosomes.
7. Vaccines are not effective against trypanosomiasis because
   a. the organisms are never exposed to the immune system.
   b. the organism is only weakly antigenic.
   c. the organism can change its protein coat and thereby evade the immune response.
   d. All of the above are correct.
8. One of the leading causes of death in AIDS patients is
   a. giardiasis.
   b. toxoplasmosis.
   c. trichomoniasis.
   d. sporotrichosis.
9. Which of the following is not caused by a protozoan?
   a. Cryptosporidiosis
   b. Cryptococcosis
   c. Both (a) and (b) are caused by a protozoan.
   d. Neither (a) nor (b) is caused by a protozoan.

TRUE/FALSE

_____ 1. Jock itch and athlete's foot are caused by the same set of fungi.
_____ 2. Subcutaneous mycoses are caused by soil-inhabiting fungi that can easily penetrate the skin.
_____ 3. *Candida albicans* is normally found in the vagina but does not usually cause disease, because the acidic pH created by *Lactobacilli* prevents its overgrowth.
_____ 4. The impact of protozoan diseases worldwide is great because there are so many species that are associated with human diseases.
_____ 5. Asymptomatic cyst shedders of *Entamoeba histolytica* do not need treatment with antibiotics.
_____ 6. Drinking water sources infected with *Entamoeba histolytica* cysts should be filtered because chlorination alone is not sufficient for destroying the cysts.
_____ 7. *Giardia lamblia* is the most common cause of epidemic waterborne diarrheal disease in the U. S.
_____ 8. Trichomoniasis is frequently asymptomatic in males, but seldom is it asymptomatic in females.
_____ 9. *Pneumocystis carinii* is now classified with the fungi instead of the protozoa.

CRITICAL THINKING

1. One of the apparent paradoxes associated with giardiasis is that outbreaks occur more frequently in the Rocky Mountain and New England states where the raw water is considered to be of fairly high quality, while outbreaks are less frequent in the south, southwest, and midwest, where quality of raw water is actually considered much poorer. Explain. (Consider the types of treatment used in the two types of areas.)

2. Describe the life cycle of *Plasmodium vivax* and discuss how it is related to the symptoms associated with malaria.

ANSWER KEY

Terms and Definitions

1. c, 2. e, 3. d, 4. b, 5. f, 6. a

Fill in the Blank

1. barbae; capitis; corporis; curis; pedis 2. opportunistic; compromised; aspergillosis; candidiasis 3. allergic; colonizing; disseminated; invasive 4. oral; thrush; vulvovaginal; balanitis 5. mucocutaneous; visceral; cutaneous

Multiple Choice

1. d, 2. b, 3. a, 4. c, 5. d, 6. c, 7. c, 8. b, 9. b

True/False

1. T, 2. F, 3. T, 4. F, 5. F, 6. T, 7. T, 8. T, 9. T

Part X MICROORGANISMS AND THE ENVIRONMENT

40 Microorganisms as Components of the Environment

CHAPTER OVERVIEW

This chapter discusses the roles of microorganisms in ecosystems. The organisms' physiological state, nutrient cycling, decomposition processes, and successional interactions are discussed. The fate and impact of genetically engineered organisms are also considered. The chapter includes a discussion of extreme environments and concludes with a discussion of the methods used in environmental studies.

CHAPTER OBJECTIVES

After reading this chapter you should be able to:

- discuss the populations and communities of microorganisms that are associated with various ecosystems environments and describe the roles they play in succession
- discuss the interactions of microorganisms with one another and with other nonmicrobial members of various ecosystems
- describe the ways in which microorganisms obtain their required nutrients from environmental sources
- discuss the fate and potential impact of the releasing genetically engineered microorganisms
- describe the microorganisms found in extreme environments and the adaptations necessary for their survival under these conditions
- discuss the various methods used to study microorganisms in the environment

CHAPTER OUTLINE

    I.   Microorganisms and the Structure of Ecosystems
         A.   Roles of organisms in an ecosystem
              1.   Primary producer—synthesizes and accumulates organic matter
              2.   Consumer—uses the accumulated organic matter as a food supply
              3.   Decomposer—cycles various nutrients from dead organisms into forms that can be used by
                   the other members of the ecosystem
         B.   Roles of microorganisms in an ecosystem
              1.   Chemoautotrophs—primary producers in limited environmental niches only, such as deep
                   oceanic hydrothermal vents
              2.   Photoautotrophs—primary producers in many aquatic ecosystems
              3.   Chemoheterotrophs—decomposers whose primary role is the mineralization of organic matter
                   to simpler inorganic forms of nutrients

4. Other roles of microorganisms
   a. Serve as food source for other organisms
   b. Produce inhibitory compounds that decrease microbial activity, and thereby limit the survival and functioning of plants and animals dependent on those microorganisms
C. Interactions of microorganisms with their environments
   1. Populations—microorganisms of the same type growing at a localized site
   2. Communities—different types of interactive populations living together within an ecosystem
   3. Microbial environments are complex and constantly changing; they are characterized by overlapping gradients of resources, toxic materials, and other limiting factors.
   4. Microenvironments—regions where resources interact (e.g., at the interface between aerobic and anaerobic environments); this sometimes gives rise to specialized groups of microorganisms suited to those unique microenvironments.
D. Microbial ecology—the term used for the study of microbes and their relationship with their particular environments

II. The Physiological State of Microorganisms in Ecosystems
A. Many environments contain low levels of nutrients, which spurs intense competition among organisms
B. Microorganisms may increase surface area to allow for more efficient uptake of limited nutrients
C. Microorganisms may attach to surfaces that may have a higher local concentration of necessary nutrients thereby creating biofilms
D. Microorganisms may sequester critical limiting nutrients, thereby making them less available to competing microorganisms
E. Natural chemicals can inhibit microbial growth in low-nutrient environments

III. Nutrient Cycling Processes—biogeochemical cycling
A. Carbon cycle—carbon can be interconverted between methane, complex organic matter, and carbon dioxide; carbon fixation can occur by the activities of cyanobacteria, the green algae, photosynthetic bacteria, and aerobic chemolithoautotrophs
B. Sulfur cycle—sulfur can be interconverted between elemental sulfur, sulfide, and sulfate forms by the actions of various microorganisms
   1. Dissimilatory sulfate reduction produces sulfide which accumulates in the environment
   2. Assimilatory sulfate reduction results in the reductin of sulfate for use in amino acid biosynthesis
C. Nitrogen cycle—a variety of microorganisms participate in this process
   1. Nitrification—aerobic oxidation of ammonium ion to nitrite, and ultimately to nitrate
   2. Denitrification—reduction of nitrate to nitrite, nitrous oxide, and gaseous molecular nitrogen
   3. Nitrogen assimilation—utilization of inorganic nitrogen and its incorporation into new microbial biomass
   4. Nitrogen fixation
      a. A series of sequential reduction steps to convert gaseous nitrogen to nitrate
      b. Requires an expenditure of energy
      c. Can be carried out by aerobes or anaerobes; the actual reduction process must be done anaerobically, even by aerobic microorganisms
      d. Anaerobic conditions can be maintained by using physical barriers, O_2-scavenging molecules, and high rates of metabolic activity
D. Other cycling processes include iron cycling from ferrous to ferric ion, phosphorus cycling between inorganic and organic forms, and manganese cycling

IV. Interactions Between Microorganisms and Metals—differing sensitivities of more complex organisms (usually less sensitive) and microorganisms (usually more sensitive) form the basis for many antiseptic procedures developed over the last 150 years; metals can be broadly characterized into three groups
A. Noble metals (silver, gold, platinum, etc.)—cannot cross the blood-brain barrier of vertebrates, but have distinct effects on microorganisms

343

B. Metals and metalloids that can be methylated (mercury, arsenic, lead, selenium, and tin are important examples)—methylation enables them to cross the blood-brain barrier and affect the central nervous system of higher organisms; also affect microorganisms; methylated mercury can be concentrated in the food chain (a process known as biomagnification)

C. Metals that occur in ionic forms (copper, zinc, cobalt, etc.) can be directly toxic to microorganisms; often required as trace elements, but excess is toxic

V. Interactions in Substrate Utilization

A. Commensalism—waste products accumulated by some microorganisms are used as nutrients by other microorganisms; microbial succession

B. Microbial succession also occurs when mixtures of electron acceptors are present and are used in a defined order

VI. Organic Substrate Use by Microorganisms—different organisms have different abilities to utilize various types of organic molecules as nutrient sources; most organic compounds can be degraded with or without oxygen, except some hydrocarbons and lignin

A. Factors influencing degradation include:
1. Elemental composition
2. Structure of basic repeating units
3. Linkages between repeating units
4. Nutrients present in the environment
5. Abiotic conditions (pH, oxidation-reduction potential, O_2, osmotic conditions)
6. Microbial community present

B. Hydrocarbon degradation usually requires oxygen: the first step involves addition of molecular oxygen; recently however, slow anaerobic digestion in the presence of sulfate or nitrate has been observed

C. Lignin—filamentous fungi that are major lignin degraders require oxygen; important in construction because wood pilings can be used below the water table where anaerobic conditions are maintained; however, if the water table drops, degradation can take place, thereby weakening the structure

D. Other organics, such as starch, cellulose, hemicellulose, chitin, protein, lipids, and biomass, can be used to provide most of the nutrients required by microorganisms; only previously grown microbial biomass contains all of the nutrients required for microbial growth

VII. Disease Causing Microbes—Survival and Fate

A. Allochthonous (introduced) microorganisms are constantly being added to natural environments, in human and animal wastes and in sewage materials

B. They generally do not grow well in natural environments, but may survive for long periods of time, particularly at lower temperatures

C. Detection of these organisms (particularly if pathogenic) is a major concern

D. Many occur naturally in the environment, especially in waters.

E. Many microorganisms recovered and cultured lose their ability to survive when re-introduced to their original environments

VIII. Genetically Engineered Microorganisms—Fate and Effects

A. Effects depend on survival, reproduction, and/or gene transfer to other microorganisms

B. Each of these will occur only at low probability, but the uncertainties are many

C. Microorganisms can be made less competitive for low nutrient levels and can be constructed with copy blocks that limit replication

D. Gene transfers to natural populations are a concern but are unlikely because of the low population density, nutrient levels, and growth rates

E. However, natural competent microorganisms coupled with stabilization of DNA released from dead microorganisms makes gene transfer a distinct possibility

IX. Extreme Environments—physical and chemical conditions restrict the diversity of microbial types that can maintain themselves

A. Salinity—favors extreme halophiles

344

B.   High barometric pressure (e.g. deep sea environments) favor barotolerant, moderately barophilic, and extremely barophilic bacteria

C.   High temperature (up to 113°C) favors thermophiles and extreme thermophiles

D.   Acidity—acidophiles maintain a high internal pH relative to the environment

E.   Alkalinity—alkalophiles maintain a low internal pH relative to the environment

X.   Methods Used in Environmental Studies

A.   Microscopic examination

B.   Viable cell counting

C.   Measurement of nutrient cycling

D.   Measurement of organic carbon by biochemical oxygen demand (BOD), chemical oxygen demand (COD), or total organic carbon (TOC)

E.   Nucleic acid probe technology can be used to look for specific organisms

F.   Gel array microchips (genosensors) containing a mixture of probes can detect single subunit (ssu) ribosomal RNA in mixed populations

G.   A summary of methods and their uses in various environments is given in Table 40.6 of the textbook

TERMS AND DEFINITIONS

Place the letter of each term in the space next to the definition or description that best matches it.

_____ 1.   An organism that synthesizes and accumulates organic matter

_____ 2.   An organism that uses the accumulated organic matter as a nutrient source

_____ 3.   An organism that cycles materials from dead organisms and changes them into forms that can be utilized by other living organisms

_____ 4.   The process of decomposition of organic matter that results in the production of simpler inorganic forms of nutrients

_____ 5.   A group of similar types of organisms living within an ecosystem

_____ 6.   A group of diverse types of organisms living and interacting within an ecosystem

_____ 7.   Accumulation of toxic metals or organic compounds in organisms, which are then consumed as food supplies

_____ 8.   Disease-causing organisms that are introduced into an environment that is not their natural environment

_____ 9.   Produces an accumulation of sulfide in the environment

_____ 10.   Results in the reduction of sulfate for use in amino acid biosynthesis

a.   allochthonous
b.   assimilatory sulfate reduction
c.   biomagnification
d.   community
e.   consumer
f.   decomposer
g.   dissimilatory sulfate reduction
h.   mineralization
i.   population
j.   primary producer

FILL IN THE BLANK

1. Specialized microorganisms have evolved to take advantage of unique _____ that can form at the interface regions of diverse habitats. These interface regions represent overlapping _____ of resources, toxic materials, and other limiting factors. One example of this is where aerobic and anaerobic environments intersect or when a lighted zone changes to a lightfree region. These areas create limiting factors that influence the survival and functioning of microbial populations according to Liebig's Law of the _____.

2. The microbial community plays a major role in _____ cycling, in which both _____ and _____ processes are involved in the cycling of nutrients important to microbes, plants, and animals.

3. In the carbon and nitrogen cycles, and, to a lesser extent, the sulfur cycle, some of the components are gases; these are sometimes referred to as _____ cycles. When there are no component gases such as occurs with the phosphorus and iron cycles, they are referred to as _____ cycles. In the latter case, the microbial community cannot change the overall level of the nutrient.

4. Sulfate, when used as an external electron acceptor, is reduced to sulfide, then accumulates in the environment. This is an example of a(n) _____ process. In comparison, if the sulfate is reduced for use in proteins and nucleic acids of microbial cells, it is referred to as a(n) _____ reduction process.

5. The aerobic process of ammonium ion oxidation to nitrite and then to nitrate is called _____, while the reduction of nitrate to nitrite, nitrous oxide, and gaseous molecular nitrogen is referred to as _____. The utilization of inorganic nitrogen to form new microbial biomass is called nitrogen _____, and the reduction of nitrogen gas to other forms of utilizable nitrogen is called nitrogen _____.

MULTIPLE CHOICE

For each of the questions below select the *one best* answer.

1. In terrestrial environments, vascular plants usually serve as
   a. primary producers.
   b. consumers.
   c. decomposers.
   d. top level consumers.

2. In which of the following environments do chemoautotrophs serve as primary producers?
   a. terrestrial soils
   b. shallow freshwater environments
   c. deep marine seeps of hydrogen sulfide
   d. In none of the above environments do chemoautotrophs serve as primary producers.

3. Although microorganisms can serve a variety of roles, they primarily function as
   a. primary producers.
   b. consumers.
   c. decomposers.
   d. top level consumers.

4. Which of the following is not a response of microorganisms to low nutrient levels and intense competition?
   a. an increase in cell surface area
   b. alteration of morphology to form mini or ultramicro cells
   c. an increase in attachment to certain surfaces
   d. All of the above are responses of microorganisms to low nutrient levels and intense competition.

5. Which of the following metals can be methylated and thereby rendered capable of crossing the vertebrate blood-brain barrier and cause central nervous system damage?
   a. silver
   b. mercury
   c. zinc
   d. platinum

6. What is the term that describes the process in which toxic compounds accumulate in organisms of a food chain such that the top level consumers ingest concentrations higher than those that are normally found in the environment?
   a. bioconcentration
   b. biomagnification
   c. bioamplification
   d. bioaccumulation

7. Which of the following contributes to microbial succession?
   a. Waste products from one organism or from a group of organisms stimulate the growth of other organisms that can utilize the waste products as nutrients.
   b. Mixtures of electron acceptors are used in a definite sequence by different sets of microorganisms.
   c. Both (a) and (b) contribute to microbial succession.
   d. Neither (a) nor (b) contributes to microbial succession.

8. Which of the following organic compounds cannot be degraded in the absence of oxygen?
   a. lignin
   b. hemicellulose
   c. chitin
   d. lipids

9. Which of the following reduce(s) the probability of harmful impact of genetically engineered microorganisms released into the environment?
   a. They can be modified so that they do not survive well in natural environments.
   b. Copy blocks can be incorporated so that they will not readily replicate without conditions they are only likely to find in the laboratory.
   c. The population density in the environment reduces the probability that gene transfer between microorganisms will occur.
   d. All of the above reduce the probability of harmful impact of genetically engineered microorganisms released into the environment.

10. Which of the following has been made possible because of the differing sensitivities of microorganisms and more complex organisms to metals and metalloids?
    a. biological processing of mining ores to liberate useful material
    b. many of the antiseptic procedures currently in use
    c. Both (a) and (b) are correct.
    d. Neither (a) nor (b) is correct.

11. Which of the following is(are) used by aerobic nitrogen-fixing bacteria to maintain the anaerobic conditions needed for the nitrogen reduction reactions?
    a. physical barriers to keep oxygen out of certain compartments
    b. O_2-scavenging molecules to remove oxygen
    c. high metabolic rates to utilize oxygen as quickly as it becomes available
    d. All of the above are correct.

12. Which of the following contributes to the possibility of gene transfer from genetically engineered microorganisms to natural populations?
    a. the existence of naturally competent microorganisms
    b. the stabilization of DNA on solid surfaces
    c. Both (a) and (b) contribute to the possibility of gene transfer.
    d. Neither (a) nor (b) contribute to the possibility of gene transfer.

TRUE/FALSE

_____ 1. Mineralization is the process of obtaining various minerals from ores by the action of microorganisms.

_____ 2. In addition to their usual roles as producers, consumers, and decomposers, microorganisms also play a role as food sources for other chemoheterotrophic microorganisms and for predatory and parasitic organisms.

_____ 3. Most microorganisms in natural environments are confronted with nutrient deficiencies that limit their activities.

_____ 4. Wood pilings that remain below the water table are generally not subject to biodegradation because the conditions there are anaerobic and the organisms that carry out lignin degradation are obligate aerobes.

_____ 5. Nitrogen fixation can be carried out by organisms growing aerobically or anaerobically, but the process itself must take place in an anaerobic microenvironment—even in aerobic microorganisms.

_____ 6. Metals such as selenium are toxic at extremely low levels and serve no useful purpose for microorganisms.

_____ 7. Microbial success driven by selective utilization of electron acceptors can by cyclical if external sources of the initially used acceptors are introduced into the ecosystem.

_____ 8. Allochthonous disease-causing microorganisms generally survive better at higher temperatures where metabolic processes are more efficient.

_____ 9. What we would consider an extreme environment may be considered optimum and may be required by the microorganisms found in them.

_____ 10. Degradation of organic material by microorganisms is influenced by several factors including the composition and structure of the organic material and the microbial community present.

_____ 11. Genosensors are gel array michrochips containing a mixture of probes that can detect specific microorganisms in a mixed population by detection of single subunit ribosomal RNA.

CRITICAL THINKING

1. Explain how microorganisms can survive and function well at temperatures approaching 225°C. What adaptations would you expect to find in these organisms?

2. Explain why wood structures can exist for longer periods of time in swamps and bogs with little or no structural degradation. What would happen if the water table fell or the swamp as drained?

ANSWER KEY

Terms and Definitions

1. j, 2. e, 3. f, 4. h, 5. i, 6. d, 7. c, 8. a, 9. g, 10. b

Fill in the Blank

1. microenvironments; gradients; Minimum 2. biogeochemical; biological; chemical 3. gaseous; sedimentary 4. dissimilatory; assimilatory 5. nitrification; denitrification; assimilation; fixation

Multiple Choice

1. a, 2. c, 3. c, 4. d, 5. b, 6. b, 7. c, 8. a, 9. d, 10. b, 11. d, 12. c

True/False

1. F, 2. T, 3. T, 4. T, 5. T, 6. F, 7. T, 8. F, 9. T, 10. T, 11. T

41 Marine and Freshwater Environments

CHAPTER OVERVIEW

This chapter discusses the general characteristics of microorganisms that are associated with marine and freshwater environments. Aquatic environments are dominated by the liquid phase, and the only oxygen available is that which can be dissolved in water. This dissolved oxygen can be quickly depleted, creating anaerobic conditions. The use of surface and subsurface waters for drinking water sources and for waste disposal is also discussed. The chapter concludes with a discussion of the contamination of groundwaters by domestic and industrial wastes and the use of microorganisms to restore this important resource.

CHAPTER OBJECTIVES

After reading this chapter you should be able to:

- discuss the nature of marine and freshwater environments
- describe the complex microbial communities found in these environments
- discuss hydrothermal vents and hydrocarbon seeps
- discuss biofilms and the consequences of their formation
- discuss microbial mats
- discuss the recent concern over *Pfeisteria piscicida*, a protozoan parasite of fish
- discuss the influence of microorganisms in the oceans on carbon, nitrogen, phosphorus and sulphur cycling
- describe the possible cyclic changes in dissolved oxygen content of aquatic environments because of the unbalanced nature of oxygen depletion and replenishment
- discuss sewage treatment systems and their logical design according to the predictable sequences of the biological utilization of organic wastes
- describe the use of indicator organisms to measure the microbiological quality of water and describe the desirable characteristics of such an organism
- discuss the recent concern over the protozoan, *Cyclospora,* which causes long-lasting diarrheal disease
- discuss the limitations of indicator organisms with regard to the potential presence of viruses and protozoa in water sources
- discuss the ways that pathogenic organisms may be introduced into water sources other than by contamination with human wastes
- discuss the increasing concern about contamination of groundwater sources

CHAPTER OUTLINE

    I.   The Nature of Marine and Freshwater Environments
        A.   Nutrient concentrations
            1.   Can vary from extremely low (e.g., micrograms organic matter per liter) to levels approaching those in laboratory media
            2.   Changes in nutrient levels can cause shifts between low-nutrient responsive (oligotrophic) and high-nutrient responsive (copiotrophic) microorganisms

3. Nutrient turnover rates can vary from hundreds to thousands of years (marine environments) to very rapid turnover (marsh and estuarine environments)

B. Gradients
  1. Movement of materials in aquatic environments causes concentration gradients to form
  2. Microorganisms are able to respond rapidly to these gradients to select their most suitable environment
  3. Mixing occurs more readily than in soil habitats
  4. Aquatic habitats are only marginally aerobic unless they are vigorously agitated
  5. Stagnant waters can develop distinct microbial growth zones that are dependent on the release of nutrients provided by other microorganisms that have already grown in that zone
  6. A Winogradsky column is a microcosm in which these microbial interactions can be studied

C. Surfaces and biofilms
  1. Microorganisms and nutrients tend to accumulate at surfaces rather than in the bulk phase of water
  2. Surface attachment is advantageous to microorganisms at concentrations of dissolved organic matter below 25 mg per liter
  3. Biofilms are layers of microorganisms and nutrients adhering to various surfaces
  4. Simple biofilms are monolayers
  5. Complex biofilms are multilayered with different microorganisms in different layers and may contain cell aggregates, interstitial pores and conduit channels between layers; these conduits are shaped by protozoans that graze the bacteria and allow nutrients to reach the biomass
  6. Confocal scanning laser microscopy (CSLM) is a new technique that assists in our study of biofilms by generating optical "thin sections" across some complex biofilms
  7. Computer simulations can be constructed from these "thin sections" in order to better understand how microorganisms function in these complex environments
  8. Biofilms have a wide impact in environmental and public health microbiology (e.g., biofilms may be a major source of *Legionella* in heated waters and also potable water systems)

D. Microbial Mats
  1. Microbial mats result from biofilms that become so large that they are visible and have macroscopic dimensions
  2. Bands of microbes are usually evident
  3. They are found in many freshwater and marine environments on rocks and sediments
  4. They form extreme gradients (e.g., light may only penetrate 1 mm into these communities and below this conditions are anaerobic)

II. The Microbial Community in Marine and Freshwater Environments
  A. Important microorganisms
    1. Photoautotrophs—sulfur and nonsulfur photosynthetic bacteria
    2. Chemoheterotrophs—include some unique species of sheathed, sessile, prosthecate, budding, and/or gliding bacteria
    3. Chemoautotrophs—aerobic hydrogen sulfide oxidizers (*Beggiatoa* and *Thiotrix*) are found in waterlogged zones where hydrogen sulfide is present
    4. Eucaryotic photosynthetic algae—primary producers
    5. Protozoans—increase nutrient cycling by grazing on other organisms
    6. Fungi that are specially adapted to an aquatic existence including oomycetes and chytrids
    7. Filamentous fungi that can sporulate under water play an important role in the processing of organic matter
  B. The marine environment
    1. Contains 97% of the earth's water
    2. The ocean has been called a "high pressure refrigerator" with temperatures near 3°C and pressures up to 1,000 atm
    3. Three types of bacteria

a. Barotolerant—grow at 0 to 400 atm; best growth is at atmospheric pressures
b. Moderate barophiles—grow best at 400 atm, but will grow at 1 atm
c. Extreme barophiles—grow only at higher pressures
4. Other major types of microorganisms include large numbers of viruses and high levels of archaeobacteria
5. The major source of organic matter is photosynthesis
6. Nitrogen and sulfur cycles function at an "ocean scale" with nitrogen frequently being the limiting nutrient
7. Dimethyl sulfoxide (DMS) release from algae can influence atmospheric acidity and cloud formulation on a global scale
8. Most of the nutrient cycling occurs in the upper 300 meters, which is where light penetrates
9. Organic matter falls, but only 1% of the photosynthetically derived materials reach the deep-sea floor (the rest is decomposed); conditions there are oligotrophic (low nutrient concentration)
10. Massive deposits of methane are found below 500 meters, and are trapped in latticelike cages of crystalline water
11. Ultra microbacteria that resist grazing by nanoflagellates have been found to be the dominant bacteria in these environments
12. Because of increasing urban development in coastal areas, nutrient enrichment and microbial pollution are increasing and taxing the ocean's seemingly inexhaustible ability to absorb and process pollutants
13. Shellfish beds are sensitive to contamination from urban runoff
14. *Pfeisteria piscicida*, a protozoan fish parasite, sense the presence of fish and attack them

C. The freshwater environment
1. Lakes and rivers provide microbial environments that are very different from those of oceanic systems
2. Lakes
   a. Mixing and water exchange can be limited in lakes, allowing large-scale vertical gradients
      (1) Epilimnion—warm, aerobic, upper layer
      (2) Thermocline—region of rapid temperature decrease
      (3) Hypolimnion—cold, often anaerobic (particularly in nutrient rich lakes) lower layer
   b. Stratified waters undergo seasonal turnovers because of temperature and specific gravity changes
   c. Eutrophication—stimulation of growth by addition of nutrients (enrichment) to a body of water
   d. Oligotrophic waters—nutrient poor; aerobic; seasonal temperature shifts do not result in distinct chemical and microbial stratification
   e. Eutrophic waters—nutrient rich; often stratified; anaerobic, except at the surface; layers will turn over in spring and fall, causing migration of microorganisms
   f. Cyanobacteria can fix nitrogen and accumulate nutrients if phosphorus is added to oligotrophic waters
   g. Cyanobacteria compete more effectively with eucaryotic algae at higher pH and temperatures
   h. Cyanobacteria and eucaryotic algae contribute to massive algal blooms in strongly eutrophied lakes
3. Rivers
   a. Horizontal movement minimizes vertical stratification
   b. Most of the functional biomass is attached to surfaces

c. Nutrients can be from in-stream production (autochthonous) or from out-stream sources (allochthonous)
d. The ability to process organic matter is limited
e. Organic material can be released into streams and rivers from a variety of sources
    (1) Nonpoint sources of pollution include field and feedlot run-offs
    (2) Point sources of pollution include inadequately treated municipal wastes and other materials from specific locations
f. If the amount of organic material is excessive, oxygen is used faster than it can be replenished, which causes an oxygen sag curve
g. If the amount of organic material is not excessive, algae will grow, which leads to oxygen production in the daytime and respiration at night (diurnal oxygen shifts)
h. Organic matter can be measured in several ways
    (1) Chemical oxygen demand (COD)—quantifies the amount of organic matter present by reacting organic material with a strong acid
    (2) Total organic carbon (TOC)—quantifies carbon concentration by reacting (reducing) organic compounds to produce carbon dioxide
    (3) Biochemical oxygen demand (BOD)—amount of oxygen needed to utilize organic material as growth substrates; indirectly measures the amount of organic material in a sample; can be affected by presence of ammonia so nitrogen oxygen demand (NOD) is inhibited by nitrapyrin addition to the sample
    (4) Biomass—measured by filtration, followed by dry weight measurement or specific content measurements
    (5) Chlorophyll measurements—indicate the mass of photosynthetic organisms in water
    (6) Number of microorganisms can be determined directly, by immunofluorescence, or by viable cell counts
i. Filtration may be needed to concentrate the bacteria from the water prior to making some of these measurements
j. If too much nutrient is added, the cycling cannot occur, which creates an anaerobic, foul-smelling body of water that is considered "biologically dead," although microorganisms may still be functioning; can be reversed only by major remediation procedures
k. Silicon removal (by dams) is reducing the diatom population, resulting in increases in toxic nitrate-utilizing algae

D. The Role of Microbial Communities in Water Quality
  1. Sewage treatment—controlled self-purification; can involve the use of large basins (conventional sewage treatment) where mixing and gas exchange are carefully controlled; can also involve constructed wetlands where reed and aquatic plant communities and their associated microbes facilitate the processing of dissolved nutrients
  2. Primary (physical) treatment—removal of particulates (20 to 30% of the BOD)
  3. Secondary (biological) treatment—removal of dissolved carbonaceous materials (90 to 95% of the BOD) and many bacterial pathogens; produces a sludge which must be further processed or disposed of
    a. Aerated activated sludge systems—horizontal flow, sludge recycling, utilization of organic matter by microorganisms
    b. Trickling filters—vertical flow over gravel on which microorganisms have developed in surface films
    c. Extended aeration systems—reduce the amount of sludge produced by the process of biological self-consumption (endogenous respiration)
  4. Anaerobic sludge digestion—reduces the amount of sludge for disposal and produces methane, which can be used as a fuel for generation of electrical power

5. Tertiary treatment (physical/chemical and/or biological)—removes inorganic nitrogen, phosphorus, recalcitrant organics, viruses, etc.
6. Constructed wetlands use floating emergent and/or submerged plants to provide nutrients for microbial growth in their root zone; they help remove organic matter, inorganic matter and metals from waters
7. Surface flow soil treatment is also being used to allow aerobic microbial processing of waste

III. Water and Disease Transmission
A. Water purification is a critical link in promoting public health and safety
  1. Aeration precipitates iron and manganese which must then be removed
  2. Sedimentation in a sedimentation basin removes sand and large particles
  3. Coagulation with alum, lime and/or organic polymers is followed by clarification in a settling basin
  4. Filtration
    a. Rapid sand filtration—physically traps particles (not efficient for removal of *Giardia* cysts, *Cryptosporidium* oocysts, *Cyclospora* and viruses)
    b. Slow sand filtration—biologically removes *Giardia,* which adheres to microbial layer on the sand particles
  5. Recently *Cryptosporidium* has become of even greater concern than *Giardia*
    a. It is commonly found in sewage (90%), rivers (75%), and drinking water (28%)
    b. Infection takes as few as ten oocysts
    c. Companion and domestic animals are common reservoirs
    d. It is self-limiting in healthy individuals but can be serious, and even fatal, in immunocompromised individuals
    e. Its small size makes it even more dangerous than *Giardia* because it is not readily removed by sand filters
  6. Disinfection with chlorine or ozone; formation of halomethanes (carcinogens) during chlorination is becoming an increasing concern
  7. Viruses must be removed or inactivated
  8. An Information Collection Rule (ICR) has been initiated to assess the threat of those pathogens to the waters of cities over 100,000 population
B. Microbiological Analysis of Water—primarily based on detecting human fecal contamination of water supplies; human fecal waste contains a variety of pathogens
  1. Ideal indicator organisms indicate fecal contamination, but may not themselves be pathogenic
    a. Indicator organism should be applicable to all types of water
    b. Organism should be present whenever enteric pathogens are present
    c. Organism should survive longer than the hardiest enteric pathogen
    d. Organism should not reproduce in the contaminated water
    e. Test should be highly specific for the indicator
    f. Test should be easy and sensitive
    g. Indicator should be harmless to humans (ensuring safety for laboratory personnel)
    h. Concentration of indicator should directly reflect the degree of fecal pollution
  2. Indicator organisms in use
    a. Coliforms—all of the facultative anaerobic, gram-negative, nonsporing, rod-shaped bacteria that ferment lactose with gas formation within 48 hours at 35°C
      (1) *Escherichia coli, Enterobacter aerogenes,* and *Klebsiella pneumoniae*
      (2) Most probable number (MPN)—statistical estimation
      (3) Presence-absence (P-A) test—limited quantitative estimate, but is gaining in acceptance
      (4) Membrane filtration technique—water is filtered, filter is placed on an absorptive pad containing liquid medium; this is incubated; and colonies are counted

        (5)   Defined substrate tests (e.g., Colilert) involve the production of a colored product (for total coliforms) or a fluorescent product (for *E. coli*) from a specific growth substrate

      b.   Fecal coliforms (FC)—more restrictive; test involves growing bacteria at 44.5°C, a temperature tolerated by fecal coliforms, but not by other coliforms

      c.   Fecal enterococci (more restrictive) are being used as indicators for brackish and marine waters because they survive longer than coliforms under these conditions

    3.   Potable water—indicates bacteria are not detectable in a specific volume of water

  C.   Water and Human Diseases

    1.   Bacterial—many different types have been previously discussed

    2.   Protozoans include *Giardia, Cryptosporidium,* and *Naegleria;* the latter organism is of increasing concern worldwide because it causes a fatal neurological disease called primary amebic meningoencephalitis (PAM)

IV.   Groundwater Quality and Home Treatment Systems

  A.   Microbiology and microbiological processes in groundwater are not well understood

  B.   Disease-causing organisms are removed by adsorption and trapping as they move through the subsurface

  C.   Home treatment

    1.   Anaerobic liquefaction and digestion in a septic tank

    2.   Aerobic digestion, adsorption, and filtration of organic material are accomplished by drainage through suitable soil in a leach (drain) field

    3.   If drainage is too rapid, there is little adsorption and filtration, with subsequent contamination of well waters

  D.   Contamination of groundwater can lead to nutrient enrichment of ponds, lakes, and rivers

  E.   Contamination sources other than septic tank leach fields include sites for land disposal of sewage sludges, illegal dumping of septic tank pumpage, improper toxic waste disposal, agricultural runoff, and deep-well injection of industrial wastes

  F.   Treatment procedures for groundwater to be used either in place or at the point of use are under investigation

TERMS AND DEFINITIONS

Place the letter of each term in the space next to the definition or description that best matches it.

_____ 1. Waters that are rich in nutrients
_____ 2. Waters that are poor in nutrients
_____ 3. Self-consumption of microbial biomass so that sewage treatment sludge volumes are reduced to a minimum
_____ 4. Poor settling in an activated sludge tank as a result of the development of massive amounts of filamentous organisms
_____ 5. Introduction of organic material into a stream or river by in-stream photosynthesis
_____ 6. Introduction of organic material into a stream or river from out-stream sources
_____ 7. A columnar set-up designed to resemble a naturally occurring water/sediment ecosystem
_____ 8. Single of multiple layers of bacteria and nutrients adhering to surfaces
_____ 9. Biofilms that become so large that they are visible and have macroscopic dimensions
_____ 10. Organisms that are low-nutrient responsive
_____ 11. Organisms that are high-nutrient responsive

a. allochthonous
b. autochthonous
c. biofilms
d. bulking
e. copiotrophic microorganisms
f. endogenous respiration
g. eutrophic waters
h. microbial mats
i. microcosm
j. oligotrophic waters
k. oligotrophic microorganisms

FILL IN THE BLANK

1. The amount of a strong acid that can be neutralized by the organic matter in a water sample is referred to as the _____; if the organic matter is reacted with oxygen at a high temperature to produce a measured amount of carbon dioxide, it is referred to as _____; if the amount of oxygen needed to utilize organic material as a growth substrate is measured, it is referred to as the _____. In the latter, _____ is inhibited by nitrapyrin because this process can also demand oxygen, which could otherwise interfere with the results.

2. When nutrients are added to water, _____, or nutrient enrichment can take place. This can occur very rapidly or over many _____. This can be particularly important with certain types of water bodies, particularly _____.

3. In high nutrient lakes, the bulk of the water, or _____, except for the surface water, can become anaerobic. In the spring and fall, the aerobic surface water and the anaerobic subsurface water turn over because of differences in _____ and _____.

4. If organic wastes are added to rivers, chemoheterotrophs become dominant and may use oxygen faster than it can be replenished, creating an oxygen _____ curve. If the amount of organic material added was not excessive, algae will begin to grow, which leads to oxygen production in daylight and consumption at night. These are referred to as _____ oxygen shifts. Eventually the oxygen level will approach saturation. This overall process is referred to as _____.

5. Marine environments contain _____ percent of the earth's water; the ocean has been called a _____ _____ _____ with temperatures near _____ °C and pressures up to _____ atm near the sea floor.

6. Organisms that can grow in pressures of 0 to 400 atm but that grow best at the lower end are called _____, while those that will grow at 1 atm but prefer pressures closer to 400 atm are called _____ _____; those organisms that will only grow at higher pressures are called _____ _____.

7. Most of the nutrient cycling in the ocean occurs in the upper _____ meters where _____ penetrates; only _____% of the photosynthetically derived biomass reaches the ocean floor, thereby resulting in _____ nutrient conditions in deep waters.

MULTIPLE CHOICE

For each of the questions below select the *one best* answer.

1. The solubility of oxygen in water is influenced by which of the following?
   a. concentration of the oxygen in the gas phase above the surface
   b. temperature of the water
   c. the gas pressure of the atmosphere above the surface
   d. All of the above influence the solubility of oxygen in water.
2. Which of the following can be a function of the concentration of carbon dioxide in water?
   a. the amount of organic material produced by photosynthetic microorganisms
   b. the pH of the water
   c. Both (a) and (b) are correct.
   d. Neither (a) nor (b) is correct.
3. In which of the following types of waters do seasonal climatic temperature changes result in distinct chemical and microbial stratification?
   a. oceans
   b. eutrophic lakes
   c. oligotrophic lakes
   d. fast-flowing rivers
4. If phosphorus is added to oligotrophic water, what organisms will play a major role in nutrient accumulation?
   a. cyanobacteria
   b. eubacteria
   c. archaeobacteria
   d. eucaryotic algae
5. Secondary sewage treatment processes remove organic material by which of the following?
   a. biological
   b. physical
   c. chemical
   d. All of the above are correct.
6. Sewage treatment that involves horizontal flow in an agitated aeration tank is referred to as
   a. lagooning.
   b. activated sludge treatment.
   c. trickling filter processing.
   d. endogenous respiration.

7. Which of the following is not an advantage of anaerobic sludge digestion?
   a. Most of the biomass produced aerobically is utilized for methane production.
   b. The remaining sludge can be dried easily before disposal.
   c. Heavy metals are concentrated in the sludge.
   d. All of the above are advantages to anaerobic sludge digestion.
8. Fecal coliforms are useful indicator organisms for contamination of water by
   a. enteric viruses.
   b. enteric bacteria.
   c. enteric protozoan pathogens.
   d. All of the above are correct.
9. Which of the following does not contribute to the removal of organic material in leach fields of home sewage treatment systems?
   a. aerobic digestion as the waste percolates through the soil
   b. adsorption of organic material to soil particle surfaces
   c. entrapment of microbes in the pores of the leach field
   d. All of the above contribute to the removal of organic material.
10. Which of the following is not true about microbial mats?
   a. They form on the surfaces of rocks and sediments.
   b. Anaerobic conditions prevail in the parts of the mat where light does not penetrate.
   c. Gradients are formed that are very gradual, with great distance between recognizable layers.
   d. All of the above are true about microbial mats.

11. Which of the following is true about primary amebic meningoencephalitis caused by *Naegleria fowleri?*
    a. Its occurrence is worldwide.
    b. It is usually fatal.
    c. Both (a) and (b) are correct.
    d. Neither (a) nor (b) is correct.
12. Which of the following is true about the Colilert defined substrate test?
    a. It can detect total coliforms but not *E. coli.*
    b. It can detect *E. coli* but not total coliforms.
    c. It can detect total coliforms and *E. coli* simultaneously but independently.
    d. None of the above are correct.
13. Which of the following occur in large numbers in marine environments?
    a. barotolerant/barophilic bacteria
    b. viruses
    c. archaeobacteria
    d. All of the above occur in large numbers in marine environment.

14. Constructed wetlands use plants and their associated microorganisms to remove which of the following?
    a. organic material
    b. inorganic material
    c. metals
    d. All of the above are correct.
15. Slow sand filtration effectively removes
    a. *Giardia lamblia*
    b. *Cryptosporidium parvum*
    c. Both (a) and (b) are correct.
    d. Neither (a) nor (b) are correct.
16. Bacterial pathogens are primarily removed by
    a. primary sewage treatment
    b. secondary sewage treatment
    c. Both (a) and (b) contribute equally to pathogen removal.
    d. Neither (a) nor (b) significantly remove bacterial pathogens.

TRUE/FALSE

_____ 1. Aquatic environments are only marginally aerobic.
_____ 2. The Winogradsky column is a microcosm of an aquatic environment that is used to demonstrate the microbial stratification that can occur in nonagitated waters.
_____ 3. Limits on COD and BOD levels in treated sewage have been established in order to prevent eutrophication of outfall areas.
_____ 4. Even though the water environment is different from the soil environment, the microbial communities are basically very similar.
_____ 5. Microbial grazing, or the use of microbes as a food source by protozoans, plays a role in nutrient cycling in aquatic environments.
_____ 6. Anaerobic sludge digestion resulting in methane production is dependent on the presence of carbon dioxide, hydrogen, and organic acids to initiate the reaction.
_____ 7. Because of our great dependence on groundwater as a drinking water supply, we have developed a tremendous understanding of the microorganisms and microbiological processes occurring in this environment.
_____ 8. If a leach field floods, it becomes anaerobic and effective treatment ceases.
_____ 9. The ocean is seemingly inexhaustible in its ability to absorb and process pollutants.
_____ 10. The horizontal water flow in rivers and streams minimizes vertical stratification.
_____ 11. Inadequately treated municipal waste is considered to be a nonpoint pollution source.
_____ 12. Potable water is unfit for consumption or recreation because of the high levels of microbial contaminants present in it.
_____ 13. Biofilms are becoming of increasing concern and may be a major source of such pathogens as *Legionella* in potable water systems.

_____ 14. Surface attachment is advantageous to microorganisms at concentrations of dissolved organic material above 25 mg/liter.

_____ 15. Cryptosporidium has recently become of greater concern than *Giardia* because it occurs more often in waters and is harder to remove.

_____ 16. Surface flow soil treatment allows anaerobic microbial processing of waste.

_____ 17. Silicon removal by dams decreases the level of toxic algae in the associated reservoir.

_____ 18. Despite their small size, microorganisms in the ocean can have a significant impact on the atmosphere on a global scale.

CRITICAL THINKING

1. Describe the primary, secondary, and tertiary (if any) processes used by the sewage treatment facility in your community. At each stage, describe the method used and the basis for its activity. If your residence utilizes a septic tank, then describe the functioning of that system.

2. Discuss the ideal characteristics of an indicator organism for fecal contamination. Using these characteristics, discuss the use of total coliforms, fecal coliforms, and fecal streptococci as indicators. Under what circumstances is one a better choice than the others? Why are none of these organisms particularly useful as indicators of enteric viruses or protozoans?

ANSWER KEY

Terms and Definitions

1. g, 2. j, 3. f, 4. d, 5. b, 6. a, 7. i, 8. c, 9. h, 10. k, 11. e

Fill in the Blank

1. chemical oxygen demand (COD); total organic carbon (TOC); biochemical oxygen demand (BOD); nitrification 2. eutrophication; centuries; lakes 3. hypolimnion; temperature; specific gravity 4. sag; diurnal; self-purification 5. 97; high pressure refrigerator; 3; 1,000 6. barotolerant; moderate barophiles; extreme barophiles 7. 300; light; 1; oligotrophic

Multiple Choice

1. d, 2. c, 3. b, 4. a, 5. a, 6. b, 7. c, 8. b, 9. d, 10. c, 11. d, 12. c, 13. d, 14. d, 15. a, 16. b

True/False

1. T, 2. T, 3. T, 4. F, 5. T, 6. F, 7. F, 8. T, 9. F, 10. T, 11. F, 12. F, 13. T, 14. F, 15. T, 16. F, 17. F, 18. T

42   Terrestrial Environments

CHAPTER OVERVIEW

This chapter discusses the general characteristics of microorganisms associated with soils. Soils are dominated by solid materials, and the environment is mostly aerobic. Soils respond to changes in temperature and moisture and to the effects of plowing and other disturbances. Many of these responses are reflected or mediated by the microbial populations found in the soils.

CHAPTER OBJECTIVES

After reading this chapter you should be able to:

- describe the general characteristics of soils from different climates
- discuss the reasons that soil microbial communities are not well-known
- discuss the interactions of microorganisms with the surfaces found in soils
- describe the interaction of microorganisms with insects and other soil animals, and explain how these interactions influence nutrient cycling and other processes
- discuss organic matter accumulation and decomposition in soils
- discuss the various associations between plants and microorganisms and the ways these associations increase the ability of the plants to compete for water and nutrients
- discuss the *Rhizobium*-legume symbiosis
- discuss mycorrhizal relationships
- discuss the use of composting for maintaining and increasing the fertility of soils
- discuss the degradation and degradability of pesticides in soils, and explain how this affects the function of these substances
- discuss the interactions of soil microbes with the atmosphere and their involvement in gaseous pollutant utilization and modification

CHAPTER OUTLINE

I. The Environments of Soil Microorganisms
   A. Soils consist of peds (heterogeneous aggregates of various sizes that contain complex networks of pores and surfaces) that influence nutrient availability and that affect interactions among different microbes
      1. Bacteria are found primarily in the microenvironment associated with smaller pores; this protects them from protozoa predation and gives them access to soluble nutrients
      2. Filamentous fungi form bridges between separated regions where moisture is available; they move nutrients and water over greater distances in the soil
      3. Protozoa, soil insects, and nematodes are located on the outside of the aggregates and feed on the bacteria and fungi
   B. Gradients of various environmental factors and nutrients can exist within the pores of the soil (e.g., oxygen gradients)

C. Various soil components' binding affinities for nutrients and other substances can create different microenvironments within the soil; (e.g., anaerobic microsites)

D. Physical factors (e.g., pH) can influence microorganisms that are associated with surfaces

E. Soil clays and humus (partially degraded and stabilized organic matter) attract and bind organic and inorganic substances

F. Clays can serve as genetic microchips where microbially derived enzymes, RNA and DNA allow binding and elongation of nucleic acids

II. Soil Microorganisms, Insects, and Animals—Contributions to Soils

A. Colonization takes place after geological disturbances begin to form soil

B. Nutrients such as phosphorus and iron must be present, but carbon and nitrogen can be imported by biological processes

C. Once formed, most soils are rich sources of nutrients because of the plant and animal life that lives and dies there, contributing to the nutrient resources

D. Except for the plant component, bacteria and fungi are in greatest abundance in terms of biomass carbon and nitrogen

E. Soil organisms play a major role in the decomposition of plant litter, which would otherwise accumulate in the environment

F. Many of the soil microorganisms have not been cultured and are, therefore, poorly understood; however, DNA analysis and PCR amplification are providing interesting and unexpected information about these unculturable organisms

G. Filamentous actinomycetes, primarily of the genus *Streptomyces,* produce an odor causing compound called geosmin which give soils their characteristic earthy odor

H. Soil microorganisms can be categorized on the basis of their preference for easily available or more resistant substrates

1. Zymogenous—organisms that respond rapidly to the addition of easily utilizable substrates such as sugars and amino acids

2. Autochthonous—indigenous forms that tend to utilize native organic matter

I. Microbes that can grow in oligotrophic (low nutrient) environments have developed special strategies to deal with those conditions

J. Soil insects and animals serve as decomposer-reducers, which not only decompose but also physically reduce the size of organic aggregates; this increases the surface area and makes nutrients more available for utilization by bacteria and fungi

K. Protozoa influence nutrient cycling by microbivory (feeding on "palatable" microorganisms), causing an increase in the rates of nitrogen and phosphorus mineralization

L. Free hydrolytic enzymes released from animals, plants, insects, and lysed microorganisms contribute to soil biochemistry

III. Microorganisms and the Formation of Different Soils

A. An imbalance of primary production and decomposition can lead to either an accumulation of organic matter or the production of an organic-poor, low-fertility soil

B. High temperature (greater than 25°C) favors decomposition

C. Decomposition is a complex process; plant litter, microbial content, soil temperature, water availability, and oxygen availability must be considered

IV. Soil Microorganism Associations with Plants

A. Plants are heavily colonized by microorganisms

B. The presence of microorganisms increases the rate of organic matter released from the roots (exudation)

C. Microorganisms influence plant growth through the release of morphology influencing compounds such as auxins, giberellins and cytokinins

D. The Rhizosphere and *Rhizobium*

1. The rhizosphere is the area of the soil in the immediate vicinity of the plant roots

2. Release of organic materials from the roots causes localized changes

3. In low-fertility soils (e.g., desert soils) this may have a tremendous impact on the microbial community, while in high-fertility soils this effect may not be as pronounced
4. Microorganisms in the rhizosphere serve as a labile source of nutrients and play a critical role in organic matter synthesis and degradation
5. Environment around certain types of plants promotes association with mutualistic nitrogen-fixing bacteria; the genus *Rhizobium* is a prominent member of the rhizosphere community
   a. *Rhizobium* is stimulated to produce Nod factors that promote activate the host symbiotic process necessary for root hair infection and module development
   b. An infection thread forms and grows down the root hair
   c. *Rhizobium* spreads within the infection thread into the underlying root cells
   d. Bacteria multiply and develop into swollen, branched bacteroids enclosed by a plant-derived membrane called the peribacteroid membrane
   e. Further growth and differentiation lead to the formation of nitrogen-fixing forms called symbiosomes
   f. Nitrogen fixation occurs within the symbiosome and the nitrogen is then assimilated into various organic compounds and distributed throughout the plant
   g. Associative nitrogen fixation also occurs and is mediated by free-living bacteria in the rhizosphere; however, recent evidence suggests that the major contribution of these bacteria may be the reduction of nitrate to ammonia, which is then available for use by the plant
   h. A major goal of biotechnology is to introduce nitrogen fixing genes into plants that do not normally form such associations

E. Actinorrhizae—actinomycete-root associations between members of the genus *Frankia* and woody plants; they can fix nitrogen and are important in the life of woody, shrublike plants

F. Mycorrhizae—fungus-root associations; 5 associations have been described
   1. Ectomycorrhizae—grow as an external sheath around the root with limited penetration of the fungus into the cortical regions of the root; found primarily in temperate regions; mycelia extend far into the soil forming a mycorrhizosphere and mediate nutrient transfer to the plant; process is aided by mycorhizal helper bacteria (MRB)
   2. Endomycorrhizae—fungal hyphae penetrate the outer cortical cells of the plant root, where they form characteristic structures known as arbuscules; they increase the competitiveness of the plant and increase water uptake by the plant in arid environments

G. Tripartite and Tetrapartite Association—involve some combination of plant, rhizobia, mycorrhizae, and actinorrhizae; enable plant to better cope with nutrient-deficient environments
   1. Capable of nitrogen fixation
   2. Important in the lives of these plants

H. Fungal and Bacterial Endophytes of Plants—some fungi and bacteria infect and live within plants as endophytes; can be mutualistic or parasitic (pathogenic)

I. Tumorigenicity or *Agrobacteriu*—Genetic Applications—*Agrobacterium* containing the Ti (tumor-inducing) plasmid cause the formation of galls (tumors) on the plant; recent interest centers around the use of the Ti plasmid as a vector to transfer new genetic characteristics (eg., herbicide resistance; bioluminescence) to the plant

V. Soil Organic Matter and Soil Fertility—microbial degradation must be balanced by the need for organic matter in order to retain nutrients, to maintain soil structure, and to hold water for plant use
   A. Plowing and other soil disturbances aerate soil and increase microbial degradation
   B. No-till and minimum-till practices reduce aeration but require more herbicides to control weed infestation
   C. Composting allows for controlled decomposition which produces physiologically stabilized compost material that, when added to the soil, increases the organic content without stimulating soil microorganisms to increase decomposition

VI. Degradation and Microorganisms—Fallibility and Recalcitrance
   A. Fallibility—not all organic compounds can be degraded by microorganisms

B. Recalcitrance—the degree to which an organic compound can resist microbial degradation
C. Degradation is promoted by the presence of easily usable energy sources—cometabolism
D. A balance is needed such that a pesticide will remain active in the soil long enough to do its job, but degrade quickly enough so that it will not linger beyond its useful period and thereby have an adverse effect
E. Degradation usually takes place in several stages
F. Degradation processes can be harnessed to degrade agricultural wastes and waste water by land farming, where the waste is incorporated into the soil or allowed to flow over the soil where degradation occurs
G. Phytoremediation involves planting fast-growing plants in contaminated soil; stimulates microbial growth through release of organic nutrients; this allows cometabolic processes that stimulate degradation

VII. Soil Microorganism Interactions with the Atmosphere
A. Ice-nucleating bacteria serve as nucleation centers for snow and precipitation formation
B. "Ice-minus" strains that have been genetically engineered seem to protect strawberries from frost damage when sprayed on the plant; however, the release of such genetically engineered microorganisms (GEMs) into the environment is of continuing concern
C. Soil microorganisms can influence the global fluxes of both relatively stable and reactive gases
   1. Relatively stable gases include carbon dioxide, nitrous oxide, nitric oxide, and methane
   2. Reactive gases include ammonia, hydrogen sulfide, and dimethylsulfide
   3. Methane has become of increasing concern and it is questionable if consumption can keep up with production rates; this may lead to atmospheric accumulation with a rise in average atmospheric temperature (greenhouse effect)

TERMS AND DEFINITIONS

Place the letter of each term in the space next to the definition or description that best matches it.

_____ 1. Soil microorganisms that respond rapidly to the addition of easily utilizable substrates, such as sugars and amino acids

_____ 2. Soil microorganisms that largely utilize native organic matter

_____ 3. Incorporation of agricultural waste into soil (or flow over soil) to allow degradation of organic material

_____ 4. The use of microorganisms as a food source by protozoans

_____ 5. The release of organic matter from the roots of plants

_____ 6. A microenvironment in the immediate vicinity of the roots of plants that may be markedly different from the soil as a whole

_____ 7. The inability of microorganisms to degrade all organic compounds

_____ 8. The ability of an organic compound to resist microbial degradation

_____ 9. Bacteria and fungi that infect and live within plants

_____ 10. Heterogeneous soil aggregates of various sizes that contain a complex network of pores and surfaces

_____ 11. Promotion of pesticide degradation by the presence of easily usable energy sources

_____ 12. Substance that promotes binding of *Rhizobium* to plant root hairs

_____ 13. Substance produced by adinomycetes that gives soil its characteristic earthy odor

_____ 14. Nodules of bacteria enclosed by a plant membrane that are not yet capable of nitrogen fixation

_____ 15. Differentiated nodules that are capable of nitrogen fixation

a. autochthonous
b. bacteroids
c. cometabolism
d. endophyte
e. exudation
f. fallibility
g. geosmin
h. land farming
i. microbivory
j. Nod factor
k. peds
l. recalcitrance
m. rhizosphere
n. symbiosome
o. zymogenous

FILL IN THE BLANK

1. Soil microorganisms are often categorized according to their preference for substrates: those that respond rapidly to the addition of easily utilizable substrates, such as sugars and amino acids, are referred to as _____ microbes; those that tend to utilize native organic matter to a greater extent are called _____ microbes; and those that can be maintained on media containing less than 15 mg/liter of organic matter are referred to as _____ microbes.

2. The processes of organic matter production and decomposition are influenced by soil _____ and _____.

3. Nitrogen fixation carried out by members of the genus *Rhizobium* takes place in nodules which are specifically attached to the roots of the plant. This is called _____ nitrogen fixation. Alternatively, other genera, such as *Azotobacter* and *Azospirillum,* do not specifically attach to the roots of plants but are dependent upon nutrients in the rhizosphere, and this process is referred to as _____ nitrogen fixation. However, recent evidence suggests that these latter organisms may be important not only in nitrogen fixation but also in nitrogen _____.

4. Associations between fungi and plant roots are referred to as _____: if this association involves the growth of the fungi as a sheath around the root tip with limited penetration of the hyphae into the cortical region of the root, it is called _____; however, if extensive penetration of the fungal hyphae into the cortical regions of the plant root occurs, then it is referred to as a(n) _____. A characteristic structure exhibited by the latter type is called an _____.

5. The inability of some pesticides and herbicides to be degraded by microorganisms is referred to as _____. The existence of these nonbiodegradable substances demonstrates the _____ of microbes (i.e., their being less than perfect).

6. Plants that form associations with mutualistic nitrogen-fixing bacteria are called _____; the bacteria infect the root hairs, stimulating the plant to form an _____ _____ down which the bacteria travels, infecting neighboring cells. The bacteria multiply within these to form swollen, branched structures called _____; further differentiation results in the formation of _____, where nitrogen fixation occurs.

7. Clays can serve as _____ _____ where microbially derived enzymes, RNA and DNA allow _____ and _____ of nucleic acids.

8. _____ involves planting fast-growing plants in contaminated soil. This _____ microbial growth through release of _____ material which allows _____ processes that stimulate degradation.

MULTIPLE CHOICE

For each of the questions below select the *one best* answer.

1. Which of the following is not used by soil microorganisms to support their growth?
   a. moisture contained in soil pores
   b. dissolved nutrients associated with soil particles or in pores
   c. organic matter released from the plant roots
   d. All of the above are used to support growth of soil microorganisms.

2. Which of the following play(s) an active role in pioneer-stage nutrient accumulation, particularly of carbon and nitrogen?
   a. cyanobacteria
   b. chemoautotrophic bacteria
   c. vascular plants
   d. soil nematodes

3. Which of the following is(are) the largest living component(s) in terms of soil biomass carbon and nitrogen?
   a. fungi
   b. bacteria
   c. plants
   d. soil nematodes

4. Which of the following is not a contribution of soil insects and animals?
   a. decomposition (mineralization) of organic material
   b. reduction in size of organic substrate particles, such as plant litter
   c. release of certain organic materials (e.g., methane) produced by the actions of insect and animal microbiota (e.g., intestinal microflora)
   d. All of the above are contributions of soil insects and animals.

5. Which of the following is not a benefit to the plant of the mycorrhizal associations?
   a. increased nutrient availability to plants in moist environments
   b. increased water availability to plants in arid environments
   c. increased utilization of plant photosynthate by the fungus
   d. All of the above are benefits to the plant of mycorrhizal associations.

6. Which of the following increases soil organic matter decomposition?
   a. composting
   b. irrigation
   c. minimum-till agriculture
   d. All of the above increase soil organic matter decomposition.

7. Given time and the almost infinite variety of microbes, all organic compounds would be degraded. This concept, referred to as _____, is in serious question because of industrially synthesized compounds.
   a. microbial infallibility
   b. microbial ubiquity
   c. microbial recalcitrance
   d. universal biodegradability

8. Which of the following influences soil nutrient cycling?
   a. protozoan microbivory
   b. released enzymes from lysed microorganisms
   c. Both (a) and (b) are correct.
   d. Neither (a) nor (b) is correct.

9. Which of the following is a potential use of soil microorganisms that interact with the atmosphere?
   a. protection of strawberries from frost damage by spraying them with bacteria resistant to ice crystal formation
   b. air purification in closed buildings by potted plants whose soil contains microorganisms capable of degrading airborne pollutants
   c. Both (a) and (b) are correct.
   d. Neither (a) nor (b) is correct.

10. Mycorrhizal helper bacteria (MRBs) help with which of the following?
    a. ectomycorrhizal development
    b. endomycorrhizal development
    c. Both (a) and (b) are correct.
    d. Neither (a) nor (b) is correct.

11. Which of the following types of compounds are released by microorganisms and influence plant growth?
    a. auxins
    b. giberellins
    c. cytokinins
    d. All of the above

TRUE/FALSE

_____ 1. Soils that are very wet tend to be anaerobic.

_____ 2. When new soils are formed by geological weathering, carbon and nitrogen must be imported by biological processes.

_____ 3. The compound geosmin, which gives soil its characteristic earthy odor, is produced by soil nematodes.

_____ 4. There is increasing interest in the oligotrophic microbes, primarily because they appear to be present in higher concentrations than the easily cultured microbes which heretofore have received most of the attention.

_____ 5. Microbivory, or the use of microbes as a food source by protozoans, removes nutrients which would otherwise be available for plant growth.

_____ 6. Excess water in the soil inhibits decomposition of organic matter by decreasing the amount of available oxygen.

_____ 7. Changes in the microbial population of the rhizosphere are less pronounced in low-fertility soils because of the general lack of nutrients to support microbial growth.

_____ 8. Rainfall can change an aerobic soil with anaerobic microsites into a primarily anaerobic soil.

_____ 9. Endomycorrhizae are detrimental to desert plants because they limit water uptake by the plant, which is growing in an already arid environment.

_____ 10. *Agrobacterium,* a tumor-producing bacterium that infects plants, has been of recent interest because of its possible use as a vector for genetically engineering plants.

_____ 11. Microbial consumption of methane easily keeps up with production so that this gas poses no problem on a global scale.

_____ 12. The vast majority of microorganisms found in soils have been cultivated in the laboratory.

CRITICAL THINKING

1. Explain how soil temperature and water content interact to produce various high-fertility and low-fertility soils. Include in your explanation a specific discussion of tropical, grassland, forest, and bog regions.

2. Describe the ideal characteristics of an herbicide with respect to recalcitrance to microbial degradation. In your description, consider a field that you planted with corn last year. You want to plant soybeans this year, and corn next year. You want to use an herbicide specifically directed against corn this year when soybeans are planted.

ANSWER KEY

Terms and Definitions

1. o, 2. a, 3. h, 4. i, 5. e, 6. m, 7. f, 8. l, 9. d, 10. k, 11. c, 12. j, 13. g, 14. b, 15. n

Fill in the Blank

1. zymogenous; autochthonous; oligotrophic 2. temperature; water content 3. mutualistic; associative; reduction 4. mycorrhizae; ectomycorrhizae; endomycorrhizae; arbuscule 5. recalcitrance; fallibility 6. legumes; infection thread; bacteroids; symbiosome 7. genetic microchips; binding; elongation 8. Phytoremediation; stimulates; organic, cometabolic

Multiple Choice

1. d, 2. a, 3. c, 4. d, 5. c, 6. b, 7. a, 8. c, 9. c, 10. a, 11. d

True/False

1. T, 2. T, 3. F, 4. T, 5. F, 6. T, 7. F, 8. T, 9. F, 10. T, 11. F, 12. F

43 Microbiology of Food

CHAPTER OVERVIEW

This chapter discusses the microorganisms associated with foods. Some of these microorganisms are used in the production of certain foods. Others are disease-causing organisms that are transmitted via foods. Still others are associated with food spoilage. Therefore, the entire sequence of food handling—from the producer to the final consumer—must be monitored carefully for the presence and activity of microorganisms.

CHAPTER OBJECTIVES

After reading this chapter you should be able to:

- discuss the interaction of intrinsic (food-related) and extrinsic (environmental) factors with the microbial communities associated with foods and the successional changes this interaction produces
- describe the various physical, chemical, and biological processes used to preserve foods
- discuss sterilization processes used in the production of foods and how these processes are monitored
- discuss the various diseases that can be transmitted to humans by foods
- differentiate between food infections and food intoxications
- discuss the detection of disease-causing organisms in foods
- describe the fermentation of dairy products, grains, meats, fruits, and vegetables
- discuss the disease-causing chemicals produced by fungi growing in moist corn and grain products
- discuss wine production by alcoholic fermentation of fruit juices or musts
- discuss the direct use of microbial cells as nutrients (single-cell protein) by humans and animals

CHAPTER OUTLINE

    I.   Microorganisms and Food Spoilage
        A.   Intrinsic Factors
           1.   Food composition
              a.   Carbohydrates—does not result in major odors
              b.   Proteins and/or fats result in a variety of foul odors
           2.   pH—low pH allows yeasts and molds to become dominant; higher pH allows bacteria to become dominant; higher pH favors putrefaction (the anaerobic breakdown of proteins that releases foul-smelling amine compounds)
           3.   Water presence and water availability
              a.   Drying (removal of water) controls or eliminates food spoilage

       b.   Addition of salt or sugar decreases water availability and thereby helps reduce microbial spoilage

       c.   Even under these conditions spoilage can occur by certain kinds of microorganisms
          (1)   Osmophilic—prefer high osmotic pressure
          (2)   Xerophilic—prefer low water availability

    4.   Oxidation-reduction potential can be affected (lowered) by cooking, making foods more susceptible to anaerobic spoilage

    5.   Physical structure—grinding and mixing increases surface area and distributes microorganisms, resulting in more rapid spoilage; vegetables and fruits have outer skins that protect them from spoilage

    6.   Some foods, particularly spices, may contain antimicrobial substances

B.   Extrinsic factors

    1.   Temperature—lower temperatures reduce microbial growth and prolong storage life

    2.   Relative humidity—lower moisture content helps prolong storage life

    3.   Atmosphere—oxygen usually promotes growth and spoilage even in shrink-wrapped foods since oxygen can diffuse through the plastic; high CO_2 tends to decrease pH and reduces spoilage

C.   Food Spoilage

    1.   Food spoilage by microorganisms alters food, which then has undesirable appearance, contains hallucinogens, or contains carcinogens

    2.   Proteolysis (aerobic) and putrefaction (anaerobic) are processes that decompose proteins in meats and dairy products

    3.   Spoilage often occurs in a successional relationship: the growth of one type of organism creates conditions conducive to the growth of a different type of organism, which eventually leads to the spoilage of the food; this is seen in the spoilage of unpasteurized milk

    4.   Fungal-derived carcinogens include the aflatoxins and fumonisins that are produced on moist corn, grain, and nut products

    5.   Aflatoxin has also been observed in milk, beer, cocoa, raisins, and soybean meal

    6.   Meats and dairy products are ideal environments for spoilage by microorganisms because of their high nutritional value and the presence of easily utilizable carbohydrates, fats, and proteins

    7.   Canned foods can undergo spoilage
       a.   Prior to canning
       b.   As a result of underprocessing during canning
       c.   Leakage of contaminated water through can seams during cooling

II.   Food Preservation Alternatives

A.   Filtration—Filtration of water, wine, beer juices, soft drinks and other liquids can keep bacterial populations low or eliminate them entirely

B.   Low or High Temperature

    1.   Refrigeration and/or freezing retards microbial growth but does not prevent spoilage

    2.   Canned food is heated in special containers called retorts to 115°C for 25-100 minutes to kill spoilage microorganisms

    3.   Pasteurization—kills disease-causing organisms; substantially reduces the number of spoilage organisms
       a.   Low-temperature holding (LTH)—62.8°C for 30 minutes
       b.   High-temperature short-time (HTST)—71°C for 15 seconds
       c.   Ultra-high temperature (UHT)—141°C for 2 seconds
       d.   Shorter times result in improved flavor and extended product shelf life

    4.   Heat treatments are based on a statistical process involving the probability that the number of remaining viable microorganisms will be below a certain level after a specified time at a specified temperature

    5.   Dehydration procedures (e.g., freeze-drying) remove water and increase solute concentration

C. Chemicals and Radiation
   1. Chemical preservatives
       a. Regulated by the U.S. Food and Drug Administration
       b. Include simple organic acids, sulfite, ethylene oxide as a gaseous sterilant, sodium nitrite, and ethyl formate
       c. Affect microorganisms by disrupting a critical factor
       d. Effectiveness depends on pH
       e. Nitrites protect against *Clostridium botulinum*, but are of some concern; however, because of their potential to form carcinogenic nitrosamines during the cooking of meats preserved with them
       f. Nisin is a polypeptide agent that inhibits peptidoglycan synthesis and can be used to preserve low-acid foods during canning
   2. Radiation
       a. Nonionizing (ultraviolet or UV) radiation—used for surfaces of food-handling utensils; does not penetrate foods
       b. Ionizing (gamma radiation)—penetrates well, but must be used with moist foods to produce peroxides from water, resulting in oxidation of sensitive cellular constituents (radappertization); used for seafoods, fruits, and vegetables; can be used to sterilize meats
III. Diseases and Foods
   A. Diseases transmitted by foods
       1. Food-borne infections
           a. Ingestion of microorganisms, followed by growth, tissue, invasion and/or release of toxins
           b. Includes infections caused by *Salmonella, Campylobacter, Listeria,* and *Escherichia coli,* particularly *E. coli* 0157:H7 which causes hemorrhagic colitis
       2. Food intoxications
           a. Ingestion of toxins in foods in which microorganisms have grown
           b. The organism need no longer be viable and need not grow after ingestion
           c. Induces staphylococcal food poisoning, botulism, *perfringens* food poisoning, and *Bacillus cereus* food poisoning
   B. Detection of Disease-Causing Microorganisms
       1. Methods need to be rapid; therefore, traditional culture methods that might take days to weeks to complete are too slow
       2. Methods need to be sensitive; therefore, traditional methods that are not sensitive enough to detect a low concentration of pathogens against a high background of normal microflora complicate matters
       3. Molecular methods overcome these limitations and are currently being used wherever feasible; includes DNA:RNA probe technology and immunochemical procedures; most recently, the use of immunomagnetic reagents allows for both detection and removal for further study using a magnetic collector
       4. Molecular methods are valuable for three reasons
           a. They can detect the presence of a single, specific pathogen
           b. They can detect viruses that cannot be conveniently cultured
           c. They can identify slow-growing or non-culturable pathogens
       5. Polymerase chain reaction (PCR) can be used to detect low level contaminants (e.g., as few as 10 toxin-producing *E. coli* cells in a population of 100,000 cells isolated from soft cheese samples)
       6. Recently bioluminescence methods (e.g., monitoring the presence of ATP) have been used to detect potential pathogens on utensils and other surfaces

IV. Microbiology of Fermented Foods
   A. Dairy products—acid produced from microbial activity causes protein denaturation; organism growth often is sequential; frequently involves *Lactococcus* and *Lactobacillus* species, among others
      1. Cultured buttermilk—fermentation of skim milk
      2. Sour cream—same fermentation but of cream
      3. Yogurt—two microorganisms (*S. thermophilus* and *L. bulgaricus*) are used for acid and aroma production, respectively
      4. Acidophilus milk contains *L. acidophilus;* improves general health by altering intestinal microflora; may help control colon cancer
      5. Bifid-amended fermented milk products (containing *Bifidobacterium* spp.) improve lactose tolerance, possess anticancer activity, help reduce serum cholesterol levels, assist calcium absorption, and promote the synthesis of B-complex vitamins; may also reduce or prevent the excretion of rotaviruses, a cause of diarrhea among children
      6. Cheeses—produced by coagulation of curd, expression of whey, and ripening by microbial fermentation; cheese can be internally inoculated or surface ripened
   B. Meat and Fish
      1. Meat products—sausages, country-cured hams, bologna, salami, etc.; frequently involves *Pediococcus cerevisiae* and *Lactobacillus plantarum*
      2. Fish products include izushi (fresh fish, rice, and vegetables incubated with *Lactobacillus* spp.) and katsuobushi (tuna incubated with *Aspergillus glaucus*)
   C. Wine, Beer, and Other Fermented Alcoholic Beverages
      1. Wines and champagnes
         a. Grapes are crushed and liquids that contains fermentable substrates (musts) are separated
         b. Must is sterilized with sulfur dioxide fumigant or by pasteurization
         c. Mashing involves incubation with water; insoluble material is removed to yield a wort, a clear liquid containing fermentable sugars and other simple molecules
         d. All grapes, regardless of color, have white juices; to make a red wine, the skins of a red grape are left in contact with the must before the fermentation process
         e. Desired strain of *Saccharomyces cerevisiae* or *S. ellipsoideus* is added, and the mixture fermented (10 to 18% alcohol)
         f. Dry wine (no free sugar)—amount of sugar is limited so that all sugar is fermented before alcohol level increases to a level that inhibits further yeast activity
         g. Sweet wine (free sugar present)—fermentation is inhibited by alcohol accumulation before all sugar is used up
         h. Aging—final fermentation to accumulate flavoring compounds
         i. Racking—removal of sediments accumulated during the fermentation process
         j. Champagnes—fermentation is continued in bottles to produce a naturally sparkling wine
         k. Brandy (burned wine)—distilled to increase alcohol concentration
         l. Wine vinegar—controlled microbial oxidation (by *Acetobacter* or *Gluconobacter*) to produce acetic acid from ethanol
      2. Beers and ales
         a. Malt is produced by germination of the barley grains and the activation of their enzymes
         b. Mash is produced by enzymatic starch hydrolysis to accumulate utilizable carbohydrates
         c. Mash is heated with hops (dried flowers of the female vine *Humulus lupulis*) to provide flavor and clarify the wort (hydrolyzed proteins and carbohydrates); inactivates hydrolytic enzymes so that wort can be pitched (inoculated with yeast)
         d. Beer—produced with a bottom yeast, such as *S. carlsbergensis*
         e. Ale—produced with a top yeast, such as *S. cerevisiae*

372

f. Freshly fermented (green) beers are lagered (aged), bottled, and carbonated
g. Pasteurized or filtered to remove microorganisms
3. Distilled spirits—beerlike fermented liquid is distilled to concentrate alcohol; type of liquor depends on composition of starting mash; flavorings can also be added; a sour mash involving *Lactobacills delbrueckii* mediated fermentation is often used
D. Bread and Other Fermented Plant Products
1. Aerobic yeast fermentation is used to produce carbon dioxide with minimal alcohol production
2. Other fermentation products add flavors
3. Other microorganisms make special breads, such as sourdough
4. Other products
a. Susu fermentation of tofu, a chemically coagulated soybean milk product
b. Sauerkraut—fermented cabbage; involves a microbial succession mediated by *Leuconostoc mesenteroides* and *Lactobacillus plantarum*
c. Pickles—involves a complex microbial succession
d. Silages—animal feeds produced by anaerobic, lactic-type mixed fermentation of grass, corn and other fresh animal feeds
VI. Microorganisms as Sources of Food—bacteria, yeasts, and other fungi have been used as food sources for animal and human consumption; referred to as single-cell protein (SCP); can be used directly as a food source or as a supplement to other foods; one of the more popular microbial food supplements is the cyanobacterium, *Spirulina*

373

TERMS AND DEFINITIONS

Place the letter of each term in the space next to the definition or description that best matches it.

_____ 1. The food spoilage process that results in the breakdown of proteins

_____ 2. The food spoilage process that results in the release of foul-smelling amine compounds

_____ 3. A process that reduces the total microbial population and that usually eliminates all disease-causing microorganisms

_____ 4. A process that eliminates all living microorganisms

_____ 5. The process of irradiating moist foods to kill contaminating microorganisms

_____ 6. The process by which complex polysaccharides found in cereals and other starchy materials are depolymerized to form fermentable substrates

_____ 7. A plant product that contains readily fermentable substrates

_____ 8. The fermentable product formed after hydrolysis of proteins and carbohydrates

_____ 9. The process by which the sediments that were formed during microbial growth and fermentation procedures are removed

_____ 10. Microorganisms that grow in media with a high salt content

_____ 11. Microorganisms that grow in media with low water activity

_____ 12. Sequential grow of microorganisms in which growth of one organism produces environmental conditions suitable for the subsequent growth of another microorganism

a. mashing
b. must
c. osmophilic organisms
d. pasteurization
e. proteolysis
f. putrefaction
g. racking
h. radappertization
i. sterilization
j. succession
k. wort
l. xerophilic organisms

FILL IN THE BLANK

1. Food spoilage is dependent on food-related or _____ factors, such as pH, moisture, water activity or availability, oxidation-reduction potential, physical structure of the food, available nutrients, and the possible presence of antimicrobial compounds, and on environmental, or _____, factors such as temperature, relative humidity, gases present (particularly _____ and _____), and the types and levels of microorganisms added to the food.

2. The process of heating foods to eliminate disease-causing microorganisms and to reduce the level of food-spoilage microorganisms is called _____; if the process involves heating to 62.8°C for 30 minutes, it is referred to as the _____ process; if the temperature is 71°C for 15 seconds, it is referred to as the _____ process; and if it involves heating to 141°C for 2 seconds, it is called the _____ process.

3. If food-borne disease transmission requires ingestion of the pathogen, growth of the organism, and release of toxins in the intestine, it is referred to as a food-borne _____; if no growth of the microorganism is required after ingestion because the toxic substances are already present in the food as a result of previous growth, then it is referred to as a food-borne _____.

4. Infection of grains by the ascomycete *Claviceps purpura* can cause _____, in which hallucinogenic alkaloids produced by the fungus can lead to altered behavior, abortion, and death after their consumption. Another group of fungi that can develop in grains and nuts produces powerful cancer-causing compounds (carcinogens) known as _____, which particularly affect the _____ where they are converted to unstable derivatives.

5. The liquid extracted from crushed grapes has readily fermentable substrates and is referred to as _____. However, if cereals and other starchy materials are used, the complex carbohydrates must be depolymerized by a process known as _____ to produce a liquid that has fermentable substrates. In this case, the liquid is referred to as a _____.

6. In the making of beer and ale, the enzymes in the grain are activated to hydrolyze the starch and produce utilizable carbohydrates. This mixture is called a _____ and can then be _____ (inoculated) with the desired yeast. If a bottom yeast is used, the product is a(n) _____; however, if a top yeast is used, the product is a(n) _____.

7. In breadmaking, yeast growth is usually carried out under _____ conditions, which results in more _____ production and less _____ accumulation.

8. The use of microorganisms directly as food sources or indirectly as a food supplements has been referred to as _____. A popular microorganism for this use is the _____, *Spirulina*.

9. Large quantities of sugar or salt added to foods cause microorganisms to _____ because of the hypertonic conditions; however, food spoilage can still result from the action of _____ microorganisms that grow best at high osmotic concentration, and from _____ microorganisms that grow best in a low water activity environment.

10. Recently _____ methods have been used to detect potential pathogens on _____ and other surfaces.

MULTIPLE CHOICE

For each of the questions below select the *one best* answer.

1. Which of the following helps preserve foods by controlling microbial access to water?
   a. drying
   b. addition of salt
   c. addition of sugar
   d. All of the above help to preserve foods by controlling microbial access to water.

2. Which of the following reduces spoilage of the foods involved?
   a. grinding and mixing of foods such as sausage and hamburger
   b. peeling off the skins of fruits and vegetables
   c. Both (a) and (b) are correct.
   d. Neither (a) nor (b) is correct.

3. Which of the following products seems to aid a variety of intestinal processes and promote general health?
   a. acidophilus milk
   b. bifid-amended fermented milk products
   c. Both (a) and (b) are correct.
   d. Neither (a) nor (b) is correct.

4. Irradiation has been used to sterilize certain types of foods. There are, however, certain limitations on its effectiveness. One of these is that it will only work on
   a. moist foods.
   b. dry foods.
   c. foods that are not in metal cans.
   d. Actually, it will work well on all foods.

5. Microbial fermentations are used in the production of cheeses in order to do which of the following steps?
   a. coagulate the milk solids to form a curd
   b. ripen the cheese to give a characteristic texture and flavor
   c. Microbial fermentations are used to accomplish both (a) and (b).
   d. Microbial fermentations are used to accomplish neither (a) nor (b).

6. Which of the following determines whether or not a wine is considered to be dry?
   a. the percentage of alcohol after fermentation
   b. the amount of free sugar after fermentation
   c. the amount of carbon dioxide after fermentation
   d. the type of grape used for starting material

7. Which of the following is the reason that molecular methods are being used instead of traditional culture methods to detect disease-causing organisms in foods?
   a. Culture methods are too slow.
   b. Culture methods are not sensitive enough to detect low levels of pathogens against a high background of normal microflora.
   c. Both (a) and (b) are correct.
   d. Neither (a) nor (b) is correct.

8. Which of the following is true about the use of immunomagnetic reagents in the food industry?
   a. They can be used to remove and concentrate microorganisms for further study.
   b. They can be used to remove microorganisms to reduce spoilage.
   c. Both (a) and (b) are correct.
   d. Neither (a) nor (b) is correct.

9. Which of the following is true about molecular methods of microorganism detection in foods?
   a. They can detect the presence of a single specific pathogen.
   b. They can be used to detect viruses that are not easily cultured.
   c. They can be used to detect slow growing pathogens.
   d. All of the above are true about the use of molecular detection methods.

10. Which of the following represents ways in which canned foods may undergo spoilage?
    a. spoilage before canning
    b. as a result of underprocessing during canning
    c. leakage of contaminated water through can seams during cooling
    d. All of the above are ways in which canned foods may undergo spoilage.

TRUE/FALSE

_____ 1. Food grinding and mixing tend to reduce spoilage by causing mechanical damage to the contaminating microorganisms.

_____ 2. Even at freezer temperatures, some microorganisms may grow and cause food spoilage, but the process is greatly slowed.

_____ 3. Microorganisms can be removed from some foods by filtration, resulting in sterilized food preparations.

_____ 4. Cooking can lead to increased anaerobic spoilage by lowering the food's oxidation-reduction potential.

_____ 5. Nitrites are used to protect against microorganisms but their use is of some concern because of the formation of carcinogenic nitrosamines upon cooking preserved foods.

_____ 6. Sour cream and cultured buttermilk are produced by the same microbial fermentation processes, but the starting material is different, which accounts for the different final product.

_____ 7. Some fermented dairy products have been suggested to have beneficial effects, including a reduction in the incidence of colon cancer and the minimization of lactose intolerance.

_____ 8. White wines are produced from white (or green) grapes. If a red grape is used, the result will always be a red wine.

_____ 9. Brandy is produced by distilling wine in order to increase the concentration of alcohol.

_____ 10. Vinegar is produced by microbial oxidation of the ethanol in wine to acetic acid.

_____ 11. Atmosphere is important in food spoilage; high O_2 tends to reduce spoilage while high CO_2 tends to increase spoilage.

_____ 12. The polymerase chain reaction can detect as few as 10 toxin-producing _E. coli_ in a population of 100,000 microorganisms isolated from soft-cheese samples.

CRITICAL THINKING

1. It is very important to test the efficiency of sterilization procedures in commercial food preparation operations. The two organisms most often used are _Bacillus stearothermophilus_ and _Clostridium_ PA 3679. Based on your knowledge of microbiology, why are these two organisms good indicators for sterilization efficiency? Why use both, rather than just one of these organisms?

2. Discuss the making of wines, champagnes, and other fermented grape products. In your discussion give particular consideration to the actions of the microorganisms involved. What steps are normally taken to control the quality of the final product?

ANSWER KEY

Terms and Definitions

1. e, 2. f, 3. d, 4. i, 5. h, 6. a, 7. b, 8. k, 9. g, 10. c, 11. l, 12. j

Fill in the Blank

1. intrinsic; extrinsic; oxygen; carbon dioxide 2. pasteurization; low-temperature holding (LTH); high-temperature short-time (HTST); ultra-high temperature (UHT) 3. infection; intoxication 4. ergotism; aflatoxins; liver 5. must; mashing; wort 6. mash; pitched; beer; ale 7. aerobic; carbon dioxide; alcohol 8. single cell protein; cyanobacterium 9. dehydrate; osmophilic; xerophilic 10. bioluminescence, utensils

Multiple Choice

1. d, 2. d, 3. c, 4. a, 5. c, 6. b. 7. c, 8. a, 9. d, 10. d

True/False

1. F, 2. T, 3. T, 4. T, 5. T, 6. T, 7. T, 8. F, 9. T, 10. T, 11. F, 12. T

44 Industrial Microbiology and Biotechnology

CHAPTER OVERVIEW

This chapter discusses the uses of microorganisms in controlled processes that are grouped under the heading of industrial microbiology and biotechnology. In these situations it is possible to choose the particular microorganism or microbial community used, and it is also possible to define and control the environment in which the activity will take place. The goals of the process will determine whether the environmental conditions will be held constant or changed. The use of genetically engineered microorganisms to increase the efficiency of the processes and to produce new or modified products is discussed, as is the integration of biological and chemical processes to achieve a desired objective.

CHAPTER OBJECTIVES

After reading this chapter you should be able to:

- discuss the use of microorganisms and the management of microbiological processes in industrial microbiology and biotechnology
- discuss the use of genetic engineering methodologies in constructing microorganisms with specific genetic characteristics to meet desired objectives
- describe the design of suitable environments in which to carry out the desired processes
- discuss the management of growth characteristics to produce the desired product
- discuss the major products of industrial microbiology and biotechnology
- discuss the use of recombinant DNA techniques in molecular biology
- discuss hypermutation and forced evolution as mechanisms of creating new microbes and microbial products—evolutionary biotechnology
- discuss protein engineering
- discuss the use of DNA chips for rapid sequencing of DNA
- discuss natural attenuation (i.e., bioremediation without the addition of specific microorganisms)
- discuss the use of biopolymers and cyclodextrins in medical, industrial and consumer products
- discuss the concepts of bioenhancement and bioremediation, and their uses in industrial microbiology and biotechnology
- discuss the use of microbial immobilization and biocatalysts
- discuss the linkage of microorganisms with electronics to produce biosensors
- discuss reductive dehalogenation as a mechanism for managing the degradation of polychlorinated biphenyls (PCBs)

CHAPTER OUTLINE

I. Industrial Microbiology and the New Biotechnology
   A. Development of microorganisms with usable metabolic activities
      1. Mutation and screening from the available gene pool is the traditional approach; it is still important

2. Intentional transfer of genetic information by recombinant DNA technology is a relatively recent development that complements the traditional approach
3. Producing a desired product in usable amounts is the goal
   B. Industrial fermentation—mass culture (aerobic or anaerobic) to produce a desired product in the desired quantities
II. Microbial Growth Processes
   A. Microbial culture
      1. Culture tubes, shake flasks, and stirred fermenters of various sizes are used
      2. All steps in growth and harvesting must be carried out aseptically
      3. Computers are often used to monitor microbial biomass, levels of critical metabolic products, pH, input and exhaust gas composition, and other parameters
      4. Newer methods include air-lift fermenters, solid-state media, and surface-attached microorganisms (biofilms) in fixed and fluidized bed reactors, where the media flows around the suspended particles
      5. Dialysis culture systems allow toxic wastes to diffuse away from microorganisms and nutrients to diffuse toward microorganisms
      6. Continuous culture techniques frequently improve output rates, but they are not always desirable for other reasons
   B. Medium development and growth conditions
      1. Low-cost crude materials are frequently used as sources of carbon, nitrogen, and phosphorus; these include crude plant hydrolysates, whey from cheese processing, molasses, and by-products of beer and whiskey processing
      2. The balance of minerals (especially iron) and growth factors may be critical; it may be desirable to supply some critical nutrient in limiting amounts to cause a programmed shift from growth to production of desired metabolites
      3. Continuous feed of a critical nutrient may be necessary to prevent excess utilization, which could lead to production and accumulation of undesirable metabolic waste products
      4. Physical environment must be defined: agitation, cooling, pH, oxygenation changes; oxygenation can be a particular problem with filamentous organisms, which limit the ability to stir and aerate the medium, thereby creating what is called a non-Newtonian broth (viscous)
      5. Attention must be focused on these physical factors to ensure that they are not limiting microbial growth; they can be very different from small-scale laboratory operations as scaleup procedures are employed
   C. Strain selection, improvement, and preservation
      1. Look for new strains with desired characteristics, which usually involve increased product formation
      2. Traditional mutation and screening procedures can then be used to further increase efficiency
      3. Protoplast fusion procedures can be used to introduce desirable genetic characteristics
      4. Molecular techniques such as the use of the "Gene Gun" or artificial chromosomes are now being used to introduce desirable genetic characteristics
      5. Preservation and strain stability are of concern; methods available include:
         a. Lyophilization (freeze-drying) and storage in liquid nitrogen
         b. Maintenance in water agar and common growth media
III. Major Products of Industrial Microbiology
   A. Primary metabolites are related to the synthesis of microbial cells and are usually produced during trophophase (exponential growth); they include: amino acids, nucleotides, fermentation end products, and exoenzymes
   B. Secondary metabolites have no direct relationship to synthesis of cell materials and natural growth and are usually produced in idiophase (a period after active growth); they include: antibiotics and mycotoxins

C. Microorganisms in balanced growth generally do not accumulate cellular products in amounts beyond those needed for growth; therefore, in industrial fermentations the cell is often tricked into producing large excesses of the desired compounds

D. Antibiotics
1. Penicillin
   a. Produced best when growth is not too rapid
   b. Sidechain precursors can be added to stimulate production of a particular penicillin derivative
   c. Product can then be modified chemically to produce a variety of semisynthetic penicillins
2. Streptomycin is a secondary metabolite that is produced after microorganism growth has slowed if the nitrogen concentration is limited

E. Amino acids—lysine and glutamic acid are needed in large amounts; they are primary metabolites whose accumulation can be increased through the use of regulatory mutants that cannot limit their production and that have increased membrane permeability, which facilitates release of the accumulated products

F. Organic acids
1. These include: citric, acetic, lactic, fumaric, and gluconic acids
2. Citric acid, which is used in large quantities by the food and beverage industry, is produced largely by *Aspergillus niger* fermentation in which trace metals are limited to regulate glycolysis and the TCA cycle, thereby producing excess citric acid
3. Gluconic acid is also produced in large quantities by *A. niger,* but only under conditions of nitrogen limitation

G. Bioconversion processes (microbial transformations or biotransformations) are frequently used instead of chemical transformation because of the high specificity of enzyme-mediated (biocatalyst) conversions in producing the appropriate stereoisomer and the mild conditions in which they can be carried out
1. Freely suspended cells are usually used only once and then discarded
2. Immobilized biocatalysts (cells or enzymes) are attached to particulates so that they can be easily recovered and used again
3. Immobilized biocatalysts are used in the bioconversion of steroids, degradation of phenol, and production of antibiotics, organic acids, and metabolic intermediates
4. Biocatalysts are also used to recover precious metals from dilute-process streams

IV. Molecular Biology and Biotechnology
A. Biochemical markers such as nutrient requirements or antibiotic resistance are often used to identify mutant or recombinant organisms
B. Modification of gene expression involves making specific changes in control regions so that desired enzymes are expressed constitutively, which leads to increases in product formation
C. Design of proteins and peptides
1. Site-directed mutagenesis and chemical synthesis of DNA can be used to tailor specific proteins (protein engineering); yeasts are becoming of increasing importance in the production of medically important proteins
2. It is possible to create enzymes that modify "unnatural substrates" (e.g., improve transformation of previously recalcitrant molecules)
3. Metabolic engineering involves the restructuring of pathways with engineered proteins to control production of end products and other small metabolites
4. Synthetic medical peptides that promote healing and blood coagulation, assist in treating cancer and AIDS, influence sexual dysfunction, and influence other processes have been developed
D. Hypermutation and Evolutionary Biotechnology
1. Forced mutation (evolution) is used to produce microbes with new degradative capabilities or which can produce compounds with new and unique properties

    2.   Optimizes production of new macromolecules at faster rates than normally occur in nature

E.   DNA Chips—computer chips designed with a layer of probes which allow detection of specific sequences and their statistical processing; used for sequencing and the detection of mutations; may lead to the development of chip-based technologies

F.   Vectors for product expression
    1.   Vectors include artificial chromosomes
    2.   The gene for foot-and-mouth disease virus antigen has been incorporated into *E. coli;* the gene product can then be used for vaccine production
    3.   Other substances can be produced in other organisms, either in larger quantities than those produced by the original organism, or in some specifically modified form
    4.   Transgenic plants have been used for large-scale production of a variety of desired metabolic products and have been produced by shooting the DNA into the plant using the "Gene Gun"

V.   Other Microbial Products

A.   Bioinsecticides—bacteria (e.g., *Bacillus thuringiensis*), viruses, and fungi are processed to form a wettable powder that can be applied to agricultural crops, where they are ingested by the insects, releasing toxins into the gut or otherwise killing the insect
    1.   *B. thuringiensis* releases a protoxin that is activated by a protease enzyme
    2.   Six of the active toxin units integrate into the membrane of a midgut cell forming a hexagonal-shaped pore leading to loss of osmotic balance and to cell lysis
    3.   This will only happen if a specific plasmid is present in the bacterium
    4.   The gene for the toxin has been successfully introduced into tomato plants that have subsequently become pest-resistant
    5.   Efforts are underway to make such transgenic plant modifications more cost-effective for routine production of commercial crops
    6.   Viruses such as nuclear polyhedrosis viruses (NPV), granulosis viruses (GV), and cytoplasmic polyhedrosis viruses (CPV) have potential as bioinsecticides
    7.   An important commercial viral pesticide (Elcar) is being used to control the cotton bollworm, *Heliothis zea*
    8.   Fungal biopesticides are increasingly being used in agriculture
    9.   The modification of baculoviruses to produce a potent scorpion toxin that is active against insect larvae is one of the most exciting advances in this field

B.   Biopolymers include microorganism-produced polysaccharides that can be used to modify the flow characteristics of liquids and to serve as gelling agents, particulate dispersing agents, and film-forming agents; they also can be used to maintain texture in ice cream and to enhance oil recovery from drilling muds by increasing water contact and displacement of oil; cyclodextrins can modify the solubility of pharmaceuticals, reduce bitterness, mask odors, selectively remove cholesterol and protect spices from oxidation

C.   Biosurfactants
    1.   Biosurfactants may replace chemically synthesized surfactants because of increased biodegradability, which thereby creates better safety for environmental applications
    2.   The most widely used biosurfactants are glycolipids which are excellent dispersing agents
    3.   Biosurfactants have been used with the Exxon-Valdez oil spill
    4.   In bioremediation programs, the biosurfactant-producing microorganism can be added directly to an oil-containing geological structure; the biosurfactant will then be produced *in situ;* this may improve oil recovery, particularly from older fields

VI.   Biodegradation and its management—microbial-mediated destruction of paper, paint, metals, textiles, concrete, and other materials

A.   Biodegradation stimulation—application of biodegradation to bioremediation efforts (removal of environmental pollutants using microorganisms); early work involved modification of the environment

1. Stimulation of an oil spill degradation—genetically engineered *Pseudomonas* species have augmented hydrocarbon degradation plasmids; clean up is enhanced by the addition of certain nutrients to stimulate a hydrocarbon metabolism; use of this technology was part of a multiple approach strategy in the clean-up of the Exxon-Valdez spill in 1989
2. GEMs, although potentially beneficial, have generally not proven to be effective
   a. They are used as a food source by protozoan predators
   b. They frequently do not come in contact with the compounds to be degraded
   c. They are unable to survive and compete with indigenous microbes
3. Biodegradation enhancement has recently taken advantage of Natural Attenuation; the environment is modified to promote biodegradation by indigenous microbes
4. Prolonged contact with "non degradable" materials such as PCBs may lead to the development of mutated organisms that can now degrade that material through a process called reductive dehalogenation
5. Subsurface biodegradation enhancement—involves similar principles as the oil-spill clean-up, but is complicated by the limited permeability of subsurface geological structures; frequently involves stimulation of naturally occurring microbial communities
6. Bioleaching of metals—involves the use of *Thiobacillus ferroxidans* to solubilize copper in low-grade ores; soluble copper sulfate is formed and is then precipitated by reaction with elemental iron; the same process can be used to solubilize uranium

B. Contol of Biodeterioration—limits undesired degradation
1. Jet fuels—fungi will grow at the water-hydrocarbon interface; this can cause pump clogging; this is alleviated by growth inhibitors, filtering, and frequent cleaning of components
2. Paper—microbial growth causes loss of paper strength, discoloration, and increased deterioration; it is controlled by use of biocides and by washing equipment with hot alkali
3. Computer chips—microbial contamination can decrease service life; the use of ultrapure water during the manufacturing of chips helps to alleviate the problem
4. Paints—controlled by metal biocides as well as other biocides; common in limiting microbial growth in paints
5. Textiles and leather—controlled by the use of fungicides
6. Metals—microbial-mediated corrosion is particularly problematic in iron pipes; control is under investigation but not readily available yet
7. Concrete—can be dissolved by *Thiobacillus* spp. that produces sulfuric acid; this is controlled by organic inhibitors and the use of cements that do not contain appreciable levels of oxidizable sulfur compounds

VII. Biosensors
A. Biosensors involve the use of microorganisms or microbial enzymes linked to electrodes in order to detect specific substances by converting biological reactions to electric currents which can then be measured
B. Biosensors have been developed to measure specific components in beer, to monitor pollutants, to detect flavor compounds in foods, and to detect glucose and other metabolites in medical situations
C. New biosensors using monoclonal and polyclonal antibodies with ssDNA-binding proteins are used to detect pathogens, herbicides, toxins, proteins and DNA

VIII. Impacts of Microbial Biotechnology—ethical and ecological considerations are important in the use of biotechnology

TERMS AND DEFINITIONS

Place the letter of each term in the space next to the definition or description that best matches it.

_____ 1. The use of microorganisms to produce organic chemicals, antibiotics, other pharmaceuticals, and important food supplements

_____ 2. The manipulation and use of microorganisms to recover metals and to improve and maintain environmental quality

_____ 3. A process whereby a critical nutrient is added periodically so that the organism will not have excess substrate available to it at any time

_____ 4. The physical conditions in the immediate vicinity of a microbe

_____ 5. The process whereby water is removed from a preparation at freezing temperatures and under a vacuum

_____ 6. Molecules produced by a microorganism that are directly related to its synthesis and that are often involved in the growth phase

_____ 7. The growth phase of a culture

_____ 8. Molecules produced by a cell that are not directly related to the synthesis of cell materials and that are produced after the growth phase has ended

_____ 9. The phase of a culture after growth has ended

_____ 10. Minor modifications carried out by a microbe on a compound not required for microbial growth

_____ 11. Microbial cells or cell products that are used to modify chemical compounds or to carry out chemical reactions

_____ 12. Undesired microbial-mediated destruction of paper, paint, metals, textiles, concrete, and other materials

_____ 13. The use of microorganisms as insecticides

_____ 14. Living microorganisms that are linked with electrodes to convert biological reactions to electrical currents

_____ 15. Use of site-directed mutagenesis and/or chemical synthesis of DNA to produce enzymes and bioactive peptides with various desired characteristics

_____ 16. Use of heterologous proteins to change pathway distributions and rates, thereby altering end product and metabolite production

_____ 17. Bacteria, viruses or fungi that have the potential to control insect pests

_____ 18. Microbially produced glycolipids that can be used as dispersing agents or emulsifiers

_____ 19. Plants into which genes from other organisms have been introduced and are being expressed

_____ 20. Removal of environmental pollutants using microorganisms

_____ 21. Forced mutation to produce microbes with new degradative capabilities or which can produce compounds with new and unique properties

_____ 22. Biodegradation mediated by indigenous microorganisms

a. biocatalysts
b. biocontrol
c. bioconversions
d. biodegradation
e. bioinsecticides
f. bioremediation
g. biosensors
h. biosurfactants
i. biotechnology
j. continuous feed
k. evolutionary biotechnology
l. idiophase
m. industrial microbiology
n. lyophilization
o. metabolic engineering
p. microenvironment
q. natural attenuation
r. primary metabolites
s. protein engineering
t. secondary metabolites
u. transgenic plants
v. trophophase

FILL IN THE BLANK

1. The growth of _____ organisms results in a viscous, plastic medium called a non-Newtonian broth, which offers great resistance to _____ and _____.

2. To assure that physical factors are not limiting microbial growth, physical conditions must be considered at the level of the individual microbe or _____ where these important biological reactions actually occur. This is most critical in _____, where a process developed in a small shake flask, if successful, must be carried out in a large fermenter.

3. Compounds related to the synthesis of microbial cells and often involved in the growth phase or _____ are called _____ _____; those that have no direct relationship to the synthesis of cell materials and that are produced after the growth phase (_____) are referred to as _____.

4. Commercial scale production of amino acids is typically carried out using _____ mutants because the normal organism avoids overprotection of biochemical intermediates by its careful control over cellular metabolism. In addition, another critical factor in the success of these fermentations is the creation of an intentional increase in _____.

5. A rapidly developing area of biotechnology concerns the linking of microorganisms to electronic components to create _____ that convert _____ into _____. This new field is referred to as _____. They are normally used to detect the presence of small amounts of specific substances.

6. *Aspergillus niger* can be used to produce large quantities of citric acid if _____ _____ are limited in order to regulate glycolysis and the TCA cycle; alternatively, conditions of _____ limitation lead to the production of gluconic acid by the same organism.

7. Effects to limit undesired degradation mediated by microorganisms is called biodegradation _____; the application of biodegradation to bioremediation efforts is referred to as biodegradation _____; one example of the latter process is the solubilization of copper from low grade ores; this process is called _____.

MULTIPLE CHOICE

For each of the questions below select the *one best* answer.

1. Which of the following is not an accepted meaning for the term fermentation?
   a. any process involving the mass culture of microorganisms, either anaerobic or aerobic
   b. the production of alcoholic beverages
   c. food spoilage
   d. All of the above are accepted meanings for the term fermentation.

2. A culture system that allows toxic waste products to diffuse away from a microbial culture through a membrane, and that also allows substrates to diffuse toward the microbial culture is referred to as a
   a. fixed bed reactor.
   b. dialysis culture system.
   c. fluidized bed reactor.
   d. continuous feed reactor.

3. Agitation is used to maintain proper oxygen availability. However, this can be a major problem when trying to culture which of the following organisms?
   a. bacteria
   b. filamentous fungi
   c. yeasts
   d. It is a major problem with all of the above organisms.

4. Which of the following methods is(are) used to improve the strains used to carry out microbial fermentations?
   a. mutation and selection
   b. recombinant DNA modification of gene expression
   c. protoplast fusion
   d. All of the above are used to improve strains used for microbial fermentations.

385

5. Which of the following would not be considered a secondary metabolite?
    a.   ethanol or some other fermentation end product
    b.   mycotoxins
    c.   antibiotics
    d.   All of the above are considered secondary metabolites.
6. Minor microbial modifications to chemical compounds that are not usually used for growth are referred to as
    a.   bioconversions.
    b.   microbial transformations.
    c.   biotransformations.
    d.   All of the above are correct.
7. Which of the following is not normally subject to biodegradation?
    a.   jet fuel
    b.   computer chips
    c.   concrete
    d.   All of the above are subject to biodeterioration.
8. Which of the following is not used as a microbial insecticide?
    a.   protozoa
    b.   bacteria
    c.   viruses
    d.   fungi
9. Which of the following best describes the production of semisynthetic penicillins?
    a.   addition of a side-chain precursor molecule to a *Penicillium* culture to produce a desired modification
    b.   chemical modification of a penicillin product after it has been released from the *Penicillium* culture
    c.   Both (a) and (b) are correct.
    d.   Neither (a) nor (b) is correct.
10. Which of the following is a reason why bioconversions are replacing chemical synthesis of certain compounds?
    a.   stereoisomer specificity
    b.   mild reaction conditions
    c.   Both (a) and (b) are correct.
    d.   Neither (a) nor (b) is correct.
11. Which of the following are being used as biopesticides?
    a.   viruses
    b.   fungi
    c.   Both (a) and (b) are correct.
    d.   Neither (a) nor (b) is correct.
12. Which of the following is a reason why genetically engineered microorganisms have not been as effective as was originally anticipated?
    a.   They are used as a food source by protozoa.
    b.   They frequently do not come in contact with the materials to be degraded.
    c.   They do not survive well once released and cannot compete with indigenous microbes.
    d.   All of the above are correct.

TRUE/FALSE

_____ 1. Industrial processes involving microorganisms require that the culture be maintained in an active growth phase in order to maximize product formation.
_____ 2. As with laboratory scale operations, industrial fermentations rely primarily on purified media components in order to maintain better control of the process and of the final product.
_____ 3. Continuous feed processes involve periodic additions of a critical nutrient so that the organism will not have excess substrate available at any time, because an excess of substrate can lead to the production and accumulation of undesirable metabolic waste products.
_____ 4. Protoplast fusion can be used only between members of the same species.

5. Environmental conditions often have to be adjusted to switch from those conducive to microbial growth to those conducive to product formation. For example, *Streptomyces griseus* is grown in a high nitrogen medium, but the availability of nitrogen must be limited before the antibiotic streptomycin will accumulate.

6. Biocatalysts are often immobilized by attaching them to ion exchange resins or by entrapping them in a polymerized matrix to enable their recovery and repeated use.

7. The first patented microorganism was one that could be used to clean up oil spills.

8. Copper can be recovered by bioleaching, in which microorganisms solubilize copper from ores that are not rich enough to be processed by conventional smelting.

9. Engineered proteins can be used to degrade previously recalcitrant molecules that were not considered to be amenable to biological processing.

10. Plants have been made pest-resistant by introducing a bacterial gene for an insect toxin into the plant's DNA.

11. Non-newtonian broths are created when filamentous microorganisms grow and prevent adequate mixing of the growth media.

CRITICAL THINKING

1. Discuss how regulatory mutants are used (and why they are necessary) in the production of amino acids. Cite a specific example and describe the nature of the regulatory changes. Why is it desirable to increase membrane permeability in these organisms as well?

2. Genetically engineered microorganisms have been developed for use as pesticides. What steps should be taken to test the safety of these organisms before releasing them for use in a natural environment? Justify your choices. If you do not think they should ever be used, state your reasons.

3. Discuss at least two ethical issues raised by the topics covered in this chapter.

ANSWER KEY

Terms and Definitions

1. m, 2. i, 3. j, 4. p, 5. n, 6. r, 7. v, 8. t, 9. l, 10. c, 11. a, 12. d, 13. b, 14. g, 15. s, 16. o, 17. e, 18. h, 19. u, 20. f, 21. k, 22. q

Fill in the Blank

1. filamentous; stirring; aeration 2. microenvironment; scale-up 3. trophophase; primary metabolites; idiophase; secondary metabolites 4. regulatory; cell permeability 5. biosensors; biochemical reactions; electrical currents; bioelectronics 6. trace metals; nitrogen 7. control; stimulation; bioleading

Multiple Choice

1. d, 2. c, 3. b, 4. d, 5. a, 6. d, 7. d, 8. a, 9. b, 10. c, 11. c, 12. d

True/False

1. F, 2. F, 3. T, 4. F, 5. T, 6. T, 7. T, 8. T, 9. T, 10. T, 11. T